Psychoanalysis: A Theory in Crisis

Books by Marshall Edelson

The Termination of Intensive Psychotherapy (1963)

Ego Psychology, Group Dynamics and the Therapeutic Community (1964)

Sociotherapy and Psychotherapy (1970)

The Practice of Sociotherapy: A Case Study (1970)

The Idea of a Mental Illness (1971)

Language and Interpretation in Psychoanalysis (1975)

Hypothesis and Evidence in Psychoanalysis (1984)

Marshall Edelson

Psychoanalysis

A Theory in Crisis

University of Chicago Press

Chicago and London

Contents

Acknowledgments

Jess Ghannam, Ken Marcus, and Morton Reiser (each of whom, in his own way, has taught me what friendship is all about) stayed with me through many months of frequent discussions about some of the chapters in this book. They were excited by the ideas, encouraged me, and through their comments and questions provoked me to think. As I wrote the chapters on the case study method, Adolf Grunbaum's eagerness for the fray, intense disagreements, indefatigable challenging, and insistence on reasoned argument stimulated and energized me. David Berg, Jonathan Lear, and Daniel Levinson were helpful in considering some specific issues. Those who have worked with me in carrying out their psychoanalyses have also been my teachers.

My wife Zelda read everything; she contributed editorial skills, and much besides. Jon and Rachel, Dave and Debra, and Bec, what they are and the lives they lead, provide a background of joy for my labors.

I am grateful to the following:

Macmillan Publishing Company, for permission to use material in Chapters 1 and 4, which previously appeared in my chapter, "Toward a Study of Interpretation in Psychoanalysis," in J. Loubser, R. Baum, A. Effrat, V. Lidz, eds., *Explorations in General Theory in Social Science: Essays in Honor of Talcott Parsons*, pp. 151–181, copyright © 1976 by The Free Press, a Division of Macmillan, Inc.

Albert Solnit, Times Books, and Yale University Press, for permission to use material in Chapter 2, which previously appeared in my paper, "Language and Dreams: *The Interpretation of Dreams* Revisited," *Psychoanalytic Study of the Child* 27 (1972), 203–282.

Williams and Wilkins, for permission to use material in Chapter 3, which previously appeared in my paper, "Psychoanalysis as Science," *Journal of Nervous and Mental Disease* 165, no. 1 (1977), 1–28, copyright by Williams and Wilkins.

Joseph H. Smith, Forum on Psychiatry and the Humanities, and

Yale University Press, for permission to use material in Chapter 4, which previously appeared in my chapter, "What is the Psychoanalyst Talking About?" in J. H. Smith, ed.. *Psychoanalysis and Language* (*Psychiatry and the Humanities*, vol. 3) (1978), 99–170.

Lawrence Erlbaum Associates, for permission to use material in Chapter 5, which previously appeared in my chapter, "Psychoanalysis, Anxiety and the Anxiety Disorders," in A. Tuma & J. Maser, eds., *Anxiety and the Anxiety Disorders* (1985), 633–644.

Melvin Bornstein and Lawrence Erlbaum Associates, for permission to use material in Chapter 6, which previously appeared in my paper, "Heinz Hartmann's Influence on Psychoanalysis as a Science," *Psychoanalytic Inquiry* 6, no. 4 (1986), 575–600.

William Alanson White Institute, for permission to use material in Chapter 7, which previously appeared in my paper, "The Convergence of Psychoanalysis and Neuroscience: Illusion and Reality," *Contemporary Psychoanalysis* 22, no. 4 (1986), 479–519.

Sage Publications, Inc., for permission to use material in Chapter 11, which previously appeared in my chapter, "The Hermeneutic Turn and the Single Case Study in Psychoanalysis," in D. Berg & K. White, eds., *Exploring Clinical Methods for Social Research*, pp. 71–104 (1985), copyright © by Sage Publications.

Leo Goldberger, Psychoanalysis and Contemporary Sciences, Inc., and International Universities Press, Inc., for permission to use material in Chapter 11, which previously appeared in my paper, "The Hermeneutic Turn and the Single Case Study in Psychoanalysis," *Psychoanalysis and Contemporary Thought* 8, no. 4 (1985), 567–614.

Cambridge University Press, for permission to use material in Chapter 12, which previously appeared in my commentary, "The Evidential Value of the Psychoanalyst's Clinical Data," *Behavioral and Brain Sciences* 9, no. 2 (1986), 232–234.

Albert Solnit and Yale University Press, for permission to use material in Chapter 13, which previously appeared in my paper, "Causal Explanation in Science and in Psychoanalysis: Implications for Writing a Case Study," *Psychoanalytic Study of the Child* 41 (1986), 89–127.

Introduction: Core Theory and Case Study in Psychoanalysis

I

This is a book on the conceptual foundations of psychoanalysis. It might equally be described as a book on the philosophy of psychoanalysis. It is written for those in the intellectual community who are interested in assessing the current status of psychoanalysis as a body of knowledge.

II

Why such an assessment now? Why do I feel, as I do, that the assessment is urgent? Why do I write in particular about the conceptual foundations of psychoanalysis?

Something strange, even disturbing, is happening to psychoanalysis in both the humanities and the sciences, and in the popular culture as well. Psychoanalysis, as a body of knowledge about human beings or the human mind, has become the object of a dismissive, disillusioned, and frequently derogatory polemic. But—and this is what I find strange—these attitudes do not seem to bear a clear relation to, or be justified by, any actual advances in our knowledge of the human mind brought about by other modes of inquiry or within other conceptual frames of reference.

I first became aware of psychoanalysis as a teenager in the forties—at the movies. The movie *Spellbound* begins with a quotation from Shakespeare--"The fault . . . . . is not in our stars, but in ourselves . . . . ." The dedicatory statement continues on a note of authoritative confidence and awe.

> Our story deals with psychoanalysis, the method by which modern science treats the emotional problems of the sane. The analyst seeks only to induce the patient to talk about his hidden problems, to open the locked doors of his mind. Once the complexes that have been disturbing the patient are uncovered and interpreted, the illness and confusion disappear . . . . . and the evils of unreason are driven from the human soul.

I suppose the hopes were too high, the expectations too great. The pendulum was bound to swing, the reactive disillusionment bound to come. In the eighties, if a psychoanalyst appears in a movie, one may be sure that he or she is lacking in integrity, or is criminal, mad, or both.[1]

In any event, I went on to have my own psychoanalysis, and eventually I read Freud. Both experiences were revelatory. The excitement I felt observing the powerful effects wrought upon me by psychoanalytic interpretations of my own mind's workings, and the exhilaration I felt retracing Freud's explorations in the human mind have never waned. If anything, these feelings have been augmented by the astonishing phenomena I am privileged to observe as a psychoanalyst in the psychoanalytic situation—phenomena, I might add, that do not seem to figure in any important way in the writings of the detractors of psychoanalysis who, gleefully or regretfully, send it on its way to the dustheap of dubious speculations.

So I became a psychoanalyst. I had every expectation that following this calling, for as such I regarded it, would lead me, as a practicing psychoanalyst and scholar, down a long and honorable path to a lifetime's achievement of which I might someday be proud.

Imagine my feelings as, rather well along in life, I see scorn heaped upon my life's love, or notice the gradual decline in the attention paid to it, or, what is worst, become increasingly aware myself of shortcomings in the psychoanalytic community to which I belong, and in the work that issues forth from it—shortcomings that seem to increase in degree with every passing year, with no apparent prospect of improvement. If I were not as fascinated as ever by the phenomena that appear in the psychoanalytic situation and that psychoanalysis alone continues to be capable of bringing to our attention, and if I did not think that psychoanalysis still provides the most interesting and indeed the best explanations of those processes, states, and products of the mind that truly belong in its province, this would be a melancholy rather than a hopeful book.

What is it that troubles me so?

In the humanities, the writings of Freud are currently subjected to hermeneutic disquisitions, which seem to have as their aim "to sabotage the text"—that is, to disregard Freud's avowed intentions and to find "between the lines" that which contradicts and overturns him. Despite his own explicitly avowed commitments to science, he is regarded as a cryptohumanist hermeneuticist and litterateur. Freud,

1. This particular change may be a manifestation of a shift in popular attitudes toward "psychiatry" in general—"psychiatry" is a rather fuzzy category for moviemakers—which has been documented by Gabbard and Gabbard (1987).

the culture hero, the scientist whose insights have shaped our world, has become the object of denigratory gossip about his personal life. His behavior as the leader of a "movement" is deplored. His intellectual integrity is questioned.

Where psychoanalysis apparently continues to be of interest to the humanities, the interest is in one or another often obscurantist offshoot of it. This interest and the choice of these particular variants of psychoanalysis are motivated by a view of psychoanalysis as a (helpful, misguided, or malevolent) hermeneutics—rather than as a body of scientific knowledge that is used to make causal inferences about the mind or to give causal explanations of products of the mind.

Feminists certainly are, on the whole, at the least wary of Freud, whose ideas about and attitudes toward women are suspect. That he was wrong about some things (and not just about women) seems to me to be unquestionably true. I would be surprised if this were not the case. Scientific knowledge is always provisional and subject to revision, a fact about science not readily recognized by those who take it too seriously in the wrong way. Freud repeatedly described himself, somewhat overmodestly perhaps, as mainly ignorant about such matters as "female sexuality," and what he put forward he put forward explicitly as based on work with relatively few cases and needing to be checked and investigated further.

In the sciences, there seems to be some conviction that psychoanalysis as a body of knowledge has been superseded by cognitive science or neuroscience. That psychoanalysis has been so superseded is surely a premature conclusion. Neuroscience, for example, is barely able to explain for us how the mind works to make possible some of its comparatively simple achievements (in such realms, for example, as perception, memory, and language). Furthermore, when members of these other disciplines argue that they have produced evidence relevant to some psychoanalytic conjecture or hypothesis—about anxiety, or the unconscious, for example—they usually misdepict what psychoanalysis has to say on the matter, or use the concept itself in a way that bears at most a vague resemblance to the way in which psychoanalysis uses it.

Quantitative/experimental research has little to say about matters of central import for psychoanalysis, and when it tries to say something, the way the individual studies are carried out departs so far from standards for such research that one is advised to contemplate the conclusions with caution.

The most telling outside challenge to psychoanalysis has been the questions raised by Adolf Grunbaum about its empirical foundations and mode of inquiry. The most telling manifestations of the difficulties in which psychoanalysis now finds itself are both the lack, for the most part, of a cogent response to these questions (I except my own work, of course), and the various kinds of responses these questions have received instead.

The attacks on Freud's character and stature, the misdepictions of his work, the social-political debates into which this work is injected, and the unwarranted claims that scientific work in other disciplines requires us to reject or drastically revise psychoanalytic ideas are not what bother me most. For all these, I believe, are symptoms of a more profound malaise, and that malaise lies within psychoanalysis itself.

None of these animadversions, no amount of neglect, would make any difference at all, nor would they so preoccupy us, if psychoanalysis itself were making steady scientific advances, if its knowledge of the mind were growing year by year, if it were progressively finding reason in evidence to reject some of its ideas and to regard others as becoming justifiably more firmly entrenched.

But what do we have instead? We have the incorporation into psychoanalysis of all of developmental psychology, all of cognitive psychology, all of every kind of psychology, as if such addition represented advance—rather than paring (rejecting some conjectures), selecting (choosing among conjectures), and increasing precision (in stating and testing conjectures). The decision at some point in the history of psychoanalysis to expand the boundaries of its core beyond its interest in the unconscious and sexuality to include "ego psychology" has not proved, whatever the gains to the therapy, an unmixed blessing to the body of knowledge.

We also have a preoccupation with new psychoanalytic theories—each one put forth with the claim that it is in toto better than, and should replace, Freud's. We do not have what instead we want: new evidence being gathered, sifted, and weighed; new observations being made; a new generation of investigators eagerly adding a new brick here, a new brick there, to an edifice that stands upon an empirically and conceptually sound, and generally accepted, foundation.

We have—not entirely, not everywhere, but mainly—scientific stagnation. We do not have new findings about the human mind. We have instead turgid theorizing, divorced from description of specific phenomena; and, too often, when phenomena are presented to document theoretical conjectures, they are only loosely connected—and not by scientific argument or reasoning—to the conjectures they purport to support.

And how, then, are theoretical disputes to be settled? How are rival theoretical claims to be resolved? As it is now, they are frequently not settled or resolved, but simply presented and re-presented, as if the mere statement of them in more and more powerful rhetorical terms would settle the matter; or they are settled and resolved locally by social-political rather than by scientific process.

What a falling off from a time of excitement, passionate curiosity, discovery. Why should it have occurred? I make two interpretations.

The first interpretation is peripheral with respect to the objectives of this book, and I offer no evidence to support it. Psychoanalysis as a therapy now has hundreds of rivals, each one claiming to do what psychoanalysis does, but faster, less expensively, and for more kinds of illnesses. I do not want to expand much upon this circumstance because, for the most part, I do not talk about or attempt to assess psychoanalysis as a therapy in this book, except in passing or when a question about or a phenomenon bearing upon psychoanalytic knowledge are at issue. I do discuss the implications of what I have to say about psychoanalysis as a body of knowledge for understanding the therapeutic action of psychoanalysis, for example, when I discuss the causal mechanisms that might be postulated to account for its effects. And I do discuss also, in considering the mode of inquiry of psychoanalysis, rational grounds for believing clinical causal inferences made in the psychoanalytic situation.

Nevertheless I will say here that it is my impression that, facing an ever-growing number of competitors in a society that favors "fast food" solutions and action over reflection and contemplation, psychoanalysis, seeking to survive economically as a treatment, has tried to increase its scope—the number of different kinds of patients or ailments for which it might claim to be helpful. What has followed is that psychoanalysts, somewhat naturally and understandably, working with many more kinds of patients and ailments, have tried to make theory over so that it applies to and is useful in understanding these patients and conditions to which it might not otherwise apply.

But psychoanalysts in carrying out a psychoanalytic therapy, even a "classical" psychoanalysis, necessarily do make, have always made, and will continue to make use of a great deal of knowledge besides psychoanalytic knowledge. It is a common, understandable mistake—but a mistake having fateful consequences with respect to the orderly development of psychoanalysis as a body of knowledge—to regard any knowledge that is useful in conducting psychoanalytic therapy as therefore *psychoanalytic* knowledge. Extensions of theory so arrived at, however well-motivated by therapeutic aims, seem to me to make for a hodgepodge of theoretical accretions. Theory becomes increasingly incoherent and unwieldy. Testing theory, choosing among rival inferences and explanations, according to scientific canons, becomes increasingly difficult.

That psychoanalysis is both a therapy and a body of knowledge has always made for some problems. More than one commentator has mentioned as crucial for understanding the problems facing psychoanalysis as a body of knowledge the way in which the tail of clinical practice wags the dog of scientific work, and specifically the way institutes devoted to training for practice are isolated from the intellectual challenges and give-and-take of a setting like the university. I am inclined to believe, however, that the clinical practice, because it produces strong effects, which are not only intensely moving but cry out for study, and because it is the source of otherwise inaccessible phenomena, is vitally needed to nourish and further the scientific work. Under circumstances different from those I have been describing, it will do so.

My second—and, for this book, more important—interpretation is that the current plight of psychoanalysis surely owes something to a failure to clarify its conceptual foundations and mode of inquiry. We need to simplify, simplify, simplify, not endlessly complicate, just what it is that psychoanalysis is about, what demarcates it from other disciplines, including other branches of psychology, and other branches of a psychology of mind.

Then, from the mass of shapeless thought, the specific distinctive contributions of psychoanalysis will emerge sharp, clear, in bold relief.

Then, these contributions to an understanding of mind can be conveyed to others who have a real interest in them.

Then, a program of research can begin to tackle the task of systematically raising and investigating one question after another—the answer to each question leading to still another question.

Then, these answers—conjectures and hypotheses—can be subjected to fair, insightful, and rigorous tests of their scientific credibility; rival explanatory claims can be adjudicated by a process of scientific reasoning.

Weeding the garden is the way to end overgrowth, choking, and stagnation. If conceptual foundations and mode of inquiry were to be clarified, the body of knowledge would have the chance to grow in a systematic and orderly way, and psychoanalysis would move, and would be perceived as moving, forward.

I shall know things are better than they are now when, for example, we cease to speak about "the psychoanalytic psychology of women." Psychoanalysis does not have a "psychology of women." Indeed, it does not have a complete account of female sexuality. It can make specific contributions to understanding some things about female sexuality. Similarly, I shall know things are better when we cease to speak about "the psychoanalytic theory of homosexuality," or indeed about "homosexuality" (as if homosexuality were not a final common pathway of any number of causal processes), and speak instead about what specifically psychoanalysis is able to contribute to an understanding of *some* of the *homosexualities.* (Other disciplines will also make their specific contributions.)

That kind of change will come about only if we can answer this question. What is psychoanalysis specifically able to contribute as a body of knowledge to our understanding of the mind's working that is not likely to be contributed and perhaps cannot be contributed by other "schools of thought," given the conceptual frameworks within which they work that determine their interests, questions, and program of research?

It is this second interpretation of mine that provides the impulse to write this book. I write it because I believe that, despite all of the problems and failings of psychoanalysis, what is of value in it is an intellectual treasure worth saving. I am convinced that what is of value in psychoanalysis is—when it comes to a way of understanding certain processes, certain states, and certain products of the human mind—the best we have.

III

Anyone who sets out to assess psychoanalysis as a body of knowledge presumably has an idea about what the body of knowledge is. What are its boundaries? How does it differ from other bodies of knowledge? What is it about and what is it not about?

Unfortunately, there is no consensus even among psychoanalysts about the answers to these questions. This lack of agreement creates problems for those who wish to understand the therapeutic action of psychoanalysis, to apply psychoanalysis in solving various practical or conceptual problems, to teach psychoanalysis to others, to subject psychoanalytic theory to tests of its scientific credibility, or to extend and develop it.

Therefore, this book, like any book attempting to assess psychoanalytic knowledge, must take some position in answering the following questions.

What is the subject matter of psychoanalysis? Is it about behavior, the personality system, interpersonal relations, the mind? An answer such as "Why not all of them?" misses what is comprehended by an analysis or explication of the conceptual foundations of psychoanalysis, and the hoped-for results of such an endeavor—simplicity and systematization, and all the benefits that derive from them.

What is distinctive about the concepts and causal explanations of psychoanalysis? Which of its concepts and conjectures are central, and which are peripheral? If it sticks to its preferred mode of inquiry, which includes free association and the case study method, how successfully can it argue that there are rational grounds for believing its clinical causal inferences and causal explanations?

It is clear from the nature of these questions that any assessment of the status of psychoanalysis as a body of knowledge must depend upon, be governed by, or presuppose the assessor's views on two matters: the content of the core theory of psychoanalysis; and the defining characteristics of science, scientific method, and scientific explanation. About these two matters, there is at this juncture of history plenty of room for disagreement. Views about them that are held by anyone who now makes such an assessment of psychoanalysis must be controversial and, furthermore, the controversy, at least for now, cannot be resolved.

If I disagree with your views about either of these matters, I am also likely to disagree with your assessment of the current status of psychoanalysis—the scientific credibility of its explanations or the directions its research needs to take. Similarly, if I disagree with your assessment of the current status of psychoanalysis, it is quite probable that the source of the disagreement, when it is not merely a matter of a difference in what facts we each have at our disposal, is a disagreement between us about what constitutes the core theory of psychoanalysis or about the nature of science, scientific method, and scientific explanation. Therefore, I take pains in this book to be as explicit and as clear as possible about my own still developing views about these two controversial matters.

IV

In Part 1 of the book, I attempt to explicate and clarify what I have called the core theory of psychoanalysis.

In Part 2, I attempt to characterize and justify the mode of inquiry—free association and the case study method—psychoanalysis uses when it argues that, given the evidence, belief in its clinical causal inferences and causal explanations is warranted.

V

The decisions about what should be taken as the fundamental core, or the conceptual foundations, of psychoanalysis, which I present in

Part 1, are my own. Someone else might take different works of Freud, or different phenomena, as starting points—and make quite different decisions about what domain to study and what concepts to choose in beginning such a study.

What will lead to acceptance of one set of decisions over another is its perceived usefulness, its effects upon psychoanalytic research, its fruitfulness—not its "truth." One set of decisions cannot be "true" and another "false." Such properties as truth and falsity are not properties of a conceptual analysis, or of decisions about what domain to study, or what concepts to use in studying it. Truth does enter in, however, when assessing the usefulness of a choice of domain and of initial descriptive and theoretical concepts—demonstrating, for example, that some set of most distinctive, essential, and central conjectures, causal inferences, or explanations about an intended domain, resulting from these initial choices, is scientifically credible.

My decisions about what constitutes the conceptual foundations of psychoanalysis are based upon my clinical experience as a practicing psychoanalyst, my scholarly studies, and my efforts as an educator and member of an intellectual community to convey to others, either to students or to colleagues in other disciplines, what psychoanalysis is all about and what it adds to an understanding of the mind.

But most of all, in making these decisions, I have over and over taken as a starting point the empirical findings and the theoretical formulations Freud gave us, first and primarily, in his great work *The Interpretation of Dreams* and its two companion works, *The Psychopathology of Everyday Life* and *Jokes and Their Relation to the Unconscious,* and, second but probably not less important, in *Three Essays on the Theory of Sexuality,* "Formulations on the Two Principles of Mental Functioning," and his case studies, especially "Notes Upon a Case of Obsessional Neurosis," "From the History of an Infantile Neurosis," and "'A Child Is Being Beaten': A Contribution to the Study of the Origin of Sexual Perversions."

It is in these works that Freud makes clear what the domain he studies is, and what his core descriptive and theoretical concepts are, and demonstrates the explanatory fruitfulness of the approach that he has taken—and that I have taken in this book in explicating the conceptual foundations of psychoanalysis.

VI

Before going on to provide a cognitive map of my discussion in this book of the core theory of psychoanalysis, I want to define a few terms.

When, engaged in a process of making decisions about what the essential core of psychoanalytic knowledge is, I mention the *intended domain* of psychoanalysis, I am addressing the following questions. To what in the world does psychoanalysis apply, if it applies to anything? About what are its conjectures, hypotheses, inferences, explanations true, if they are true about anything?

When I mention *distinctive* concepts, questions, and answers, I mean by distinctive that other disciplines do not employ such concepts, raise such questions, or give such answers (come up with such conjectures, inferences, or explanations).

What do I mean by *the core theory of psychoanalysis?* The concepts that belong to this theory are those distinctive *descriptive concepts* that are applicable to entities in the intended domain of psychoanalysis and that inspire or enable psychoanalysis to raise distinctive questions about this domain. The distinctive *theoretical concepts* that belong to this core theory are used in formulating distinctive answers to these questions. This set of distinctive answers—*theoretical conjectures, clinical causal inferences,* or *causal explanations*—are constituents of the core theory, insofar as psychoanalysis is confident that if any of its generalizations, conjectures, hypotheses, inferences, or explanations are true of its intended domain, these certainly must be among them.

If they are in fact demonstrated to be true in the intended domain, then the confidence this demonstration engenders in the fruitfulness of the conceptual framework will lead to some attempts to apply it in some systematic way to other domains as well. On the other hand, if just these distinctive generalizations, conjectures, inferences, or explanations were to be demonstrated to be false in the intended domain of psychoanalysis, then, at the least, serious doubt would be raised about the fruitfulness of this conceptual framework as a way of looking at things.

For a conceptual framework to survive and hold the interest of the intellectual community, some core concepts must prove useful in formulating questions about a domain, questions that otherwise are unlikely to be asked. The answers to these questions should be the kind that lead to still other interesting questions. Other core concepts must prove useful in formulating answers to these questions. Some answers are constituents of the core theory just because, for the theory to survive, at least these answers must be true of some sorts of things in the world, no matter how limited or circumscribed that set of things is. If these desiderata were not true of psychoanalysis, there would be little interest in it or in extending it beyond its intended domain to other domains.

An assessment of the current status of psychoanalysis must respond to those who believe that such an apocalyptic end of a failed and exhausted paradigm is occurring now. They have concluded that psychoanalysis has failed to demonstrate that its core concepts have been fruitful, and has failed to establish the scientific credibility of its distinctive, most central, essential, characteristic theoretical conjectures, clinical causal inferences, or causal explanations.

VII

My decisions about what is distinctive about psychoanalysis are, in some cases, controversial. It will be clear in Part 1 that I have decided to give primacy, for example, to the causal force of psychic reality over that of actual events (including those that are traumatic), the causal force of fantasy over that of memory, the causal force of the quest for sexual pleasure over that of the quest for the object, the causal force of relations to one's body and to internal objects in fantasy over that of relations with an actual mother or other significant external objects, and the causal force of sexual wishes over that of aggressive (and nonsexual) wishes.

When I say I have given primacy to one causal factor over another in explicating the conceptual foundations of psychoanalysis, I do not mean that in each case the first factor has a causal role and the second factor is denied a causal role in psychoanalytic theory. (Freud, in his concept of a complemental series, emphasizes the causal interactions between, and the varying causal weights to be given in different cases to, both circumstances and constitution, for example.) Rather I am identifying what causal factors I believe are prior and primary in an explication of the conceptual foundations of psychoanalysis. I am identifying what causal factors should be especially associated with psychoanalysis as a body of knowledge—in contrast to other psychologies—because it has something distinctive to say about these factors and because it is, given its mode of inquiry, in the best position to investigate and elucidate their causal roles.

In all cases, I have tried to give reasons for my choices, but I have not picked out for rebuttal specific works of those who think otherwise. (That omission should not be taken to indicate that I am unfamiliar with such works.) Attempts to refute do not seem appropriate when I am claiming just that the core I identify is distinctive, that it characterizes what is unique and essential about psychoanalysis.

I expect others, addressing the same problem, might come to formulations different from mine. Such formulations might begin with behavior or interpersonal interactions, rather than mental states, for example, as the fundamental entities with which a theory like psychoanalysis concerns itself, or regard the entire personality system (not only mental states, but skills, every kind of psychological capacity, etc., as well) as the intended domain of psychoanalysis. About the formulations of those who have made other conceptual choices, I might well concede that their formulations turn out to yield scientifically credible conjectures, and even that these conjectures have a place in psychoanalytic theory. But I might also still maintain that their place is peripheral rather than central.

I have taken other controversial positions. In Part 1, I reject the conception of psychoanalysis as a general psychology. I argue for psychoanalysis as a commonsense intentional psychology (that is, as a psychology about the nature and causal interrelations of such entities as wishes and beliefs). I tend to doubt the cogency of much of psychoanalytic metapsychology (although some version of the structural theory seems to me to be necessary). I am especially dubious about the place of ego psychology in, and its effect upon, psychoanalytic theory.

With regard to the relation of psychoanalysis to other disciplines, I have argued in Part 1 that psychoanalysis cannot be reduced by neuroscience. I have also argued that much experimental/quantitative research purporting to test psychoanalytic theory, and much work in applied psychoanalysis, misfires, because it involves a misdepiction of psychoanalysis.

All these positions follow from my decisions about what is core theory in psychoanalysis.

VIII

Part 1 records then the development of my thinking about the following questions. What is the *domain* of psychoanalysis: the organism, the person, the personality system, interpersonal relations, the mind? What sorts of entities belong to the class of individuals—things or objects, states of affairs, or events—psychoanalysis investigates? Does it study behaviors, interpersonal relations, or mental states? What sorts of properties possessed by these entities, and what sorts of relations among them, does psychoanalysis identify as of special interest?

What is distinctive about psychoanalytic *concepts* (the way psychoanalysis groups and characterizes objects, as it attempts to cut nature at its joints)? What is distinctive about the questions these concepts inspire or enable psychoanalysis to raise about its intended domain? What is distinctive about the answers to these questions, which psychoanalysis, using these concepts, comes up with?

I begin my answers to these questions by rejecting as grandiose and infertile the idea that psychoanalysis is a general psychology. Here is how I go about then progressively delimiting the domain of psychoanalysis. (In Part 1, I do not carry out this project in the orderly sequence that now unfolds in my introduction.)[2]

Psychoanalysis is, most generally, a psychology of mind, and not, for example, a psychology of behavior, or a psychology of interpersonal interactions or relations.

Psychoanalysis as a psychology of mind is, more specifically, an intentional psychology. That is to say, it is concerned with the nature and causal interrelations of such entities as wishes and beliefs, whether these be mental dispositions or occurrent mental states. That it is an intentional psychology distinguishes it from those bodies of knowledge (other branches of a psychology of mind) concerned with belief-and-wish-impenetrable mental capacities and processes. ("Belief-and-wish-impenetrable" means that these particular mental capacities and processes—some aspects of reality-accommodated perception, for example—are not influenced by wishes and beliefs, and contrasts with "belief-and-wish-penetrable," which is used to describe capacities and processes that are influenced by wishes and beliefs.)

If psychoanalysis is an intentional psychology, then mental states are entities in its domain. Constituents of mental states include a person; that person's mental representation of an object, state of affairs, or event (the contents of a mental state); and a relation between the person and the person's mental representation. *Wishes, believes, perceives, remembers, imagines* are examples of such relations.

The word "intentional" in "intentional psychology" alludes to the fact that wishing, believing, perceiving, remembering, and imagining, for example, are *directed to* mental representations.

Psychoanalysis as an intentional psychology is also, more specifically, a science of the symbolizing activity of the mind, because it is interested in how the capacity for symbolization is manifested in constructing mental representations. Mental representations are symbolic representations. They symbolize objects, states of affairs, or events. The mind operates upon such symbolic representations, to

2. I have been influenced in what follows by the work of Noam Chomsky, Jerry Fodor, Talcott Parsons, Rom Harre, and Richard Wollheim, among others, whose specific contributions are documented in various chapters of this book.

transform them, linking one causally to another. Operations that operate upon symbols are symbolic operations.

This interest in the construction of mental representations, and in the symbolic operations that form and transform such representations, identifies psychoanalysis as a representational and computational psychology of mind.

However, psychoanalysis does not have things to say about every kind of mental state, representation, and operation. Psychoanalysis accords causal priority to psychic reality. (It has been called a science of illusions.) Psychoanalysis as a science of the symbolizing activity of the mind, a representational and computational psychology of mind, is, more specifically, a science of imagination. Even more specifically, psychoanalysis is a science of imagination as imagination is implicated in the vicissitudes of sexual wish-fulfillments.

It is the special relation between wishes, and in particular sexual wishes, and imagination that is of central importance to psychoanalysis.

In contrast, it is the relation between conjunctions of instrumental or reality-accommodated beliefs and wishes that are possible to gratify in actuality, on the one hand, and action, on the other, that is of central importance to a nonpsychoanalytic but representational psychology of mind. In a nonpsychoanalytic but representational psychology of mind, interest is focused on rational action (caused by belief and desire acting together) and irrational action (beliefs are in error or relevant beliefs are missing).

When I describe psychoanalysis as a science of imagination, I am not talking about reality-oriented anticipatory imagining. I am talking about imagination functioning under special conditions. More specifically, I am talking about unconscious fantasy.

Fantasies are mental representations of events, or more usually sequences of events; these mental representations are formed and transformed by certain kinds of mental operations. Psychoanalysis focuses on the causal efficacy of unconscious fantasies as mental dispositions. The manifestations of such a mental disposition (occurrent mental states, for example) may be evoked by an external stimulus, or by the subject's own willful mental activity, which brings the manifestation about or stirs it up. Such a mental disposition may also manifest itself spontaneously and periodically, because to manifest itself in this way is a property of the disposition.

When I talk here of unconscious fantasies, I talk not only of fantasies about external objects and their doings, but of fantasies about internal objects and their doings. Internal objects are imagined to inhabit, to be located inside, the mind. I talk then about imagination when it is dominated—this is one of the special conditions—by a conception of the mind as spatial or even as corporeal, and about all of the consequences of the domination of imagination by such a conception of the mind.

Psychoanalysis is distinctive in according causal priority to unconscious fantasies. Psychoanalysis is also distinctive in according causal priority to the links among the contents of various mental states forged by such symbolic operations as condensation, displacement, translation into imagery, and iconic or metaphoric symbolization. (Psychoanalysis has been called a science of tropes.) This distinguishes it from cognitive science, which accords causal priority to the links among the contents of various mental states created by truth-conserving symbolic operations, as these operations transform mental representations in processes of cognitive inference.

IX

What is distinctive about the questions psychoanalysis raises about its domain as just defined? Psychoanalysis raises questions about the *contents* of mental states—the objects, states of affairs, or events symbolized by a mental representation. It especially raises questions about *causal gaps.*

That is, psychoanalysis raises questions about the inexplicability of the contents of particular mental states. These contents are inexplicable, sometimes because they do not seem to have any causal links to the contents of other mental states to which a subject has access (for example, they do not seem to be caused by any wishes and beliefs the subject is aware of); sometimes because the objects, states of affairs, or events in external reality, which are their putative cause, or which would be the right kind of cause of them, do not seem to exist or ever to have existed (for example, dreams, hallucinations, nonveridical memories); and sometimes because they do not seem to produce the actions one might expect, given the nature of the contents and the particular mental state to which they belong (for example, a subject is unable to form the right kind of intention to act, or the intention, once formed, does not result in action). In some cases, it is the absence of certain contents when these contents might be expected to be present, given a particular causal constellation, that calls for explanation (for example, a subject whose memory, as a psychological capacity, is unimpaired tries but is unable to remember certain contents).

Neurotic symptoms, dreams, and parapraxes are paradigmatic of such causal gaps. (Although I tentatively add jokes to this list in various places in this book, my guess is that Freud's primary interest in jokes was in the opportunity they afforded him to demonstrate the existence and nature of certain causal processes and mechanisms or mental operations, as these were active in the "joke work," and so to justify his postulation of such causal processes and mechanisms in explaining neurotic symptoms, dreams, and parapraxes.)

Neurotic symptoms, dreams, and parapraxes are the phenomena to which the core theory of psychoanalysis must apply, if it is to apply to anything, and must be able to explain, if it is to be able to explain anything. But the core theory of psychoanalysis does not take as a focus of inquiry, nor is it its task to explain the nature of, psychological capacities such as cognition, perception, memory, and linguistic competence. In general, psychological capacities that it is *not* the business of psychoanalysis to explain have one or both of the following characteristics.

First, at least one crucial component of the capacity (and possibly more than one) is not and cannot be influenced by wishes and beliefs.

Second, an *external* actual object, state of affairs, or event is a *necessary cause* of a mental state in which such a capacity (e.g., memory) is realized. (The capacity is realized in relating a person and that person's mental representation; "remembers" is such a relation, as in "this person *remembers* such-and-such." A property of the mental representation is that it represents itself as having been caused by an external actual object, state of affairs, or event.)

I do not mean to say, of course, that psychoanalysis does not or cannot make use of knowledge of such psychological capacities, only that it is unlikely to generate such knowledge. I would expect psychoanalysis to contribute something new to our knowledge of such capacities as symbolization and imagination. However, I would not expect it to generate a complete account of even these psychological capacities.

X

The core theory of psychoanalysis explains neurotic symptoms, inexplicable dreams (not all dreams are inexplicable), and inexplicable parapraxes (not all parapraxes are inexplicable) by postulating mental dispositions such as unconscious sexual wishes, which produce occurrent mental states—those whose contents are to be explained. These mental dispositions also produce other mental dispositions, rather than reality-adapted action. These other mental dispositions (unconscious fantasies, for example) have the power in turn, by virtue of the causal mechanisms or processes they evoke, to produce occurrent mental states—those whose contents are to be explained. Such contents may be explained, for example, as more or less disguised sexual wish-fulfillments.

I insist not only upon the causal role of unconscious mental contents and processes, but especially upon those having to do with sexual wishes and counterwishes, and especially sexual wishes and counterwishes arising in childhood. I consider the psychoanalytic theory of sexuality to be in danger of dilution and displacement to the periphery by current preoccupation with "the self," "identity," "object-relations," "interpersonal interactions," "the importance of the mother-infant relation and the pre-oedipal experiences of the very young infant," and "aggression." I want to restore it to its central position in the body of psychoanalytic knowledge.

An unconscious fantasy is not ordinarily a realization of pure and simple wish-fulfillment. Its scenario is complicated by the beliefs and anticipations attributed in imagination to the hero or heroine of the fantasy, whose wishes are being fulfilled. Its scenario is also complicated by the attribution in imagination to other actors in the fantasy of unfavorable responses to the hero or heroine. These other actors are written into the screenplay because they are necessary for the hero or heroine's wish-fulfillment.

The beliefs and anticipations of the hero and heroine, and the unfavorable responses of other actors, are likely to have the result that the hero or heroine of the fantasy experiences anxiety, shame, or guilt as well as excitement or pleasure. So manifestations of the fantasy are likely to include the experience of anxiety, shame, and guilt, as well as excitement and pleasure. The problem for the writer of the screenplay is to achieve wish-fulfillment under conditions of threat, to create a compromise in which some lessening or alteration of wish-fulfillment pays for some mitigation of anxiety, shame, or guilt.

The neurotic symptom, dream, or parapraxis the core theory of psychoanalysis seeks to explain is often explained by it as a solution of, or a more or less successful attempt to solve, such a problem. The particular results of any such problem solving depend upon the use of certain kinds of mental (symbolic) operations—designated variously as primary process, dream work, defense mechanism—as these form and transform mental (symbolic) representations. Such operations are the causal processes or mechanisms that link the cause (the unconscious fantasy) to its manifestations (inexplicable occurrent mental states, inexplicable enactments).

The explanatory power of psychoanalysis, and I expect its therapeutic action as well, resides pretty much in its steady attention to, and explication of, these causal processes and mechanisms—how the mind "works" under these circumstances.

I begin in this book to use a cinematic model of the mind to explain the psychic force or causal efficacy of such mental dispositions as unconscious fantasies. Further explication and development of the implications of this model are future work.

XI

Part 2 of this book focuses on the case study method and its use by psychoanalysis.

The characteristic mode of inquiry of psychoanalysis includes free association and the case study method.

Free association is a method for obtaining relevant data that using other means are hard to come by. From such data, the psychoanalyst makes clinical causal inferences and formulates causal explanations. The psychoanalyst also uses such data as evidence in arguing that belief in his clinical causal inferences and causal explanations is warranted. How good is this evidence? In Part 2, I shall suggest that the evidence is better than it has sometimes been thought to be.

The case study method is used to argue that, given the data obtained in the psychoanalytic situation, the clinical causal inferences and causal explanations of psychoanalysis are scientifically credible. How good are these arguments? In Part 2, I propose canons for the case study method in psychoanalysis that, if adopted by psychoanalytic investigators, would make such arguments better than they have sometimes been. I also suggest some kinds of arguments that might be especially convincing but have not been much used in the case studies of psychoanalysis.

XII

A shift in my thinking about science, scientific method, and scientific explanation is evident in Part 2. Increasingly, I favor:

1. a *generative or powers account of causality* over a successionist or covariation account of causality;

2. identification of *causes that are enough* rather than necessary or sufficient to bring about an effect;

3. a view of theory that foregrounds hypotheses about the existence and nature of *theoretical (nonobservable) entities* rather than empirical generalizations about relations among variables;

4. a view of explanation as making assertions about the existence and nature of *actual causal processes and causal mechanisms* rather than subsuming the less general under the more general "covering" law; and

5. the use of *analogy, consilience* (convergence of different evidence upon a conclusion), and *inference to the most likely cause* in

arguing for the scientific credibility of clinical causal inferences and causal explanations in psychoanalysis, especially when experiment is not possible.

The importance of analogies or models—in understanding the relation between psychoanalysis and other bodies of knowledge or disciplines, and in arguing for the existence and nature of theoretical entities, causal processes, and causal mechanisms—forms a theme that runs through both Parts 1 and 2.

XIII

Why do I encourage use of the case study method in Part 2 rather than enjoin psychoanalysis to turn to other modes of inquiry?

First, because it seems to me that the case study method seems especially appropriate for psychoanalysis, given that psychoanalysis uses a generative or powers account of causality, and that it is interested in explanations involving the postulation of theoretical (nonobservable) entities and the invocation of causal processes and mechanisms.

Second, because I am not persuaded by the various kinds of denigration of the case study method.

A conception of science as a search for "proof," for dogmatic certainty, will result in denigration of the case study method. Freud may have been substantively wrong in some of his case studies. But science always assumes that work of scientific value yields tentative, provisional conclusions that are scientifically credible at a given time against the background of available accepted knowledge. Later, as new things are discovered to be true, previous conclusions will be rejected in toto or revised. When I discuss Freud's case studies in this book, in explicating the methodology of the case study, I do not go into the things found out later about the Rat Man or Wolf Man. For I take it for granted that the inferences drawn from a case study, like the results of any scientific mode of inquiry (including highly controlled experiments), are vulnerable to revision with new knowledge. We all build on a swamp. Does that mean we should give up and sink?

The mistaken assumptions that tend to creep into case studies also lead to denigration of the case study method. Typical examples of such assumptions: that co-occurrence or correlation implies causality; that enumerating even large numbers of positive instances of a hypothesis makes it scientifically credible. That Freud fell into such errors on occasion does not mean the case study method must inevitably be flawed.

Dogmatic uncertainty—complaints that we can never know everything, that we cannot reproduce the richness of experience in our case studies, that whatever one tries to communicate is just too complicated to communicate—is acknowledgment of difficulty at best, and misconception of the objectives of science or a refusal to accept limits at worst. But it can be, for those who wish to use the case study method in a quest for knowledge, simply demoralizing.

That psychoanalysts disagree in the inferences they reach from the same clinical data has been thought to discredit the case study method. But psychoanalysts may disagree for reasons that are not intrinsic to this method. They may disagree because they have not specified rigorously enough what is to be explained; they have not agreed upon a particular question that all, making use of the same data, will then try to answer. They may disagree because they do not adequately distinguish what is fact from what is inference. They may disagree because of the over-general or over-abstract level at which they carry on their discourse; concepts are used like counters, moved around, without much regard for their relation to specific phenomena. They may disagree because they are looking primarily for confirmations, each of his own conjecture or hypothesis, rather than joining together in a search for data that will decide among conjectures or hypotheses, eliminating some, and favoring some over others. But these are problems in the sociology of psychoanalytic discourse, not problems in the case study method.

XIV

In Part 2, I argue that the case study method is not simply a heuristic for generating conjectures or hypotheses, but is capable of testing conjectures or hypotheses, of making a difference in deciding whether or not they are scientifically credible.

I am not ready to adopt statistical inference in all cases as a superior mode of testing a conjecture or hypothesis—superior because it supposedly subjects the conjecture or hypothesis to *neutral* standards of judgment. I have come to doubt that there are neutral tests of conjectures or hypotheses, which do not presuppose anything about the world—about the subject matter addressed in the conjecture or hypothesis. For example, much statistical inference presupposes that the domain to which the conjecture or hypothesis applies has the same structure as mathematics.

The very notion of what is meant by a "test" needs to be examined. Testing is not equivalent to using statistical inference or experimenting. What testing requires is available in an exemplary use of the case study method. Testing requires that the situation of investigation has encouraged and permitted data to decide among rival conjectures or hypotheses. It also requires that when data are favorable to some conjecture or hypothesis, there is some way to control, to estimate the influence of, or to eliminate extraneous factors that might be held responsible for obtaining such favorable data—or, at least, there is some way to render implausible the supposition that these particular extraneous factors did so influence the outcome of the investigation.

My interest in Part 2 of this book is to explicate how a case study should be carried out and written, if the objective of the case study is to make a contribution to scientific knowledge. I have no interest in either abandoning, or demanding exclusive allegiance to, the case study method. I simply want to know how to use it better than in most cases it has been used.

XV

I argue in Part 2 that the case study method is capable of providing evidential support for the credibility of psychoanalytic clinical causal inferences and causal explanations, and that as a method it has some advantages for psychoanalysis.

For example, this method does provide—as other methods do not—information needed to explain what psychoanalysis intends to explain.

It is possible in a case study that different kinds of information will converge upon the same conclusion.

It is possible that the information obtained in a case study favors one psychoanalytic clinical causal inference or causal explanation over a rival psychoanalytic inference or explanation.

It is possible in a case study to argue effectively that extraneous factors, such as suggestion, that might have been responsible for the investigator's obtaining evidence favoring his conjectures or hypotheses have been, if not eliminated, rendered implausible.

I do not argue that psychoanalysis should limit itself to the case study method. Whether or not it should is a false issue. Every method for testing conjectures or hypotheses has advantages or disadvantages. In choosing a method, one recognizes dilemmas. There are always trade-offs, maximizing something desired (explicating complicated causal processes and mechanisms, for example) at the expense of something else desired (formulating empirical generalizations about relations between precisely measured variables, for example), when it is not possible to maximize both desiderata. Similarly, one may decide that making it possible to generalize to other subjects or settings conclusions reached in the situation of investigation shall take precedence over controlling as many extraneous factors in the situation of investigation as possible. There is no such thing as an ideal method. Any investigator does well to use more than one, and see if the outcomes converge upon the same conclusion.

XVI

But there still remains the question of whether, or how, one can generalize the findings achieved by the case study method. Since I do not say much about this problem in Part 2, and since doubts about generalizability may indeed stand in the way of the reader's interest in the method, I shall make a few comments about this question here.

The case study has been depicted as only good for dealing with singular conjectures or hypotheses. I think that the importance of singular conjectures or hypotheses tends to be underestimated. We may regard that a single person or mind is a domain of a certain kind we wish to explore. To demonstrate that some conjecture or hypothesis about a particular person or mind is scientifically credible is not a small achievement. And having found that the conjecture or hypothesis holds in one domain, we face a new task.

The task—not an unfamiliar one in scientific work—is to extend systematically this same conceptual framework to other domains (other persons, other minds). We show that the conjecture or hypothesis holds in those domains as well, or we discover in what way the conjecture or hypothesis must be modified to hold in more than one such domain, or we find that the conjecture or hypothesis turns out to hold only in the one domain with which we began.

The belief that no knowledge of a single subject can be legitimately generalized to other subjects is simply fallacious.

What is the essential meaning of "generalization?" It should not be confused with a specific method, such as using statistical reasoning to justify inferences from a sample to the population from which it was drawn.

"Generalization" has to do with the scope of a conjecture or hypothesis. The kind of question raised is: "Of how many entities in a domain, in how many settings, or on how many occasions, or in how many time-slices is this true?" Another kind of question, not often distinguished from the first, is: "Of how many similar and of how many apparently very different domains is this true?" Scientific work may legitimately concern itself primarily at times with expanding or limiting the scope of a conjecture or hypothesis.

Generalization may occur within a single case as well as across cases. We may want to know, for example, over how many entities in the domain of a single mind—how many mental states and how many kinds of mental states, and in how many time-slices—does a generalization about a single mind hold.

What kinds of arguments are available to justify generalization from one person or mind to others? Here are a few.

Other persons or minds may be similar in crucial ways, leading to an argument for generalization by analogy. (Such an argument is not disreputable; it is used when animals are subjects in experiments.)

"If this were true only of this person or mind, that would be a rare event, and a rare event is unlikely" is an argument by probability.

Direct replication of a case study leading to the same results tends to support the credibility of these results. But, also, systematic replication across cases when crucial characteristics are varied one at a time is not beyond the capability of psychoanalysis. The results of such systematic replication suggest what the scope of a conjecture or hypothesis is.

XVII

Of the fifteen chapters in this book, ten (Chapters 1–7 and 11–13) make use of previously published material. Of these ten chapters, all but two (Chapters 5 and 11) have been extensively revised. Chapters 8, 9, 10, 14, and 15 have not been previously published.

Chapters 1–4 issue from the same impulse that gave rise to my books *The Idea of a Mental Illness* and *Language and Interpretation in Psychoanalysis.* These chapters are all based on papers published in the 1970s, but I have been careful in revising them to bring them into accord with, or at least relate them to, the thinking I have done in the rest of this book, and to excise from them what are abandoned projects (e.g., the formalization of psychoanalytic theory) or projects that do not fit well into this book (e.g., explication of the ideas of Noam Chomsky and Talcott Parsons).

The remaining eleven chapters were written after the publication of my 1984 book *Hypothesis and Evidence in Psychoanalysis,* and may be considered sequel to that book. They continue and develop lines of thought stated there, and in some cases diverge from positions taken there.

Chapter 1 is a commentary on Freud's distinction between psychic reality, on the one hand, and external, material, practical, or historical reality, on the other.

I may seem to be inconsistent in emphasizing in this chapter that external reality is symbolically constructed, and adopting a naive realism about external reality in Part 2. However, what I point out in

Chapter 1 is that we know there is an external reality when we realize from our experience that there is something that does not accommodate itself to our wishes. But, granting its existence, how we conceive of it is influenced by the symbolizing activity of the mind.

I mention ideas of Talcott Parsons and quote lines from the poetry of Wallace Stevens in Chapter 1. The mentioning may seem to the reader digressive and the quotations merely a rhetorical flourish.

Talcott Parsons interpreted a *systems-subsystems* model in sociology. His work with the model is so intellectually rigorous that I believe it to be instructive for psychoanalysis, because the same model is an important part of the conceptual foundation of psychoanalysis. It underlies the psychoanalytic conception of the structure of mind (the structural theory). Furthermore, Parsons's thinking about the model is important in clarifying certain conceptual issues arising from Freud's distinction between psychic reality and external reality. Therefore, although it was no part of the project of this book to explicate Parsons's ideas, or to show their connection to psychoanalytic ideas, the occasional mention of Parsons is appropriate and deserved. The reader who wants such an explication or who wants to explore the connection more thoroughly may consult my original paper.

Despite Wallace Stevens's personal reservations about psychoanalysis, his poetry is a prolonged and profound meditation on the relation between psychic reality and external reality. The quotations, as those who know that poetry will recognize, belong in the chapter.

The conclusion of Chapter 1 is that psychoanalysis is a science of the symbolizing activity of the mind, a definition foregrounded in Chapters 1–4 but a subtext of other chapters as well, and finding full expression in Chapters 9 and 10.

In Chapter 1, I begin to explore a *mental representations and symbolic operations* model of the mind. That exploration continues in Chapter 2, in relation to the work of Noam Chomsky, who made use of the model in linguistics. Again, his work with the model is so intellectually rigorous that I believe it to be instructive to psychoanalysis, because the same model is an important part of the conceptual foundation of psychoanalysis. There, its content includes primary process, dream work, and defensive mechanisms, all of which can be conceived as kinds of mental operations upon mental representations. Again, I do not in this book explicate the ideas of Chomsky, or trace in any systematic way their connection to ideas in psychoanalysis. The reader who wants that explication or who wants to explore that connection may consult my original paper.

Chapter 3 begins my examination of the relation of psychoanalysis to other disciplines. In this chapter, my focus is on the use by psychoanalysis of other scientific bodies of knowledge, and the use by other scientific disciplines of psychoanalytic knowledge.

This theme continues in my discussion of applied psychoanalysis in Chapter 8 where I am, however, more involved with the relations between psychoanalysis and humanistic disciplines. In Chapter 8, I present a set of canons for work in applied psychoanalysis, which I fear some readers may find over-severe.

In Chapter 3, I also examine some conceptual relations between psychoanalytic theories and theories in other scientific disciplines. Discussions of the role of homologies, analogies, and models in analyzing the conceptual relations between sciences—also, in formulating the conceptual foundations of psychoanalysis or in developing and testing causal explanations—occur in Chapters 1, 2, 3, 6, 9, 13, and 15.

The examination of such conceptual relations continues in Chapter 7 where, in arguing against the possibility of the reduction of psychoanalysis by neuroscience, I carry out the most exhaustive delineation in this book of the differences between not only psychoanalysis and neuroscience but between psychoanalysis and other branches of psychology, including other branches of psychology of mind.

The relation between psychoanalysis and other disciplines comes up again in Chapter 5 where, in the course of a discussion of the psychoanalytic theory of anxiety and the anxiety disorders, I consider problems arising from the use of animal experimentation to test psychoanalytic ideas, and the misdepiction of the psychoanalytic concept of anxiety that is rife in such work.

In Chapters 9 and 10, the reader will find my fullest treatment in the book of the role of imagination and fantasy in psychoanalysis (imagination and fantasy is already a major theme in Chapter 1). In Chapter 9, I consider especially the implications of this role for the psychoanalytic theory of defense. In Chapter 9, I also consider problems in the use of quantitative/experimental methods of research to test such psychoanalytic ideas, as well as, given such ideas, the misdepiction in such research of psychoanalytic theory, especially by learning theorists and cognitive scientists but also by some psychoanalytic investigators.

It is in Chapter 9 that I begin to develop a *cinematic* model of the mind.

Chapter 4 discusses psychoanalytic interpretation from the point of view of the symbolizing activity of the mind, and in the course of this discussion makes use of some logical-linguistic ideas.

In Chapter 6, in order to avoid redundancy, I draw together ideas about psychoanalytic theory as an intentional psychology and as a representational and computational theory of mind, about the domain of psychoanalysis, and about the distinctive questions it raises about that domain and the distinctive answers it gives to those questions. These ideas are part of a sequence of thought in Chapters 7 and 13, and I have indicated in those chapters where the reader may want to refer to Chapter 6.

It is in Chapters 6 and 7 that I express most explicitly my rejection of the conception of psychoanalysis as a general psychology.

In Chapter 10, I discuss the specific contributions psychoanalysis makes to a theory of sexuality. I also state here five major substantive nomothetic themes in psychoanalytic theory. In these themes, unconscious sexual fantasy plays the primary causal role. Psychoanalytic explanations make use of these themes and it is a major part of the research program of psychoanalysis to provide evidence for such explanations.

In Chapter 11, I point out the pernicious effects of using appeals to subjectivity, meaning, and uniqueness to justify abandonment of ordinary canons of scientific work by those who do case studies, and indicate in what ways a case study can be a scientific argument. The shift in my thinking about science to which I have alluded in Section XII above begins here, continues in Chapter 13 and reaches its fullest expression in Chapters 10 and especially 15.

In Chapter 12, I respond to Adolf Grunbaum's critique of psychoanalysis. In Chapter 14, I reflect upon the various responses his critique has evoked in others.

In Chapter 13, I discuss the way in which a conception of the nature of causal explanation may influence our reading of the Wolf Man case and enhance our understanding of what Freud was attempting to do in this case study and others as well. In this chapter, I state canons for the case study method in psychoanalysis, both for the case study in which the goal is to argue that an empirical generalization is scientifically credible, and for the case study in which the goal is to argue that a complicated causal explanation, a causal story, is scientifically credible.

In Chapter 15, I indicate how a folk or commonsense psychology (an intentional psychology) functions as a model of the mind in psychoanalysis.

In this final chapter, I have come much farther than ever before in presenting what I believe are *rational grounds for believing clinical causal inferences in psychoanalysis.* It contains my latest thinking about the nature of science and about the kind of scientific work, the mode of inquiry, and the kinds of scientific argument that best suit the objectives and subject matter of psychoanalysis.

PART 1
The Core Theory of Psychoanalysis

Psychic Reality[1]

I

Psychic Reality and External Reality

Freud's discovery of psychic reality is described reluctantly in relatively few passages throughout his writings; yet it is the foundation of all his major achievements. Rarely has any discovery been made so contrary to the intentions and predilections of its discoverer.

It was in the *Project for a Scientific Psychology*, in the midst of his earliest efforts to ground explanation in psychology in an external, empirically observable neurophysiology, that Freud indicated he had already made the distinction between thought-reality and external reality—notably, in a sentence referring to symbolic activity: "Indications of discharge through speech are also in a sense indications of reality—but of thought-reality not of external reality" (1895, p. 373).

Subsequently, Freud wrote to Fliess "the great secret" that he no longer believed in the traumatic theory of neuroses. The expected "feeling of shame" was mitigated by the "feeling of a victory" in the recognition that "honest and forcible intellectual work" had led him not only to criticize his previous assumptions about the role of experience in the etiology of the neuroses but to realize, however tentatively, that the world of fantasies has a force, a causal efficacy, akin to that of external reality (1892–1899, pp. 259–260).

In *Totem and Taboo*, he wrote confidently: "What lie behind the sense of guilt of neurotics are always *psychical* realities and never *factual* ones." He added, however, in a tone that suggests a depreciation of psychical

1. This chapter makes use of material, here extensively revised, that appeared in Edelson (1970c).

reality: "What characterizes neurotics is that they prefer psychical to factual reality and react just as seriously to thoughts as normal people do to realities" (1913, p. 159). By 1919, he had added and revised the following sentence in *The Interpretation of Dreams:* "If we look at unconscious wishes reduced to their most fundamental and truest shape, we shall have to conclude, no doubt, that *psychical* reality is a particular form of existence not to be confused with *material* reality" (1900, p. 620).

It was not until *On the History of the Psychoanalytic Movement* that Freud finally made explicit the despair and recovery that accompanied his discovery of psychical reality.

> The firm ground of reality was gone. At that time I would gladly have given up the whole work, just as my esteemed predecessor, Breuer, had done when he made his unwelcome discovery. . . . At last came the reflection that, after all, one had no right to despair because one has been deceived in one's expectations; one must revise those expectations. If hysterical subjects trace back their symptoms to traumas that are fictitious, then the new fact which emerges is precisely that they create such scenes in *phantasy,* and this psychical reality requires to be taken into account alongside practical reality. (1914a, pp. 17–18)

Freud's despair and even antipathy were not simply a rejection of the sexual content of psychical reality (the seduction by parents of children), although it may be noted in that connection that he deferred until late in his life the revelation that not only the father but also the mother was the parent in these fantasies (1931). His anguish is that of the utilitarian rationalist who, wishing the cause of psychopathology to be "out there," is confronted by the obdurately nonrational and subjective. In his account in *The Introductory Lectures on Psychoanalysis,* he wrote somewhat plaintively:

> If the infantile experiences brought to light by analysis were invariably real, we should feel that we were standing on firm ground; if they were regularly falsified and revealed as inventions, as phantasies of the patient, we should be obliged to abandon this shaky ground and look for salvation elsewhere. But neither of these things is the case: the position can be shown to be that the childhood experiences constructed or remembered in analysis are sometimes indisputably false and sometimes equally certainly correct, and in most cases are compounded of truth and falsehood. Sometimes . . . symptoms represent phantasies of the patient's *which are not, of course, suited to playing an aetiological role.* (1916–1917, p. 367; italics mine)

In a vivid passage, he went on to depict the dilemmas of thought and technique that the distinction between physical reality and external reality continued to present. The psychoanalyst is perplexed by "the low valuation of reality, the neglect of the distinction between it and phantasy," and is "tempted to feel offended at the patient's having taken up . . . time with invented stories." The psychoanalyst is also in doubt, when the patient "brings up the material which leads from behind his symptoms to the wishful situations modelled on his infantile experiences," whether these experiences are reality or fantasy. When the psychoanalyst is "enabled by certain indications to come to a decision" and is "faced by the task of conveying it to the patient," difficulties arise. If the patient is told "that he is now engaged in bringing to light the phantasies with which he has disguised the history of his childhood . . ., his interest in pursuing the subject further suddenly diminishes in an undesirable fashion." He agrees with the psychoanalyst in wanting reality. He "despises everything that is merely 'imaginary.'" However, if the patient is left in the belief that he and the psychoanalyst "are occupied in investigating the real events of his childhood," he may eventually accuse the psychoanalyst of having made a mistake and laugh at him for being credulous.

> It will be a long time before he can take in our proposal that we should equate phantasy and reality and not bother to begin with whether the childhood experiences under examination are the one or the other. Yet this is clearly the only correct attitude to adopt towards these mental productions. They too possess a reality of a sort. It remains a fact that the patient has created these phantasies for himself and this fact is of scarcely less importance than if he had really experienced what the phantasies contain. The phantasies possess *psychical* as contrasted with *material* reality, and we gradually learn to understand that *in the world of the neuroses it is psychical reality which is the decisive kind.* (1916–1917, pp. 367–368)

Freud's attempted resolution of his ambivalence toward his own conceptual offspring is further adumbrated in a passage in which he described himself as compelled to enter into "the origin and significance of the mental activity which is described as 'phantasy' [or 'imagination']." It is through fantasy that "human beings continue to enjoy the freedom from external compulsion which they have long since renounced in reality." They "cannot subsist on the scanty satisfaction which they can extort from reality," but contrive "to alternate between remaining an animal of pleasure and being once more a creature of reason." The mental realm of fantasy is like the "reservation," which preserves nature in "its original state which everywhere else has to our regret been sacrificed to necessity."

> Everything, including what is useless and even what is noxious, can grow and proliferate there as it pleases. The mental realm of phantasy is just such a reservation withdrawn from the reality principle. (1916–1917, p. 372)

In his last works, Freud still manifested traces of his complex attitude toward psychical reality. In *Moses and Monotheism,* he attributed the "compulsive quality" of such phenomena as symptoms, ego-restrictions, and stable character-changes to their "great psychical intensity" and their "far-reaching independence of the organization of the other mental processes, which are adjusted to the demands of the real external world and obey the laws of logical thinking." Such phenomena were said to constitute "a State within a State, an inaccessible party, with which co-operation is impossible, but which may succeed in overcoming what is known as the normal party and forcing it into its service." He warned that "the domination by an internal psychical reality over the reality of the external world" opens the path to psychosis (1939, p. 76).

At the same time, he sought to redeem psychical reality by ascribing some truth-value to its manifestations, which were seen as not merely distorting but conserving traces of some past reality. Having made, in a number of his writings, a distinction between material truth and historical truth in reference both to religious belief and psychotic delusions, he declared in *Constructions in Analysis:* ". . . there is not only *method* in madness, as the poet has already perceived, but also a fragment of *historical truth"* (1937, p. 267). In a final *Postscript to an Autobiographical Study,* he recanted *The Future of an Illusion's* "essentially negative valuation of religion." "Later I found a formula which did better justice to it; while granting that its power lies in the truth which it contains, I showed that the truth was not a material but a historical truth" (1925, p. 72).

External Reality

The distinction between psychic reality and external reality is at the center of Freud's inveterate dualism. This distinction is paralleled by the proposal of sociologist Talcott Parsons that the theory of action be built upon the distinction between actor and object. Similarly, Jean Piaget seeks to understand symbolic functioning in terms of the balance between processes of assimilation of reality to internal schema and processes of accommodation to external reality.

The poet Wallace Stevens represents the same view in a symbolic form different from that used by scientist or philosopher. His work is a "variations on a theme" of surrender to the pressure of the world of things-as-they-are and the imagination's sometimes triumphant, sometimes impotent struggle to resist mere surrender by creating a world of "necessary fictions."[2] "From this the poem springs: that we live in a place / That is not our own and, much more, not ourselves / And hard it is in spite of blazoned days" (Stevens 1961, p. 383; poem titled "Notes toward a Supreme Fiction").

We have seen that Freud had trouble with "psychic reality." But judging from the variety of adjectives preceding "reality"—external, factual, material, practical—we may conclude that the conceptual status of "external reality" offered as much difficulty. Freud avoided philosophical questions as much as possible in his work in the interest of creating an empirical science, but here an ontological specter seems impossible to evade. In his distinction between a material and a psychical reality, was Freud merely contrasting a naively accepted "real" reality with an unreal, spiritual, psychical reality? There seems to be good reason to think not.

He had certainly thought about such questions. That he knew and admired the work of Kant and was aware that our knowledge of external reality was shaped by the character of our minds is evident from Jones's biography (Jones 1953–1957, 1:367, 2:415, 3:466), and from the passages in which Freud refers to Kant, especially the following (1915b, p. 171):

> The psycho-analytic assumption of unconscious mental activity appears to us, on the one hand, as a further expansion of the primitive animism which caused us to see copies of our own consciousness all around us, and on the other hand, as an extension of the corrections undertaken by Kant of our views on external perception. Just as Kant warned us not to overlook the fact that our perceptions are subjectively conditioned and must not be identical with what is perceived though unknowable, so psycho-analysis warns us not to equate perceptions by means of consciousness with the unconscious mental processes which are their object. Like the physical, the psychical is not necessarily in reality what it appears to us to be.

2. Imagination as a capacity and imagining something as a mental state, wish-fulfillment as effective imagining, and fantasy become increasingly important in the repeated attempts made throughout these chapters to define the core theory of psychoanalysis. See especially Chapters 9 and 10.

He concludes, perhaps somewhat surprisingly: "We shall be glad to learn, however, that the correction of internal perception will turn out not to offer such great difficulties as the correction of external perception—that internal objects are less unknowable than the external world."

About his own speculations, Freud could declare as emphatically as the most radical of empiricists, ". . . these ideas are not the foundation of science, upon which everything rests: that foundation is observation alone" (1914b, p. 77). However, it is clear from the passage with which he begins *Instincts and Their Vicissitudes* that Freud was not a naive empiricist. He did not suppose the scientist simply extracts his ideas from a close observation of phenomena, although, to be sure, the "true beginning of scientific activity consists . . . in describing phenomena and then in proceeding to group, classify and correlate them." He was aware that theory or ideas about reality and phenomena as experienced are interdependent, that observations are always made in a conceptual frame of reference, that indeed ideas determine what is perceived as "real." Ideas are conventions that fit, that order experience.

> Even at the stage of description it is not possible to avoid applying certain abstract ideas to the material at hand, ideas derived from somewhere or other but certainly not from the new observations alone. Such ideas . . . must at first necessarily possess some degree of indefiniteness; there can be no question of any clear delimitation of their content. So long as they remain in this condition, we come to an understanding about their meaning by making repeated references to the material of observation from which they appear to have been derived, but upon which, in fact, they have been imposed. Thus, strictly speaking, they are in the nature of conventions—although everything depends on their not being arbitrarily chosen but determined by their having significant relations to the empirical material, relations that we seem to sense before we can clearly recognize and demonstrate them. (1915a, p. 117)

Freud wrote that the "human ego is . . . slowly educated by the pressure of external necessity to appreciate reality and obey the reality principle" (1916–1917, p. 371). Piaget wrote that for adapted thought there must be accommodation to reality. In both cases "reality" refers to an *imaginative* conception: of a something "out there" that exists independently of the activity of the mind that knows it, a something that is indifferent to the wishes, desires, intentions, anticipations, attitudes, value-preferences, and interests constituting such activity. Stevens remarks that even "the absence of the imagination [has] / Itself to be imagined" (Stevens 1961, pp. 502–503; poem titled "The Plain Sense of Things").

"External reality" or "material reality" represents an imaginative conception of the chaotic and meaningless, of the experience of a resistance to man's rage for order—a resistance located "out there."[3] It is the pressure of that unknowable reality that, so man imagines, threatens to defeat him, oppresses him, and, finally, should the activity of his mind fail, overwhelms him.

External reality is an irresistible inference, but it is not directly known, even by the heroic effort of abstinence from anthropomorphizing exerted by Stevens's snow man, who is "not to think / Of any misery in the sound of the wind," but who "listens in the snow, / And, nothing himself, beholds / Nothing that is not there and the nothing that is" (Stevens 1961, pp. 9–10; "The Snow Man"). We construct our conception of external reality by the activity of our minds. We know of Stevens's singer "that there never was a world for her / Except the one she sang and, singing, made" (Stevens 1961, pp. 129–130; "The Idea of Order at Key West").

The Conceptual Foundation of Psychoanalysis

Three Distinctions

If external reality is constructed, as is psychic reality, by an activity of the mind, then the distinction between external reality and psychic reality is actually three distinctions.

One distinction is between the object world and the subject's own aims and activities.

The second distinction is between kinds of functions that mental representations are designed to serve. Mental representations are symbolic representations of conceptions of outer or inner objects or states of affairs. These symbolic representations are constituents of mental states. Some mental representations are constructed in such a way that considerations of veridicality, truth, and adaptation are given primacy. Constructing these mental representations, in other words, is regulated by the reality principle. Other mental representations are constructed in such a way that considerations of expressiveness and pleasure are given primacy. Constructing these mental representations, in other words, is regulated by the pleasure principle.

3. "Man," "he," "his," "him," and "himself," are used in their generic sense unless the context indicates otherwise in order to avoid the repeated use of such awkward phrases as "he or she."

Talk of "functions" is talk of what the mind is able to do. The various kinds of things a mind is able to do are conceived by psychoanalytic theory to be allocated among subsystems of the mind (id, ego, superego, ego-ideal). The distinction between external reality and psychic reality leads, then, to a conception of the structure of the mind—how subsystems of the mind differ and what relations between them are possible.

The third distinction is between the kinds of materials and the kinds of rules for combining these that are used in constructing mental representations, and between the kinds of transformational operations the mind carries out on mental representations. The distinction between external reality and psychic reality requires a conception of what psychoanalytic theory refers to as primary and secondary processes.

The three distinctions implicated in the distinction between external reality and psychic reality are at the very core of psychoanalytic theory, which is to say that these distinctions form an essential part of the conceptual foundation of psychoanalysis.

1. Object World and Instinctual Aim

The distinction between external reality and psychic reality is in part the distinction between the object world and the subject's instinctual aims. In the broadest terms, this is the distinction between the object-situation of the subject and the subject's own aims and activities. The subject's activity includes wishes, impulses, desires; intentions and intentional acts; choices or value-preferences; attitudes; interests. The states the subject seeks to attain or maintain may be desirable in and of themselves—without reference to their status as means to any other end. That is, there is no incentive to cease such activity; the subject repeats or maintains it.

Interference with the subject's activity instigates wish-fulfilling fantasies or symbolic evocations of a future in which the activity will actually occur and realize its ends. If it were possible for the subject's activity to continue uninterrupted and without interference, or if wish-fulfillment were sufficient, no conception of external reality could arise. As it is, the subject's activity comes also to be organized around orientations to, anticipations of, and reality-accommodated beliefs about possible future states of affairs.

In Freud's instinct theory, certain activities of the subject are conceived to be independent of the external object world; the ends sought are conceived to be ends in themselves without regard to what objects are used to achieve these ends. An instinctual aim is a kind of activity—such as sucking, touching, expelling, holding in, looking, penetrating—that affords sensual pleasure. Objects are interchangeable with respect to the gratification of an instinctual aim. Any one of a variety of objects may be a means to the realization of a particular instinctual aim. No object is committed a priori to the gratification of one and only one instinctual aim.

What is the relation between instinct and body? From the point of view of the mind, the subject's body is part of his situation. The subject's body, strange as it may seem to conceive it so, is from the point of view of the mind part of the external object world. Different subsystems of the mind have different conceptions of this object (the subject's body) and seek different kinds of relations to it.

The self, as the term is commonly used, includes all that has the feeling of "me," and all that is located inside rather than outside "me." The body may be felt to be part of the self. One's capacities and one's aims both may be felt to be part of the self.

However, it is useful to distinguish between conceptions of the self as situational object and conceptions of the self as the locus or source of the subject's activities or aims. The self conceived as situational object may be the locus of obstacles to the gratification of the subject's instinctual aims, the locus of means used to realize the subject's instinctual aims, or the locus of occasions for the gratification of the subject's instinctual aims. The self, including both body and mind, may be conceived as possessing or as failing to possess, by virtue of inheritance or past achievements, capacities for performance required if instinctual aims are to be gratified, or qualities that interfere with, facilitate, or afford opportunities for the gratification of the subject's instinctual aims.

The nature of the symbolic representation of the subject's aims, of the subject as subject or agent (not as situational object) is a theoretical problem, to which Hartmann alludes when he distinguishes between the cathexis of an object-representation (I am invested in whatever it is I am thinking about) and the cathexis of an object-directed ego function (I am invested in thinking itself). This allusion occurs, for example, in a discussion that begins with the distinction made by the child between his activity and the object toward which this activity is directed (H. Hartmann 1964, pp. 187–188).

Langer (1953, 1967) has concluded that art, or in general what she calls presentational (as distinct from discursive) symbols, is the symbolic form par excellence for the symbolic presentation (as distinct from representation) of feelings or motivational states or activity. I suggest that feelings themselves can be symbols: a subject's interiorized (in subjective experience) or exteriorized (in expressive action) symbols of his own conceptions of mental states in which he finds himself. Feelings symbolize in particular the subject's evaluative assessments (especially with regard to the gratification of instinctual aims) of mental states in which he finds himself.[4]

Clearly, it is fantasy that Freud holds to be the symbolic representation, which is a presentational symbol, of a subject's conceptions of his own instinctual aims or activities. This formulation is an amplification of Freud's view that an instinct is the "psychical representative" (I would say, a symbol or mental representation) "of the stimuli originating from within the organism and reaching the mind" (Freud 1915a, p. 122).

A subject is, however, not always able to symbolize *conceptions* of his own mental activity or states. Sometimes feelings are merely signs or indicators of such an activity or state.[5] An important distinction in psychoanalytic theory is that between anxiety or guilt as a signal—that is, as a symbol of a conception or evaluative assessment of a possible state of affairs—and anxiety or guilt as a reaction to an actual (not simply a possible) state of affairs, which overwhelms the subject. In the latter case, anxiety or guilt indicates, is a part of, is connected to an existent pressing state of affairs, for the symbolization of which the subject has inadequate resources (including time), and to which the subject merely reacts, essentially without mediation by conceptual activity.

Any object in the subject's situation (including his own body) viewed as an opportunity for instinctual gratification is an instinctual object. As such, it is potentially useful for the gratification of a variety of instinctual aims. Symbolic representations of conceptions of the significance of objects with respect to the gratification of instinctual aims are causally efficacious.

4. "Symbol" or "symbolic representation," in this book, refers to any entity that represents, designates, or stands for a conception of or idea about a possible or actual object or state of affairs. When I mean "symbol" in the narrow psychoanalytic usage (as in "freudian symbol"), I use the term "iconic symbol," unless context makes the meaning clear.

A symbolic representation, in this book, may be either presentational or discursive. I shall use the terms "presentational symbol" or "symbolic presentation" when I want to contrast discursive symbolic representations and presentational symbols.

Discursive symbols are exemplified by the language of rational discourse and by cognitively informed veridical perception. These symbols conform to the reality principle, are submitted to reality testing, and possess their efficacy by virtue of the meanings—for example, conceptions of situational exigencies—they represent.

Presentational symbols are typically used in expressive acts and art. The presentational symbol resembles or exemplifies what it symbolizes; it is quintessentially metaphoric.

A mental representation is an interiorized symbolic representation. It is also, when linguistic or governed by linguistic description, an interiorized semantic representation.

5. A sign or index, unlike a symbol, is part of, or existentially connected to, the actual object or state of affairs it announces or points to.

2. The System-Subsystems Model in Psychoanalysis

A *System-Subsystems Model* is an important part of the conceptual foundation of psychoanalysis. It is similar to the model that has also been used—albeit with different content—by Parsons in studying social systems. It underlies the psychoanalyst's conception of the structure of mind (id, ego, superego, ego-ideal).

Fantasy possesses psychic reality—and has effects—because it presents with compelling immediacy, through homologous morphology, a kind of meaning: the very qualities of the subjective activity and experience it symbolizes. As a semblance, an illusion, an apparition, it presents its meanings in the form of a virtual appearance like the virtual image in a mirror; a dream or a work of art presents its meanings.

Fantasy possesses symbolic rather than intrinsic efficacy. Its meanings are communicated in intrapsychic exchanges among subsystems of the mind. That is, one subsystem communicates a symbolic presentation of meaning to another subsystem, which responds to it by acts of its own. So, the ego and superego respond to symbolic presentations of the pleasurable and the ideal. Each kind of symbolic representation is generated by, and is a characteristic output of, some subsystem of the mind, and an input to some other subsystem of the mind, which works with it, responds to it, operates upon it.

As the realization grows that the distinction between external and psychic reality is a distinction between kinds of symbols (and kinds of symbolic processes), and that one cannot know external reality independently of any imaginative apprehension of it or valuation of it (external reality is not an antidote to the imaginary), the distinction between id and ego (and primary and secondary process) becomes less pejorative. There is a greater appreciation of the interdependence of these subsystems or processes—for example, in the work of Ernst Kris on creativity. Heinz Hartmann has repeatedly challenged the assumption that there is an intrinsic connection between fantasy and illness on the one hand and reality and health on the other. He has emphasized that rationality is not necessarily an indicator of health, and that rational action is not the only or at all times the best means to optimal adaptation (H. Hartmann 1958, 1964).

Freud's occasional intimation that illness is equivalent to subjugation to fantasy, to the nonrational, while health is recognition of brute reality should be rejected. Such a conclusion in effect equates health with primacy of one kind of symbol and denies any but dysfunctional effects to any other kind. The relation between fantasy and illness is a very complex one. Fantasies present conceptions of the desirable; such conceptions vary widely in content. The capacity to symbolize an internal world, and to live in that world, may have functional or dysfunctional effects. The world may be one in which its creator is at home or in which he suffers, in which he loves or must hate, in which he cares for or destroys others or is cared for or destroyed by them. The effects a fantasy has depend not only on its intrinsic content, but on how that content is received, evaluated, or treated by—on the significance it has for—various subsystems of the mind.

It follows from our understanding of what is involved in the distinction between external reality and psychic reality that it may be useful to think of ego, id, superego, and ego-ideal as subsystems of symbolic processes. Each such subsystem of the mind is dominated by a different conception of what kind of object and what kind of relation with that object is sought: what gives sensual pleasure; what, as a realization of an ideal, gives self-esteem; what, as a realization of the "right" pattern of relations with others, gives a feeling of goodness or community-belongingness; what, as an adaptive achievement, gives a sense of mastery of, or control over, means and resources, including knowledge. Each subsystem has its characteristic kind of symbolizations and its characteristic modes of symbolizing (e.g., primary process, secondary process). Each subsystem regulates the construction of and choice among symbolizations according to its own principles (e.g., pleasure principle, reality principle). Each subsystem requires resources (e.g., attention, effort) and must recognize and overcome obstacles (e.g., lack of access to a desired object) if it is to achieve its goals. Such subsystems of the mind may compete with each other for the allocation of these resources, conflict with each other in serving different ends, or contribute in various degrees to "multiply determined" compromise-formations (including such symbolic forms as symptoms and symptomatic enactments).

Evaluative standards provide bases for selection or rejection of symbolic representations. One value standard prescribes that the choice of a kind of symbolic representation of object or state of affairs should be made to maximize achievement of a subject's instrumental aims—adaptation to, or mastery of, a situation (means and conditions), according to a cognitive understanding of verifiable means-ends relations. A second value standard prescribes that such a choice should be made to maximize achievement of a subject's expressive aims—gratification, or attainment of ends desirable in and of themselves without regard to their status as means to other ends. A third value standard prescribes that such a choice should be made to maximize integration of a subject with others, who together are members of the same social collectivity. A fourth value standard prescribes that such a choice should be made to maximize a subject's self-esteem.

In the context of psychoanalytic theory, these value-preferences are likely to be formulated as regulatory principles of the mind, such as the reality principle and the pleasure principle. Such value-preferences are also implicit in references to the regulation of symbolizing processes through the effects of dysphoric or euphoric affects such as anxiety, pleasure, guilt, and shame.

3. The Mental Representations and Symbolic Operations Model

A *Transformational-Generative Model* or a *Mental Representations and Symbolic Operations Model* is an important part of the conceptual foundation of psychoanalysis. It is similar to the model that has also been used—albeit with different content—by Chomsky and Schenker in studying such symbolic systems as language and music. This is a model about symbolic operations, such as primary process, secondary process, dream work, defensive operations, that operate on mental representations. The transformations in mental representations wrought by symbolic operations are the subject of many of the psychoanalyst's acts of interpretation.

Langer, just before discussing the relation of Freud's dream work to a theory of art, wrote:

> The laws of combination, or "logic," of purely aesthetic forms—be they forms of visible space, audible time, living forces, or experience itself—are the fundamental laws of imagination. They were recognized long ago by poets, who praised them as the wisdom of the heart (much superior to that of the head), and by mystics who believed them to be the laws of "reality." But, like the laws of literal language, they are really just canons of symbolization; and the systematic study of them was first undertaken by Freud. (Langer 1953, p. 241)

So, we have Freud (1911a, pp. 63–65) formulating certain mechanisms of symptom-formation in terms of transformational rules for operating on verbal representations. "I (a man) love him" becomes "I do not love him—I hate him," which in turn becomes "He hates (persecutes) me, which will justify me in hating him." Here, transformational rules contradict or negate by operating on the verb, and by transposing subject and object, of an underlying semantic representation, resulting in delusions of persecution.

"I (a man) love him" becomes "I do not love him—I love her—because she loves me." Here, transformational rules contradict or negate by operating on the grammatical object in the *same* underlying semantic representation, resulting in erotomania.

"I (a man) love him" becomes "It is not I who love the man—she loves him." Here, transformational rules contradict or negate by operating on the grammatical subject of the same underlying semantic representation, resulting in delusions of jealousy.

Finally, "I (a man) love him" becomes "I do not love at all—I do not love anyone," or perhaps, then, "I love only myself." Here, transformational rules contradict or negate by operating on the entire proposition, resulting in megalomania (if the second change from "him" to "myself" occurs).

The id, in the interest of gratification, creates symbolic forms—"hallucinatory" images of the gratification of sensual wishes—using transformational operations such as displacement, condensation, translation from verbal symbols into imagery, and iconic symbolization. The "deep," underlying, or unconscious mental representations in psychoanalysis are often constituents of wishes.

The ego, in order to achieve a cognitive map of—and so master—the object world, creates verbal symbolic forms to represent that object world, using the transformational rules of natural language. The superego creates symbolic forms by using transformational operations that, for example, change verbs from indicative to imperative. The ego-ideal creates luminous images, which awe and inspire, using transformations that erase imperfections and abstract out, emphasize, and foreground just those features that exemplify the ideal.

Freud's discovery of primary process and of the differences between primary and secondary process seems essentially a discovery that different symbolic systems have different kinds of rules for operating on symbolic representations and so transforming them. Primary and secondary process both involve canons of symbolization. Both may involve in Langer's sense of the term "a logic," but different kinds of canons, and different logics.

Psychoanalysis: A Science of the Symbolizing Activity of the Mind

Memory

Freud was disturbed to discover that his patients' reported memories were fictions. Evaluated according to standards of empirical validity and instrumental logic, they were not factual, not true. However potent as symbolic representations of a patient's conceptions of reality in governing his life, these vivid mnemic images were illusions. E. Kris (1956a, 1956b) later showed that even apparently factual memories are inevitably representations of what has been abstracted from the varied, chaotic flux of experience. They are personal myths—nodes of condensation of innumerable experienced realities.

The conviction commanded by a memory is similar to the conviction commanded by a work of art. Both are virtual happenings, semblances, a seeming to be and to transpire. Both command conviction through the creation of an illusion, embodied and presented in the form of an apparition. Pieces of "real" events may be merely a kind of material used to create the illusion. Such an illusion is efficacious by virtue of its symbolizing a conception in a form that does not merely represent the conception but exemplifies it.

Stevens (1951, p. 139) suggested that Freud "might have said that in a civilization based on science there could be a science of illusions." The literary critic Lionel Trilling, in a similar vein, made the following comments:

> For, of all mental systems, the Freudian psychology is the one which makes poetry indigenous to the very constitution of the mind. . . . [It] was left to Freud to discover how, in a scientific age, we still feel and think in figurative formations, and to create, what psychoanalysis is, a science of tropes. of metaphor and its variants, synecdoche and metonymy. (1950, pp. 52–53)

The Transference Neurosis

The transference neurosis is one of Freud's great scientific discoveries. Here is a phenomenon that, to judge by complaints about the behavioral sciences, is rare indeed in their realm. It is unexpected. It is of critical theoretical significance. It is replicable under carefully controlled conditions.

The carefully controlled conditions are the features of the psychoanalytic situation, which includes a patient capable of prolonged commitment to attempts at free association and a psychoanalyst whose participation is rigorously disciplined. The patient is willing and able to devote himself to making verbal productions in a situation designed to minimize external excitants, guidance, or interference that might constrain these productions. Furthermore, he agrees to try to refrain from preventing in any way the utterance in verbal form of whatever comes to his mind; he will avoid, for example, judging and making deliberate efforts to select and order occurrent mental contents. The psychoanalyst's aims in relation to the patient's verbal productions are limited to interpreting their meaning and communicating such interpretations in a way that increases the likelihood they will be meaningful to the patient. All the psychoanalyst's skills are exercised toward these ends alone. Ideally, the psychoanalyst will not be persuaded to respond to the patient's verbal productions in any way other than acts of interpretation.

What happens under these circumstances is truly remarkable. Typically, after an initial, apparently relatively uninhibited period of expression, the patient's symptoms may suddenly disappear, gradually subside in severity, or increasingly cease to concern or preoccupy him in the psychoanalytic sessions. At the same time, he may find free association increasingly difficult. Regularly, it is ever more persistent thoughts and intense feelings about the psychoanalyst that he is reluctant to voice, despite the injunction to free association. He is absorbed by his (to him usually unacceptable) conceptions of the psychoanalyst and the psychoanalyst's attitudes toward, feelings about, and intentions in relation to himself; by his own (to him usually unacceptable) attitudes toward, feelings about, or intentions in relation to the psychoanalyst as so conceived; and by his own efforts to verify his conceptions of the psychoanalyst and to realize his aims, or to bring about some state of affairs involving a particular kind of relationship with the psychoanalyst.

It may be, moreover, that, given the conditions described above, these preoccupations of the patient will hold sway over him only or mainly for the period of the psychoanalytic hour. Astonishingly enough, after an hour of hesitation, strain, hints of passion, explicit torment, muteness, imprecations, beseechings, the patient may rise calmly from his recumbent position, perhaps indicate, however fleetingly, his recognition of the psychoanalyst as psychoanalyst, and go about his business, relatively untroubled, only to immerse himself once again in his *creation* the next hour. For the impression is irresistible that the patient creates something, something circumscribed in space and time, something made out of the materials of the psychoanalytic situation and process itself. Using methods determined or made possible by the properties of the psychoanalytic situation, he makes something with form, however strange, the shape of which at first is dim, vague, as if seen always from afar through a mist, there, lost, recovered, and lost again through many hours, but in time looming closer, and increasingly precise in outline and rich in detail and design.

Freud did not ignore this phenomenon, damn it as a nuisance, or exploit it to noninterpretive ends. His astonishing feat, of course, was instead to discover that this apparent impediment to free expression in the psychoanalysis, this apparent obstacle to the patient's unconstrained participation in the work of the psychoanalysis, was, in fact a symbolic representation of the patient's conception of his inner world (his psychic reality), a symbolic representation of the conflicts between imagined entities (of which his symptoms were still another symbolic representation), now quintessentially in the form of enactments and fantasies constituting the transference neurosis. It, no less than the patient's dreams, symptoms, and stream of verbal associations, called for interpretation.

The transference neurosis is not a revival of earlier events or relationships. It is a revival of the patient's earlier, perduring conceptions of these events and relationships. It is a presentational symbol of these conceptions. Fantasies are an inextricable part of it. Fantasies are symbols of conceptions of inner reality, visualized as states of affairs in time past, time present, or time future. Such symbols are exemplars of the conceptions of inner reality they also represent.

Psychoanalysis, despite its inclusion of a genetic or developmental frame of reference, despite the historicism of many of its theoretical formulations, is not a historical science, but is instead a science of the symbolizing activity of the mind. Psychoanalysis cannot be concerned merely with the recovery of the actual historical past. As a method it is not suitable for the study of actual events. A patient may refer to what is apparently the same event at different times and in different contexts during a psychoanalvsis. At these different times and in these different contexts, presentations of the event, its elements and properties, differ. Details, emphases, conceptions of the event and of its significance, and the attitudes and feelings aroused by or associated with the event, differ. The very history of the patient seems to change as he reconstructs it during different periods of a psychoanalysis. What the event and its elements and properties mean, what the patient, using "actualities" as material, made and makes of them, changes. If a history is revived, it is the history of the acts of the patient's mind, creating through time his symbolic representations of his conceptions of past, present, and future "reality."

We may or may not infer an actual event at that imaginary point where a patient's various symbolic representations of that event seem to converge. But that actual event as an entity is not knowable through, and cannot be investigated by, the method of psychoanalysis. The pathogens exorcised by psychoanalysis are not physiological processes or historical situations but symbolic representations of conceptions of such processes and situations and transformations of these symbolic representations: mental shades, memories, fantasies. Not reality but symbolic representations of conceptions of reality. Not organism but symbolic representations of conceptions of the body. Not object-relations but symbolic representations of conceptions of objects and kinds of relations to them. Between stimulus and response, between event and behavior, falls the act of the mind. It is the creation of the symbol, the "poem of the act of the mind" (Stevens 1961, pp. 239–240) that is the object of study in psychoanalysis.

Language and Dreams[1]

2

The Semiological Foundations of Psychoanalysis

According to Freud's great conception in *The Interpretation of Dreams* (1900), the dream is a symbolic entity. The dream derives its efficacy from its status as a symbolic representation of latent thoughts. So Freud placed psychoanalysis in the realm of semiology, making it a science of symbolic functioning, which studies particular kinds of symbolic systems (organizations of symbolic entities), their relations to each other, and their acquisition and use. Psychoanalysis is, by virtue of the nature of its conceptual foundations, a theory of symbolic functioning, rather than a general psychology or a general theory of behavior.

Freud asks about the dream as a manifest representation of meaning: out of what materials is it made? By the use of what techniques are the materials treated, combined, or transformed to make a dream? He views the dream as actively constructed. The dreamer makes the dream—out of available materials (for example, somatic stimuli, and recent but indifferent impressions), and according to definite methods or techniques (the dream work). The dreamer chooses materials that will serve his purpose. Instigated by various disturbances of sleep, he aims to compensate for these disturbances, to oppose

1. This chapter is an extensively revised version of Edelson (1972). The material presented here has its origins in a strenuous effort to investigate the analogies and disanalogies between Freud's theory of dreams and Chomsky's linguistic theory and, in general, the relation between language and dreams. The strain of the effort has left its mark, despite my efforts to expunge it. Results of the investigation are included here because of their seminal relation to my own further thoughts on many issues discussed later in this book. But the reader who wants to forego immersion in a somewhat demanding exposition and skip to Chapter 3 may do so without loss of continuity.

certain virtual intrusions, by attempting to construct in a state of sleep a virtual wish-fulfillment—a dream.

The Transformational-Generative Model

I shall refer in this chapter to Chomsky's explicit use of a transformational-generative model in his linguistic theory (Chomsky 1957, 1959, 1965, 1966a, 1966b, 1967, 1968), because I believe that Freud implicitly used such a model in *The Interpretation of Dreams,* although with some important differences, which I shall mention, and certainly giving it a very different content. (Chomsky was primarily interested in syntax.) That Chomsky, faced with the exigencies of linguistic facts, came to modify his early transformational-generative grammar in later work does not detract from the heuristic value of observing the analogies (and disanalogies) between his rigorously developed use of a transformational-generative model in one domain and Freud's rather more intuitive use of a very similar model in another domain.

The First Chapter of *The Interpretation of Dreams*

Freud's Thought Processes as a Scientist

The first chapter of *The Interpretation of Dreams* is a review of the scientific literature dealing with the problem of dreams. This review is cogent and scientifically sophisticated, and its organization and argument have great power. Not so, the obligatory ritual to which we are often subjected by our journals, in which a collection of quotations and citations descends, at its worst, to pedantry and appeal to authority. A review should rise, as Freud's does, to an awareness of science as a corpus of interlocking propositions, which requires an investigator to assess what has apparently been established by previous workers and what consequences for the whole the abandonment of old or the introduction of new findings and concepts will have.

No wonder Freud put off writing this chapter to the last, and found writing it so exhausting that he did not return to it in revising subsequent editions. For in this chapter, he exemplifies the thought processes of a creative scientist: how he surveys a congeries of often contradictory assertions; how he seizes upon the theoretically significant fact; how he marshals his material according to the questions he asks and, having asked them, holds onto them with obstinate tenacity as to the thread that will lead out of the maze; how he proposes criteria for explanation, not the least of which is the power, transcending apparent contradictions, to answer his questions (though these are not the questions of others); and with the help of such criteria, how he trenchantly dismisses explanations that have satisfied others and begins to fashion in their stead those that will satisfy himself.[2]

Freud frames his account according to topics rather than authors, raising "each dream-problem in turn," because, despite the points of value of individual studies, no one "line of advance in any particular direction can be traced," "no foundation has been laid of secure findings upon which a later investigator might build," but rather "each new writer examines the same problems afresh and begins again, as it were, from the beginning." He expresses his conviction "that in such obscure matters it will only be possible to arrive at explanations and agreed results by a series of detailed investigations" (p. 5f.).

He takes a casually noted but drastic step given the scientific milieu in which he worked. (He was to continue to accept many of the assumptions of this milieu, even while he paved the way through his own work for their abandonment.) He distinguishes his own "piece of detailed research," "predominantly psychological in character," from "a problem of physiology," such as the problem of sleep. He supposes, however, that "one of the characteristics of the state of sleep must be that it brings about modifications in the conditions of functioning of the mental apparatus" (p. 6).

The distinction Freud makes here is a subtle one, which he found difficult to adhere to, and which is by no means even today generally understood or accepted. As is apparent from the actual conceptual and investigative strategies he pursued in his own work, he regards the physiological and psychological realms as two different empirical and theoretical systems. The data of one system are not mere epiphenomena of the processes of the other, and the explanatory propositions accounting for data in one realm cannot be used to account for data in the other. Nevertheless, there is a systematic relation between the two realms, neither a simple parallelism nor a metaphysical gap. Processes in the physiological realm set the conditions for—that is, provide resources or occasions for, or constrain or limit—processes in the psychological realm, but the former pro-

2. Characteristically, in citing writers who present views compatible with his own, Freud chides one for putting himself "in the wrong by not giving any refutation of the material which contradicts his thesis" (p. 60).

cesses do not determine the direction of, control, cause, or explain the latter ones.[3]

> Dreaming has often been compared with "the ten fingers of a man who knows nothing of music wandering over the keys of a piano" . . .; and this simile shows as well as anything the sort of opinion that is usually held of dreaming by representatives of the exact sciences. On this view a dream is something wholly and completely incapable of interpretation; for how could the ten fingers of an unmusical player produce a piece of music? (P. 78)

This passage is only one of many that make clear Freud is determined to follow an unusual interest for a scientist of his time: an interest in the dream as a creatively generated symbolic entity. I emphasize "creative" because of his recurrent emphasis on the problem of accounting for the "choice" of materials used in constructing a particular dream, especially the "choice" of the dream images that ultimately appear in the manifest dream. The term "choice," if taken seriously, appears to imply rules that can be violated or imagined ends that govern what choices are made. It is unlikely that Freud has simply stumbled here into an anthropomorphic formulation of physiological or purely physical (that is, nonsymbolic or nonsemantic) processes.[4]

Freud's emphasis, diction, and many examples suggest instead that he considers the production of the dream as analogous in important ways to the production of other kinds of symbolic entities, by artists, for example, or, as we might think now, by *bricoleurs* (jack-of-all-trades), who are described by Levi-Strauss, when he describes myths as a kind of intellectual *bricolage.*

The *bricoleur* "works with his hands and uses devious means compared to those of a craftsman," and, while "adept at performing a large number of diverse tasks . . . does not subordinate each of them to the availability of raw materials and tools conceived and procured for the purpose of the project." Instead he makes do "with 'whatever is at hand,' that is to say with a set of tools and materials which is always finite and is also heterogeneous because what it contains bears no relation to the current project, or indeed to any particular project, but is the contingent result of all the occasions there have

3. That a theory like psychoanalysis cannot be replaced or reduced by neuroscience is argued in Chapter 7.

4. In Chapter 7, I describe what it means to regard psychoanalysis as a semantic science.

been to renew or enrich the stock or to maintain it with the remains of previous constructions or destructions," the elements having been "collected or retained on the principle that 'they may always come in handy' " (Levi-Strauss 1962, pp. 17f.).

Given this inevitably contingent, ad hoc, nonlawful aspect of the production of dreams, and Freud's interest in the interpretation of dreams as symbolic entities, it is clear that he will not formulate propositions of the kind, "given this set of conditions (a specification of the observable characteristics of the situation in which the dream occurs), this particular dream will certainly follow." The discovery of the meaning of a particular dream—that is, the discovery of the conceptions of objects and states of affairs the dream as a symbolic entity represents—is to be made by providing specific answers to the following questions about that dream.

What circumstances provide the occasion for the dream?

What materials does the dreamer choose as he begins the mental work of constructing the dream?

What is the nature of the system that is responsible for the ultimate transformation of these materials by processes that are universal—i.e., that are characteristic of all dreams?

What is the way these transformations operate on the contents of latent mental states—symbolic representations of objects and states of affairs—to produce the particular images that make up the manifest form of a dream?[5]

Despite the focus on the motivational and conative in Freud's work, the emphasis here, apparently on the cognitive, on mental contents (symbolic representations) and transformational operations upon these, is not inappropriate. Indeed, Freud appears to identify "impulse" as at least a type of idea, although having particular characteristics to be sure. For example: "It will be seen that the emergence of impulses which are foreign to our moral consciousness is merely analogous to what we have already learned—the fact that dreams have access to ideational material which is absent in our waking state or plays but a small part in it" (p. 71).

More than is generally appreciated, Freud's interest in meaning in *The Interpretation of Dreams* is precisely in that sense of "meaning" that has to do with conceptions or ideas, necessarily, then, involving some level of abstraction—that is, the connotative sense of "mean-

5. Hereafter, the phrase "mental or symbolic representations of objects or states of affairs" should be understood to mean "mental or symbolic representations of *conceptions of or ideas about* objects or states of affairs."

ing."[6] He is also of course interested in other senses of "meaning."[7]

He does mention in passing that affects in contrast to impulses do not appear to be transformed by the processes producing the dream images. He quotes Stricker as writing, "'Dreams do not consist solely of illusions. If, for instance, one is afraid of robbers in a dream, the robbers, it is true, are imaginary—but the fear is real.'" Freud comments, "This calls our attention to the fact that *affects* in dreams cannot be judged in the same way as the remainder of their content" (p. 74).

If this is true, either (1) the fate of a subject's *relations* to mental contents (. . . believes that . . ., . . . wishes that . . ., . . . fears that . . .) is different from the fate of the mental contents themselves, or (2) impulses and feelings may be very different kinds of mental states from the point of view of the symbolic function.

The presence of an affect in a dream appears to function as an index of an existing state—an actual reaction to or assessment of latent mental contents. But that part of the underlying impulse that survives in the manifest dream is a transformation of these mental contents.

Four Problems

Freud's review of the literature traces four main lines of thought—four problems to be investigated in *The Interpretation of Dreams.*

1. The Materials Used in the Construction of Dreams

Dreams are symbolic representations of thought. In that sense, dreaming is a "form of thought"—a form given to thought. "As a rule dream-pictures contain what the waking man already thinks" (p. 8).

Memories are important in providing the materials for construct-

6. Without choice there cannot be connotative (abstract or ideational) meaning. If a situation coerces an utterance ("say 'pencil' or I will shoot you dead"), then the utterance cannot be said to have connotative meaning. That is not to say that such a vocalization—a terrified scream, for example—would not communicate information. An index communicates information, although, in this rather uninteresting because so extended sense of "information," almost anything may be said to communicate information since from an observation of it some state of affairs may be inferred. Such considerations provide one basis for rejecting a strict stimulus-response account of verbal behavior or, indeed, of any behavior involving symbolization (Chomsky 1959).

7. The motivational sense of "meaning" has to do with the intention inferred to underlie or dictate a particular suasive use, or presentation in communicative processes, of a symbolic entity. The deictic or denotative sense of "meaning" has to do with an existent concrete object or thing, and a conception of it, to which a symbolic entity may be said to refer. "Meaning" in the sense of meaningfulness or meaningful form has to do with an order or pattern discernible in, or exemplified by, a symbolic entity such as a work of art.

ing dreams. "All the material making up the content of a dream is in some way derived from experience" (p. 11).

Childhood experience is "one of the sources from which dreams derive material for reproduction—material which is in part neither remembered nor used in the activities of waking thought."

Recent experience is another source: ". . . elements are to be found in most dreams, which are derived from the very last few days before they were dreamt."

The choice of material is the "most striking and least comprehensible characteristic of memory in dreams." "For what is found worth remembering is not, as in waking life, only what is most important, but on the contrary what is most indifferent and insignificant as well" (pp. 15, 17, 18).

Freud does not take any one of these features as *the* point of departure for his explanation of dreams, nor does he regard these features as contradicting one another. Rather, all of them must be encompassed by a theoretical account of dreams. The point of view that makes this possible is seeing all of these features as results of choosing various types of material in constructing dreams.

Freud is drawn to the paradoxical unexpected fact. He recognizes that the choice of indifferent, insignificant memory images has tended to lead to discounting or overlooking the connection between the dream and the experiences of waking life, and so to abandoning promising lines of research. One writer, who was "unquestionably right in asserting that we should be able to explain the genesis of every dream-image if we devoted enough time and trouble to tracing its origin," allowed himself to be "deterred from following the path which has this inauspicious beginning" by the "exceedingly laborious and thankless" nature of the task, thus losing the opportunity to be led "to the very heart of the explanation of dreams" (pp. 19f.). We are to conclude that the gifted scientist, confronted by the prospect of a possibly thankless labor among details of apparent triviality, does not shrink back; indeed, he may be moved to what appears to others as eccentrically persistent exertions.

Similarly, Freud seizes upon a fact that is not only odd but is of "remarkable . . . theoretical importance . . . that dreams have at their command memories which are inaccessible in waking life" (pp. 12f.). "No one who occupies himself with dreams can, I believe, fail to discover that it is a very common event for a dream to give evidence of knowledge and memories which the waking subject is unaware of possessing" (p. 14). Among his examples, he mentions that "in dreams people speak foreign languages more fluently and correctly than in waking life" (p. 11). Some theories about dreams must be judged incorrect because they ignore this crucial feature in seeking "to account for their absurdity and incoherence by a partial forgetting of what we know during the day. When we bear in mind the extraordinary efficiency . . . exhibited by memory in dreams we shall have a lively sense of the contradiction which these theories involve" (p. 20).

Some materials, then—some memory images—are available for use in symbolizing during the sleeping state that are not available during the waking state. From this, and from knowledge of the strange transformations that occur in constructing a dream, we may conjecture that two factors, at least, determine how transformational processes in a symbolic system will operate to alter and filter or screen deep, underlying, or unconscious mental contents. Both factors are apparently affected by the sleeping state. One, materials may be available for dream formation that are not available as memories during the waking state. Two, constraints imposed upon processes of symbolization during the waking state may be absent or relaxed during processes of symbolization characteristic of the sleeping state.

Dreams are generated by combining meaningful entities such as memory images, rather than meaningless entities such as the sound elements combined in speech. As Freud's later work was to make clear (see also E. Kris 1956b), memories are themselves symbolic entities, the result of a creative, constructive (not purely imitative) process that involves selecting, rejecting, combining, and altering various details. Memory images symbolically represent abstractions, conceptions of experience, not experience itself.

Langer (1953, 1957, 1962), similarly, argues persuasively in her theory of art that (like memory images in dreams) *meaningful* images (a pictorial image of someone or something, an imitation of a particular sound) are but so much raw material pressed into service by the artist. These raw materials are not to be taken as giving the meaning of a work of art. They are used to create a significant form, an apparition, a virtual reality, an illusion.

Langer's theory of art provides an analogy to a theory of the dream as an illusion of an experience of wish-fulfillment. Dance is an illusion of virtual powers, music an illusion of virtual time, plastic arts illusions of virtual space, poesis illusions of virtual events, literature an illusion of virtual history, and drama an illusion of a virtual present. The real meaning of a work of art is a conception of inner reality or the world of feeling.

Dreams use only fragments of memory images, and these are often altered in some way. In dreams, aspects of memory images are deleted; something else substitutes for a part of a memory image; some aspect of the form of a memory image is altered. Freud concludes that therefore dreaming cannot be simply a process by which memory images are reproduced as an end in itself. "Views of this sort are inherently improbable owing to the manner in which dreams deal with the material that is remembered" (pp. 20f.). (Current cognitive theories of dreaming, in which dreaming is seen as serving the function of information processing and the fixation of memories rather than wish-fulfillment, would presumably have been included by Freud as "views of this sort.")

By what kinds of processes are memory images transformed so that they may be used in constructing a dream? A memory image can be used to represent meaning by virtue of what it already represents as a memory image—a conception of a past event. But it also can be used to represent meaning by virtue of other, apparently fortuitous features—formal or sensuous properties—that in some way resemble, evoke, or allude to what it represents. In fact, the content of a memory image may not be crucial for its use at all, but only "the fact of its being 'real.' " The fact that it is a *memory* image determines its being chosen to represent a thought such as " 'I really *did* do all that yesterday' " (p. 21).

Here is a disanalogy between dreams and language. In dreams, the relation between form and content is not arbitrary or conventional, as it is in language. A dream is a presentational, apparitional, or "motivated" symbol; such symbols resemble, imitate, or exemplify what they represent.

2. The Occasions Providing an Opportunity for, or Instigating, the Construction of a Dream

One or another kind of stimuli have been taken by other investigators to be *the* cause of dreams. However, external stimuli cannot be so taken, for nothing about the intrinsic features of the stimulus accounts for one group of memories rather than another being aroused, or accounts for the dreamer making one interpretation rather than another of an external stimulus. Subjective sensory excitations cannot constitute satisfactory explanations, for they "are ready at hand, one might say, whenever they are needed as an explanation." Furthermore, "the part they play . . . is scarcely or not at all open to confirmation . . . by observation and experiment" (p. 31). Medical men may prefer organic sensations or internal somatic stimuli as an explanation of dreams, because such stimuli provide a "single aetiology for dreams and mental diseases," but "one is often faced with the awkward fact that the only thing that reveals the existence of the organic stimulus is precisely the content of the dream itself," and again nothing about a somatic stimulus accounts for the choice of one particular dream image rather than another (p. 37).[8]

Therefore, external sensory excitations, internal sensory excitations, internal somatic stimuli, or purely psychical sources of stimulation are seen by Freud as instigating dream construction but not explaining it. Such stimuli provide occasions for dream construction. These stimuli do not *determine* the dream's form or meaning. In themselves they give us no clue about the way in which a particular dream is constructed—the choice of images, for example—or what a particular dream means.

Here again we have an analogy to language. Situations requiring communication, calling for the expression of thoughts or feelings, demanding thought, may result in the use of language. These situations, because they provide an occasion for such use, do not on that account explain the characteristics of a particular piece of language—an utterance, for example—or determine the choices (phonological, syntactic, or semantic) made in generating that particular piece of language.

3. The Characteristics of the Manifest Dream

The characteristics of dreams follow from the way dreams are constructed from these materials on these occasions.

a) Some dreams are forgotten very quickly, whereas others are remembered with extraordinary persistence. Their lack of intelligibility and orderliness is especially important in accounting for their being so easily forgotten.

> If sensations, ideas, thoughts, and so on, are to attain a certain degree of susceptibility to being remembered, it is essential that they should not remain isolated but should be arranged in appropriate concatenations and groupings. If a short line of verse is divided up into its component words and these are mixed up, it becomes very hard to remember. (P. 44)

The processes by which dream images are combined are largely unknown to the waking symbolizing mind. They are so different from those used in waking acts of symbolization (rules of language, e.g.), that to remember a dream, to understand it, and to know what

8. The awkward fact to which Freud refers is similar to the one to which Chomsky (1959) refers when he points out that there is no way to tell which stimulus in a situation will be reinforcing until after reinforcement has occurred.

thoughts it represents are indeed difficult. Dreams may be remembered solely because of the contributions of waking consciousness in the form of secondary revision, which, through interpolations or by filling in gaps, forges apparent causal connections between merely juxtaposed sequences of images, and so gives a dream a degree of coherence and order.

b) There are "modifications in the process of the mind" that make the dream seem alien. "The strangeness cannot be due to the material that finds its way into their content, since that material is for the most part common to dreaming and waking life" (p. 48). In falling asleep, we notice that thoughts are represented by images, which rise involuntarily, in contrast to the voluntary flow of ideas while awake. Dreams hallucinate. They present events as actually happening. "We appear not to *think* but to *experience*; that is to say, we attach complete belief to the hallucinations," since in the state of sleep reality testing does not occur (pp. 50ff.).

The value ordinarily attached to perceptions seems detached from dream images. Dreams "are disconnected, they accept the most violent contradictions without the least objection, they admit impossibilities, they disregard knowledge which carries great weight with us in the daytime, they reveal us as ethical and moral imbeciles" (p. 54).

The judgment that the absurd, incoherent ideation of dreams is due to "lowered psychical efficiency" is based only on observable features of the manifest dream. (The radically empiricist linguist who limits himself to a description or classification of what is empirically given—the manifest representations of actual utterances—in attempting to account for language makes a similar mistake in explanatory strategy.) "Anyone who is inclined to take a low view of psychical functioning in dreams will naturally prefer to assign their source to somatic stimulation; whereas those who believe that the dreaming mind retains the greater part of its waking capacities have of course no reason for denying that the stimulus to dreaming can arise within the dreaming mind itself" (p. 64).

To refute these low views, Freud shows that the memory function in dreams can in some respects be superior to that of waking life. He also shows that the latent content of dreams are mental representations of the same kind as the mental representations generated in waking consciousness.

It is not generally appreciated that Freud held the view that the thoughts underlying both dreams and psychopathological symptomatology are of the same kind that may appear in waking consciousness. What is different are the transformational processes operating on these thoughts. In the case of waking consciousness, transformational processes yield rule-governed sequences of meaningless sounds to form *meaningful* representations. In the case of dreams or symptoms, transformational processes yield what often appear to be *meaningless* images or pictograms, or enactments.

One of Freud's tasks in *The Interpretation of Dreams* is to alter the judgment that the ideation of dreams is deficient by showing (1) that the difference between dreams and the linguistic representations of waking consciousness is not a difference in what is represented but a difference in the form and manner of construction of the representation; and (2) that the incoherence of dreams does not involve a breakdown into chaos either of thought or of its representation, but rather the nature of definable processes, which depart from the rules governing linguistic representations in waking consciousness, but which have just as specifiable distinctive features as those rules. (The rules of a natural language are in fact enormously difficult to specify, and have not been completely specified for any language.)

4. A Theory of Dreams

A theory of dreams should seek "to explain as many as possible of their observed characteristics from a particular point of view" and, at the same time, should define "the position occupied by dreams in a wider sphere of phenomena" (p. 75). From a theory of dreams, one should be able to infer a function for dreaming. So, it might be held that one ought to be able to infer from a theory of language what it is about language that makes it possible to use it in thinking and communicating; that it is so used, however, is not an explanation of it, any more than the function of dreaming explains a particular dream.

Freud rejects the idea that dreaming can be explained by theories assuming that psychic mental activity continues as usual during sleep but necessarily produces different results under such altered conditions. If the mind continues to function in all ways as usual, how derive the distinctions between dreams and mental representations in waking consciousness? Moreover, such theories suggest no function for dreaming.

Freud rejects theories that hold that dreams are the result of hypofunction of the psychic apparatus. These theories also leave "no room for assigning any function to dreaming." They are based on an a priori devaluation of dreams as unworthy to rank with the psychic processes of waking life.

Freud rejects theories that hold that dreams are the result of hyperfunction of the psychic apparatus. They are too vague, and they lack universality (i.e., they are unable to relate dreams to other phenomena). However, these theories suggest a utilitarian function—a reviving, healing, creative function—for dreaming. Freud mentions in this connection Scherner's work, which stresses the role of the dream imagination. It "makes use of recent waking memories for its building material" and "erects them into structures bearing not the slightest resemblance to those of waking life: it reveals itself in dreams as possessing not merely reproductive but *productive* powers." (Freud's attempt to refute the thesis that dreams are *productive*, in Scherner's sense, is a major theme in *The Interpretation of Dreams.*) Since dream imagination is "without the power of conceptual speech," it is "obliged to paint what it has to say pictorially, and, since there are no concepts to exercise an attenuating influence, it makes full and powerful use of the pictorial form." Freud in rebuttal notes the arbitrary character of dream imagination; its recalcitrance to scientific investigation seems to him obvious. Nevertheless, he urges that Scherner's ideas should not be rejected just because they seem fantastic. "Ganglion cells can be fantastic too" (pp. 84–87).

The missing ingredients of universality and function are supplied by two propositions, suggested by Freud in a final section of the first chapter. He does not separate these propositions, but I think they ought to be separated. The first has to do with the nature of the ideas dreams represent. The second has to do with the function served by the particular kind of symbolic representation dreams are—that they are presentational, apparitional, or motivated symbols.

The first proposition is that, among the ideas represented by dreams, there are uniquely and always ideas that are wishes. The second proposition is that dreams use memory images to create the illusion of a virtual reality (an apparently actual experience of what is in fact a hallucinated reality), and thus ideally—as presentational symbols—serve the function of representing wishes that are only "in the mind" as actually fulfilled.

Is it likely that dreams have only one function? Probably not. Another symbolic system, language, has more than one function—serves more than one kind of purpose. Both thought and communication come immediately to mind; there are others. One must be clear here, however, about the distinction between "dreams" as a set of phenomena, on the one hand, and, on the other hand, "the symbolic system that generates dreams," which may have a variety of functions, generating as it does different kinds of presentations symbols—dreams, psychopathological symptoms, the enactments of the transference neurosis, even poetry (Edelson 1971, 1975).

The Six Dimensions of Symbolic Functioning

Freud adumbrated in *The Interpretation of Dreams* six dimensions of symbolic functioning, and therefore six dimensions by which different modes of symbolization and their uses can be compared with each other.

1. The Symbolic Entity: The Manifest Dream

The manifest dream is a symbolic entity. Symbolic entities represent abstractions—classes, categories, thoughts, ideas, conceptions. Symbolic entities have significance and produce effects not by virtue of what they are as objects in and of themselves but by virtue of their meaning—what they represent. Symbolic entities are constructed by selecting from among a set of relatively discrete elements or materials some of them, and arranging or combining these according to rules, methods, or procedures characteristic of the symbolic system to which they belong.

The manifest dream is an organization of images that, according to Freud, because of the very purposiveness of all psychic activity, must take the form of a virtual or hallucinatory wish-fulfillment as the end of a process of psychic activity in a state of sleep (p. 567).

In studying natural language, Chomsky distinguished speech or performance (the use of language to make particular utterances) from linguistic competence or capacity, which is knowledge of the system that includes the rules of phonology, syntax, and semantics. Analogously, any manifest dream must be distinguished from the system (including the dream work) that is logically prior to any use of it to construct a particular dream.

2. The Materials Used in Constructing a Symbolic Entity: Somatic Stimuli; Recent and Indifferent Impressions

Freud drew an analogy between the materials used in constructing a dream and the materials used in constructing a work of art. Somatic sources of stimulation, of insufficient intensity to disturb sleep, as well as recent but indifferent impressions left over from the previous day, are

> brought in to help in the formation of a dream if they fit in appropriately with the ideational content derived from the dream's psychical sources, but otherwise not. They are treated like some cheap material always ready to hand, which is employed whenever it is needed, in contrast to a precious material which itself prescribes the way in which it shall be employed. If, to take a simile, a patron of the

> arts brings an artist some rare stone, such as a piece of onyx, and asks him to create a work of art from it, then the size of the stone, its colour and markings, help to decide what head or what scene shall be represented in it. Whereas in the case of a uniform and plentiful material such as marble or sandstone, the artist merely follows some idea that is present in his own mind. (P. 237)

Linguistic elements may also be used as cheap plentiful materials, partly by virtue of the situations or contexts in which they have appeared in experience and to which they may therefore allude, partly by virtue of their perceptible properties as objects (sound, length), and partly by virtue of their relations to other linguistic elements (these relations are defined by the linguistic system to which they jointly belong).

Freud asserts that the appearance in dreams of speech alludes to the contexts in which it occurred. Such speech is merely building material; it does not represent thoughts. He also shows how syllables and words may be used as building materials in constructing dreams. Syllables and words form links—and so allude—to other members of a class of sounds or class of meanings, by virtue of their shared membership in the same class. So, if such linguistic units belong to semantic categories, they may be substituted for one another because they are completely synonymous, antonymous, partially synonymous, or contrasting in one or more semantic features.

Syllables and words also may evoke or allude to each other because they are connectable or combinable according to the rules of syntax. If such linguistic elements are substituted for one another, rather than connected in sequence, the result is a deviant construction, which may be understandable as a metaphor. For example, starting with the symbolic representations "father wears a big hat" and "father yells at me," and substituting "big hat" for "father," which is syntactically related to it, one may derive "the big hat yells at me," which ultimately may appear in the manifest content of a dream. Since "father" and "hat" do not belong to the same semantic class, the representation "the hat yells at me" is deviant but understandable metaphorically.

Materials used in constructing a symbolic representation may be considered from the point of view of their relative availability as well as the stringencies (a term used by E. Kris 1952) they impose on the act of construction. As the earlier quoted passage by Freud about the various materials used in making different works of art suggests, some materials impose relatively little in the way of limitations on what can be represented and how it can be represented. Other materials predetermine stringently what is possible and what impossible to represent and how what is represented can be represented. In other words, some materials demand more accommodation to their intrinsic characteristics than others.

"Both [indifferent and recent] impressions satisfy the demand of the repressed for material that is still clear of associations—the indifferent ones because they have given no occasion for the formation of many ties, and the recent ones because they have not yet had time to form them" (p. 564). An advantage that recent impressions have as materials in dream construction is that they have not yet become committed or tied to certain meanings. Their recency enhances not only their availability but the relative lack of stringency their use imposes on dream construction. The dreamer can do almost anything he wants with them. The relative lack of stringency is especially important because the dream work is governed by the principle that a dream will represent a number of meanings simultaneously.

An advantage that indifferent impressions have as materials in dream construction is that they are relatively empty of significance. This indifference not only facilitates serving the end of censorship but evoking multiple meanings or generating multiple allusions as well. These materials have no claim to interest in and of themselves. Insofar as they are merely materials to be used in constructing a manifest dream, they do not act as disturbers of sleep and therefore in themselves do not instigate or provide any occasion for dreaming. Sounds used in speech have similar characteristics—their easy accessibility, and the relative lack of stringency their use imposes on the representation of thought. They are not intrinsically significant. To the extent a sound were intrinsically significant, such as a scream of pain is, that sound would lose its value as material that might be used in constructing a representation of thought. The innate expressiveness of a sound and the utility of a sound in constructing a symbolic representation are inversely related.[9]

In dreams and language, the materials used in constructing perceptible symbolic representations are different. In language, meaningless sound elements are combined according to rules to form meaningful perceptible symbolic representations. Dream construction, however, makes use of meaningful elements, memory images, and what they evoke or allude to, to create a perceptible presentation of thought. "My recollection of the monograph on the genus Cyclamen would thus serve the purpose of being an *allusion* to the conversation

9. This fact raises a question concerning the popular postulation of the origins of language in emotional or interjectional utterance.

with my friend, just as the 'smoked salmon' in the dream of the abandoned supper party . . . served as an *allusion* to the dreamer's thought of her woman friend" (p. 175).

3. The Instigators of Symbolization: Disturbers of Sleep

Instigators of dream construction incite, or provide occasions for, symbolic functioning in the state of sleep. Dream instigators include external sensory stimuli, organic somatic stimuli, psychic stimuli—any intrusion sufficiently intense to threaten a disturbance or disruption of sleep. Such dream instigators are not causes of particular dreams, any more than an incitement of, or occasion for, a particular act of speech is the cause of that particular act of speech.

One cannot understand or predict a particular dream from an examination of its instigator, or an act of speech from the most exact knowledge of the situation that is its occasion. Many different dreams may be incited by the same instigator, just as many—an indefinite number of—different utterances are possible in a given situation. An instigator or occasion does not necessarily result in a dream or in symbolic functioning of any kind, for such instigators or occasions may result in other kinds of mental states, acts, or functioning. In themselves instigators or occasions have no power to construct and provide no commitment to construct a dream, or indeed a symbolic representation of any kind. If a symbolic representation does ensue, an instigator or occasion does not determine what content or form it shall have.

Knowledge of the canons of symbolization of different symbolic systems constitutes, and defines the nature of, a capacity for symbolizing. Knowledge, a dispositional mental state, is manifested in performances—particular acts of symbolization. It is this kind of capacity (e.g., linguistic competence) that confers upon a person the power to create a particular symbolic representation: he knows how to do it.

Canons of symbolization govern the way elements of a symbolic system are used to create particular symbolic representations. Such canons do not govern relations between stimulus and response. These canons permit choices among many alternatives, usually if not always involve normative considerations, and—for example, in any instance of speech—may be violated to one degree or another, as is never true when a law of nature involves existential connections between external events that are unmediated by psychological activity.

It is important to note, with respect to the question of choices among alternatives in dream construction, that when Freud questioned whether associations to a dream, which are produced after the dreaming of it, were necessarily elements involved in its construction, he decided that they were not. A set of associations produced during the day following the dream may represent new connections "set up between thoughts which were already linked in some other way in the dream-thoughts." These new elements may form links between latent meanings and their manifest representation that are alternatives to those used in the construction of the dream. These new links are, "as it were, loop-lines or short-circuits, made possible by the existence of other and deeper-lying connecting paths" (p. 280).[10]

4. The Capacity for, or Commitment to, Symbolization: The Wish

It is a characteristic of the mind that wishes are necessary to move it to carry out mental work. Making a dream requires a commitment to achieve ends by symbolic means, and no such commitment will be forthcoming except at the behest of a wish. Freud postulated that the use of the capacity for dream construction, the commitment of resources to an attempt to make a dream, depends upon the activation or recruitment of a wish that is usually unconscious and usually sexual.

What is Freud's conception of what wishes are (pp. 565f.)? A need brings about a memory trace of excitation. A mnemic image of a "particular perception"—a perception of the need's being satisfied (a perception of receiving nourishment, for example)—becomes and remains associated with that memory trace of excitation. When the need arises again, "a psychical impulse will at once emerge." That impulse expresses itself in attempts to bring about the particular perception of the experience of satisfaction—that is, "to re-establish the situation of the original satisfaction." "An impulse of this kind is what we call a wish; the reappearance of the perception is the fulfillment of the wish; and the shortest path to the fulfillment of the wish is a path leading direct from the excitation produced by the need" to the particular perception itself. "Direct," because the shortest path does not involve taking any action in relation to the external situation. When the shortest path is taken, wishing ends not in action but in hallucinating.[11] A dream is just such an end; in constructing it, a direct path to wish-fulfillment has been taken.

10. This would form the basis of Freud's response to those, such as Glymour (1983) and Grunbaum (1984), who accuse him of a causal inversion fallacy because he appears to them to be attributing to thoughts appearing in associations *after* the dream a causal role in producing the dream. I provide a critique of this accusation in some detail in Chapter 15.

11. I will say in later chapters, and in particular in Chapters 9 and 10, not that

Thought and purposeful action aim to alter the external world in attempts to re-evoke the perception. But whenever there are circumstances where action seems or is impossible (for example, the state of sleep), when the world seems or is recalcitrant to all attempts to alter it, or when what is wished-for is by its nature impossible to bring about, then again a wish may follow the shortcut to wish-fulfillment. Wishing has the power to assist in producing a dream not simply by virtue of its intensity (its "energy") but by virtue of the fact that it is, as a state of desire, directed to a symbolic representation (a mnemic image of a perception of past satisfaction). For Freud, that's what wishing is—a commitment to re-evoke what is absent, the perception itself.

The wish clearly then cannot be a blind, inchoate, unorganized, structureless, noncognitive urge. (See Schafer 1968a; Schur 1966.) The mnemic image of the perception is a symbolic representation of a conception of an absent state of affairs. That is, the mnemic image is not merely an imitation of the perception it represents; it represents the perception "as remembered." This symbolic representation guides or governs efforts to re-evoke the perception of that state of affairs, in a way that perhaps has some similarities to the way a representation such as a map guides a process of exploration, or the way verbal representation such as inner speech regulate action.

That we speak of a mnemic image implies that what is remembered—perceiving a state of affairs—is located in the past. That wishing is directed to this symbolic representation implies that the state of affairs is also projected as a possibility in the future, and that a commitment exists to make the possibility an actuality.

Are wishes the *causes* of dreams? Although I do agree that Freud claims that wishes are causally relevant, that is, necessary for the construction of dreams, I question that he claimed that wishes are causes of dreams ("causes" in the sense of "producing" or "bringing about"). I believe that the two senses of "cause" ("causal relevance" and "power to produce"), which is often neglected in readings of Freud's work, have very different implications for the methodology of research.[12]

On the basis of passages I choose to emphasize, Freud's major hypothesis is that dreams have sense or meaning (in contradiction to

---

"wishing ends in hallucinating the experience of satisfaction," but rather, following Wollheim, that "wishing ends in *effectively imagining* the experience of satisfaction."

12. Various conceptions of causality are examined in Chapters 11, 13, and 15, especially the various implications these different conceptions have for psychoanalytic research—for dealing with the task of providing grounds for the credibility of the many kinds of causal claims that are made in the psychoanalytic situation and in psychoanalytic case studies.

the major rival hypothesis that they are the disorganized products of defective or inferior functioning by the brain during sleep). The sense or meaning of a dream is to be found in the wish or set of frustrated wishes (and ideas related to them) the dream attempts to represent or succeeds in representing as fulfilled.

This sense or meaning is sometimes difficult to discover because the dream as a representation of ideas is constructed by mental operations referred to by such terms as "condensation," "displacement," "symbolization," and "considerations of perceptual representability." These mental operations, collectively subsumed under phrases such as "dream work" or "primary process," are the way the mind works in certain states of consciousness (e.g., during sleep, or when conscious purposiveness is interfered with, as in states of varying degrees of sensory deprivation, or suspended, as in the procedure of free association).

The obstacles to the gratification of wishes represented by dreams have not necessarily arisen from intrapsychic conflict. That is, censored cravings are not the universal cause of dreams. The dream work, while it may serve motives of censorship when such conflict does exist, operates to give form to thought in sleep even when such conflict does not exist.

After many close readings of *The Interpretation of Dreams,* I myself remain somewhat perplexed by the causal status Freud has in mind for wishes in general and repressed infantile wishes in particular. He sometimes speaks as if dreams must be wish-fulfilling because the very nature of mental activity, as he defines mental activity, is purposive—somewhat as Talcott Parsons defines *action* as goal-oriented and value-governed and regards what does not have these properties as *behavior* but not *action* and therefore not covered by his theory. "No influence," writes Freud, "that we can bring to bear upon our mental processes can ever enable us to think without purposive ideas" (p. 528); ". . . it is self-evident that dreams must be wish-fulfilments, since nothing but a wish can set our mental apparatus at work" (p. 567).

Freud gives a number of arguments for the generalization that dreams are attempts at wish-fulfillment. He draws upon dreams in which sexual wishes, thirst, hunger, or the desire to change or get rid of situations that otherwise call for waking up are clearly represented as fulfilled. So, children's dreams, certainly the dreams of starving or thirsty men, the wet dreams of sex-obsessed adolescents, and such dreams as that of a student who wishes to sleep longer and also be on time to class and so dreams of being in his classroom are all clearly wish-fulfillments. Freud suggests that other dreams should then, presumably out of considerations of parsimony, be regarded as having this same property.

The difficulties in detecting attempts at wish-fulfillment in obscure dreams are to be solved theoretically—by making the distinction between manifest and latent content, and by postulating the mental operations of the dream work in order to get from one to the other. The theoretical devices used to explain these difficulties and to decipher the obscurities must have independent justification, of course. This strategy bears some similarity to Chomsky's, who formulated the rules that yield obviously grammatical sentences and then let these rules decide whether obscure sentences are grammatical or not.

However, Freud also depicts the wish as merely fuel or energy that the dreamer, whose sleep is disturbed by any kind of intense stimuli, must recruit to construct a dream. What appears to be necessary to instigate the construction of dreams is any stimulus (external, somatic, or psychic) intense enough to disturb sleep. What appears to be necessary if the attempt to construct a dream is to succeed is some frustrated wish, and this wish may be any kind of wish, preconscious or unconscious, recent or remote.

Why a wish? It provides the resources of energy needed for the effort of constructing the dream. Presumably, also, although Freud does not say so, the experience of wish-fulfillment is soothing—or, if the disturber of sleep is a frustrated wish, makes it seem unnecessary to wake up in order to gratify it.

"I am ready to admit," Freud writes, "that there is a whole class of dreams the *instigator* to which arises principally or even exclusively from the residues of daytime life" (p. 561). But he contrasts the entrepreneur, this dream instigator, with the capitalist (the wish), who has the resources, the motive-force or energy, to enable the entrepreneur to carry out his idea, that is, to construct the dream. Without capital to support the work of obtaining materials and constructing a dream out of them, the enterprise of the entrepreneur becomes difficult to carry out. Intrusion during sleep will most likely then provide an occasion for some other act than dreaming (such as waking up).

> A daytime thought may well play the part of *entrepreneur* for a dream; but the *entrepreneur,* who, as people say, has the idea and the initiative to carry it out, can do nothing without capital; he needs a *capitalist* who can afford the outlay, and the capitalist who provides the psychical outlay for the dream is invariably and indisputably, whatever may be the thought of the previous day, *a wish from the unconscious.*
>
> Sometimes the capitalist is himself the *entrepreneur,* and indeed in the case of dreams this is the commoner event: an unconscious wish is stirred up by daytime activity and proceeds to construct a dream. So, too, the other possible variations in the economic situation that I have

> taken as an analogy have their parallel in dream-processes. The *entrepreneur* may himself make a small contribution to the capital; several *entrepreneurs* may apply to the same capitalist; several capitalists may combine to put up what is necessary for the *entrepreneur*. (Freud 1900, p. 561)

Freud's choice of analogy here is felicitous. Money like the dream is a symbolic entity. Money is efficacious only by virtue of what it symbolizes, the meanings it represents. Money is a symbolic resource that may be used in a social system to attain a wide variety of ends. The nature of these ends is not determined a priori by the resource used.

According to the capitalist analogy, unconscious wishes are resources that can be used (but not used up) to construct a variety of dreams and, indeed, more generally, psychoanalytic theory proposes, to achieve a variety of ends, the exact nature of which is not predetermined by the resource used. A sine qua non of instinctual wishes is that they are displaceable from one object or state of affairs to another, when these are perceived as similar in certain respects.

In this capitalist analogy, the wish is like oxygen when a match is struck to make fire, or gasoline when someone wants to get a car moving. It is causally relevant, necessary, but it does not *cause*—that is, it does not have the power to produce or make—the fire or the car to move, unless currently evoked and activated it serves also as an instigator (entrepreneur). Indeed, Freud refers to the dream instigator (intense external, somatic, or psychic stimuli), which may not be a wish, as "looking for a wish," analogous to the way an entrepreneur looks for money.

> . . . it may even happen that a wish which is not actually a currently active one is called up for the sake of constructing a dream. A dream . . . has no alternative but to represent a wish in the situation of having been fulfilled; it is, as it were, faced with the problem of looking for a wish which can be represented as fulfilled by the currently active sensation. (P. 235)

Freud appears to have the following conception. The construction of dreams is mental work. Work takes energy. Repressed infantile wishes are sources of the energy required to construct *adult* dreams, presumably because their strength, relative to the strength of wishes of everyday conscious life, is so great, enhanced as it is by their having so long been refused gratification.

The belief that an infantile wish is represented in a dream, writes

Freud, "cannot be proved to hold universally; but it can be proven to hold frequently, even in unsuspected cases, and it cannot be *contradicted* as a general proposition" (p. 554). Here, he refers to the fact that the degree to which the evidence becomes available in free associations in a particular case depends on the degree to which the method is successfully implemented, that is, how strong the analysand's defenses are, how intense his conflict is, and so on.

Whatever instigates the construction of a dream, and to be sure in some cases the instigator may be the activation of a repressed infantile wish or any recent frustrated wish, that instigator will require the resources associated with repressed infantile wishes to carry out the task of constructing the dream. Wishes in many cases, then, are just that—neither the instigator nor the cause of an adult's dream but the necessary resources for its construction. Similarly, external stimuli, somatic stimuli, and recent indifferent mnemic images, which are not intense enough to disturb sleep, have no causal status. They may, however, become pressed into service as "cheap" materials used in the operations of the dream work to construct the dream.

Whatever one makes of these ideas and the particular passages cited, they do not seem to translate easily into the formulation that unconscious wishes *cause* (in the sense, produce) dreams. Unconscious wishes are unconscious mental states. Such states are enduring dispositions that may be, under certain circumstances, evoked. They are capable of becoming manifest in a wide variety of ways, especially in actually occurring conscious mental states. In any mental state, whether it be dispositional or occurrent, some attitude of a subject—wishing, for example—is directed to mental representations (symbolic representations) of objects or states of affairs. Such mental representations are the mental contents that are one kind of constituent of mental states.[13]

Certainly, I hold that unconscious wishes possess their efficacy by virtue of the symbolic representations that are constituents of them. These symbolic representations—of unconscious fantasies, for example—are causal processes, which, as they undergo transformations, transmit causal influence through space and time.[14]

In addition, I hold that the *cause* of dreams—what gives the dreamer the power to produce a dream when instigated to do so—is a dispositional mental state or capacity. That disposition or capacity is

13. I shall consider psychoanalysis as an intentional psychology—concerned with such intentional acts as wishing and believing—in much greater detail in Chapters 6 and 7.

14. For an explication of causality in terms of such causal processes, see especially Chapter 13.

the dreamer's possession of (one might want to say "knowledge of") the system of operations or procedures (including the dream work) for making a dream.[15]

5. The System of Operations Used in Constructing a Symbolic Entity: The Dream Work

Freud considered the dream work the essence of dreaming. His penultimate chapter on the dream work is the climax of *The Interpretation of Dreams,* and is in many ways the masterpiece of his masterpiece. He made explicit the operations by which a dream—a symptom, a joke, a myth, a work of art—is constructed. This must rank as one of his greatest contributions to psychoanalysis as a science of symbolic functioning. His discovery of the dream work is the foundation of his theory of dreams and (involving as it does the distinctions between primary and secondary process and between pleasure and reality principle) of his theory of mind. Upon this foundation these theories take their most secure position.

Operations (techniques, rules, procedures) and higher-order criteria (regulating principles) determine the form a symbolic entity will take. Operations—such as (in dreams) condensation, displacement, and giving primacy to representability, as well as, for example, the rules of language—determine the choice of materials or elements used and the way these are combined to make a symbolic entity. Regulating principles—such as reality principle and pleasure principle, as well as, for example, universal constraints on what kinds of rules may occur in a natural language—determine what operations may be used and toward what end their use is to be oriented. Another regulatory principle (in dreams): ". . . the dream-work is under some kind of necessity to combine all the sources which have acted as stimuli for the dream into a single unity in the dream itself" (p. 179).

Freud explicitly drew the analogy between the rules of language and the dream work: ". . . the dream-content seems like a transcript of the dream-thoughts into another mode of expression, whose characters and syntactic laws it is our business to discover" (p. 277). In more than one place, he suggested that the dream work operated, in part at least, through a linguistic transformation of a verbal representation of the latent dream thought into a verbal representation that is capable of manifestation in imagery.

> . . . the dream-thought that was represented was in the optative: "If only Otto were responsible for Irma's illness!" The dream repressed the

15. I elaborate this formulation in Chapters 13 and 15.

> optative and replaced it by a straightforward present: "Yes, Otto is responsible for Irma's illness!" This, then, is the first of the transformations which is brought about in the dream-thoughts even by a distortionless dream. . . . Thus dreams make use of the present tense in the same manner and by the same right as day-dreams. The present tense is the one in which wishes are represented as fulfilled. (Pp. 534f.)

The explication of the operations by which a symbolic entity is made is the most important step that can be taken toward an understanding of it. This explication is, as Chomsky has argued, logically prior to the consideration and solution of other problems: How is knowledge of a symbolic system acquired? What determines how knowledge of a symbolic system is used in performances? What are the consequences of various such performances? What kinds of impairments affect processes of acquiring and using knowledge of a symbolic system? It makes sense, before attempting to answer such questions, to try to understand first what is the nature of that which is acquired, used, or impaired.

To understand a symbolic entity is to be able to make it, to see how it is made, or to reconstruct it (E. Kris 1952). Freud declared that the easiest way of making the processes of the dream work clear and of defending his formulations about the dream work against criticism would be, if it were practical, to collect the dream thoughts he discovered through dream analysis and, starting with them, to reconstruct or synthesize the manifest dream (p. 310). Many of his statements seem to cry out for reformulation as specifying how an artisan, who wants to make a dream and who, in making it, must choose among alternative available materials, should proceed—what is possible, what required, what impermissible. The dreamer as artisan is perhaps epitomized in that kind of dreamer of whom Freud said, when he "is dissatisfied with the turn taken by a dream, he can break it off without waking up and start it again in another direction—just as a popular dramatist may under pressure give his play a happier ending" (pp. 571f.).[16]

To understand a symbolic entity is to comprehend how it is made. To comprehend how it is made is to understand the mind that made it. To discover mind through an analysis of the modes of symbolization and their products—poetry and science, mathematics and history, religion and neurotic symptoms—is the strategy of an important group of scientists and philosophers. Cassirer (1923–1929, 1944, 1946) sought to know man through the study of his works of culture.

16. The role of the dramatist or director in fantasies and in acts of effective imagining, such as wish-fulfillments are, is taken up again in Chapters 9 and 10.

Levi-Strauss (1958, 1962, 1964), who wrote that the mind of man is a mystery so long as music is a mystery, sought to find the mind of man through a study of his myths and social institutions. Chomsky reveals the mind of man through the study of his unique possession, language. They all owe something, their works are kindred, to this work of Freud.

Throughout Freud's scientific life, he did his most creative work not in trying to formulate a general psychology but in trying to explain relatively delimited symbolic structures—dreams, jokes, the psychopathology of everyday life, neurotic symptoms, works of art—by showing what goes into constructing them. He was what Piaget (1970) in his own terminology might now acknowledge him to have been: an ardent constructionist, the progenitor and prescient exemplar of constructive structuralism. Freud was quite right to announce proudly that as a result of his work "we are . . . presented with a new task which had no previous existence: the task, that is, of investigating the relations between the manifest content of dreams and the latent dream-thoughts, and of tracing out the processes by which the latter have been changed into the former" (p. 277).

A line of reasoning in Freud's later writing, especially concerning the two principles of mental functioning and the primary and secondary processes, suggests that the dream work is to be considered just one way of constructing symbolic representations of thoughts. The set of techniques, procedures, or operations used are those that will be used during sleep. One might argue, in other words, that intrusions during sleep instigate a process of thought, which may be given symbolic representation. In the conditions characteristic of the state of sleep, if thought is to be symbolically represented, it will be represented by dreams, as, in the waking state, thought is represented by speech, whether inner, written, or spoken. The first kind of representation makes use of the operations of the dream work. The second makes use of the rules of language.

In 1925, Freud added the following footnote to *The Interpretation of Dreams.*

> At bottom, dreams are nothing other than a particular *form* of thinking, made possible by the conditions of the state of sleep. It is the *dream-work* which creates that form, and it alone is the essence of dreaming—the explanation of its particular nature. . . . The fact that dreams concern themselves with attempts at solving the problems by which our mental life is faced is no more strange than that our conscious waking life should do so. (Pp. 506f.)

The phrase "form of thinking" should not be taken to mean that the thoughts represented are different from those represented in waking life, but rather that the form given to the representations of thought in sleep is different from that given to representations of thought in waking life. "[The dream work] does not think, calculate or judge in any way at all; it restricts itself to giving things a new form" (p. 507).

Dreams are a form of thought in which the representations of thought are generated by definite, distinctive canons of symbolization (Langer's laws of symbolization). The faculty of thought remains intact. The state of sleep does not alter the capacity to think, or the nature of thinking itself, but alters rather the processes by which symbolic representations of thought are generated. (Shall we conclude that secondary process and primary process may involve primarily, when thinking is at issue, differences in the symbolic representation of thought rather than differences in thought processes or contents?)

The consequence of this view, not enunciated or unequivocally accepted by Freud, is that the motive to censor cannot be a necessary cause of the use of the dream work during sleep. Disguise or dream distortion may be simply a consequence of the use of the particular kinds of operations comprising the dream work, which are the operations used—motive to censor or not—to construct symbolic representations during sleep. To be sure, this consequence is desirable from the point of view of a motive to censor, whenever such a motive is active.

Analogously, in natural language, that the possibility of recovering deep from surface structures (i.e., understanding the propositional content of a sentence) is maximized by transformational rules may be a consequence of the particular constraints natural language imposes upon what kinds of transformational rules may exist to generate surface structures from deep structures. One need not postulate, on that account, that every time such transformations are used, a motive to maximize communicative intelligibility or thought retrieval is active.

The thinking of the dreamer (or of some kinds of psychotic persons) is not different from the thinking of the nonpsychotic person in the waking state. Here, "thinking" refers to the most basic, probably universal categories and the most basic relations between these categories. What is different is the way in which thoughts of the dreamer, on the one hand, and of the nonpsychotic person in the waking state, on the other, are symbolically represented.

In psychiatry we have been preoccupied with differences in thought processes and with sometimes-called functional thought disorders. We might better be concerned with differences in the means by which thought is represented, with differences in processes of symbolization, and with the dysfunctional consequences of such differences. Some modes of symbolization, for example, might obscure the relation between a subject's thoughts and feelings, create obstacles to a subject's conscious recognition and understanding of his own thought, or interfere with a subject's communication of his thought to others.

6. The Meaning of a Symbolic Entity: Latent Thoughts and the Dream as Rebus

The meaning of the manifest dream is the set of latent thoughts it represents. We do not consider the dream work a type of thinking. It adds nothing, but only operates on symbolic representations of conceptions of objects and states of affairs in order to produce the manifest dream, which represents these same mental contents in a different way. There is no consciousness of the operations of the dream work. These operations are inferred from an examination and comparison of what they begin with and what their consequences are.

In what symbolic form is the latent dream thought clothed? How is it symbolically represented? The first answer is to suppose a naked thought waiting indifferently to be clothed in the form of one symbolic system or another. The second answer is to suppose that thought is always represented by means of some symbolic system. In this latter case, we are interested in the means by which a symbolic entity produced by means of one symbolic system comes to be translated into a symbolic entity produced by means of another symbolic system, when both symbolic entities represent the same thought.[17]

Freud seems to have chosen the first answer, when he considers thought in the unconscious to have no verbal form—how is it represented, if it is symbolically represented at all?—and to require a verbal form if consciousness of it is to be possible. But he seems to have chosen the second answer, when he describes the dream as a kind of rebus. Here, he is explicitly concerned with the translation of

17. I do not include here automatic reflex-arc-like sequences, however complex an assembly of such sequences may become, or conditioned responses, when a signal or index is inseparably attached to and an aspect of an immediately existent concrete entity that it signifies, or any problem solving unmediated by conceptual or imagist abstraction. I do include as a symbolic entity a completely interiorized event such as an image—what Piaget calls an "interiorized imitation" of a reality that is not necessarily present or existent—whether or not an exteriorization of it in a physical or objective medium has taken place.

symbolic entities generated by one symbolic system into symbolic entities generated by another symbolic system.

The latent dream thoughts and the content of the manifest dream "are presented to us like two versions of the same subject-matter in two different languages." The manifest dream content "seems like a transcript" of the latent dream thoughts "into another mode of expression, whose characters and syntactic laws it is our business to discover by comparing the original and the translation." We have no difficulty understanding the latent dream thoughts, as soon as we recover them. But the manifest dream content "is expressed as it were in a pictographic script, the characters of which have to be transposed individually into the language" representing the latent dream thoughts.

> If we attempted to read these characters according to their pictorial value instead of according to their symbolic relation, we should clearly be led into error. Suppose I have a picture-puzzle, a rebus, in front of me. It depicts a house with a boat on its roof, a single letter of the alphabet, the figure of a running man whose head has been conjured away, and so on. Now I might be misled into raising objections and declaring that the picture as a whole and its component parts are nonsensical. A boat has no business to be on the roof of a house, and a headless man cannot run. Moreover, the man is bigger than the house; and if the whole picture is intended to represent a landscape, letters of the alphabet are out of place in it since such objects do not occur in nature. But obviously we can only form a proper judgment of the rebus if we put aside criticisms such as these of the whole composition and its parts and if, instead, we try to replace each separate element by a syllable or word that can be represented by that element in some way or other. The words which are put together in this way are no longer nonsensical but may form a poetical phrase of the greatest beauty and significance. A dream is a picture-puzzle of this sort and our predecessors in the field of dream-interpretation have made the mistake of treating the rebus as a pictorial composition: and as such it has seemed to them nonsensical and worthless. (Pp. 277f.)

Language and Dreams

Manifest Dream Images and the Dream as Rebus

Freud repeatedly implies or explicitly states that knowledge about language (what Chomsky calls linguistic competence) is a resource used in constructing, as well as in interpreting, a dream. In the first chapter of *The Interpretation of Dreams,* he has already noticed the number of interpretations of dream elements made by oriental dream books that depend upon a play on words, "upon similarity of sounds and resemblances between words" (p. 99). He also wrote, "Indeed, dreams are so closely related to linguistic expression that Ferenczi . . . has truly remarked that every tongue has its own dream-language" (p. 99). No one who has read the passage in Freud (1911a), discussed in Chapter 1 of this book, in which he traces the familiar principal forms paranoia can take to linguistic transformations of the verbal proposition "I (a man) love him (a man)," can have much doubt about the felicity of the analogies I have drawn in this chapter between Chomsky's use of a transformational-generative model in linguistics and Freud's use of it in his theory of dreams.

The manifest dream is, given what the psychoanalyst actually interprets, operationally a linguistic entity. For it is the verbal description of the manifest dream that is the object of study in the psychoanalytic situation. That verbal report includes casual, apparently extraneous comments about and characterizations of the manifest dream as a whole, as well as amendments of its description.

The interpretation of free associations to the dream involves relating linguistic elements to each other. Free associations may be broken down into sentences, words, phonemes, even when reference is to memories or images. Linguistic elements in the description of the manifest dream evoke associations, and these associations in turn evoke other linguistic elements. We might well ask to what extent the evoking of free associations by verbal descriptions of manifest dream images are influenced by features of linguistic elements and by features of natural language as a symbolic system. To what extent does dream construction depend on, to what extent is it a function of, to what extent does it exemplify a use of linguistic competence?

If it is true that a dream is a rebus, then language and dreaming are intimately related. How do we understand a rebus? A picture of a building near a stream is followed by a picture of a road with trees on either side and that picture by one of a key. We take each picture to represent not a concrete word or syllable but an abstraction, a class of linguistic elements. The first picture represents in imagery the class of all syllables that sound like "mill," which includes the syllable "Mil." The second picture represents in imagery the class of all syllables that sound like "walk," which includes the syllable "wauk." The third picture represents in imagery the class of all syllables that sound like "key," which includes the syllable "kee." We select a member of each class represented, our active search governed by the constraint that the three selected syllables in sequence will themselves form a symbolic entity—a word, phrase, or sentence. "Milwaukee," then, is a symbolic representation in language of what was represented in imagery by the rebus.

A rebus, like a dream, may represent a class of words or syllables not only by means of the name or description of some object it pictures but by means of the formal characteristics of the picture. A picture of a stand over an eye may represent a conception of "position," or a class of words concerned with position, and so ultimately yield the selection of "under" to form the symbolic representation "I understand." A picture of two letters "A L" next to each other may represent a conception of two objects side-by-side or *together*, yielding "altogether" as the linguistic translation of the rebus.

The manifest dream report, a verbal description of the dream, evokes free associations, which in some cases at least represent classes of linguistic entities. The dreamer selects members from these classes to form a translation of the manifest dream in the verbal language of waking thought. Since the procedure is so much like that of translating a rebus, Freud is justified in emphasizing that the dream-interpreter must know and make use of the analysand's exact wording in reporting and describing the characteristics of the manifest dream.

Classes of symbolic entities can be formed according to the rules of language or constituting systematic deviation from the rules of language. (See the many examples and the discussions of them in *The Interpretation of Dreams* and also in *The Psychopathology of Everyday Life* and *Jokes and Their Relation to the Unconscious*—works to which we ought frequently to return to remind ourselves of the importance of language in understanding the phenomena of interest to psychoanalysis as a science of symbolic functioning.)

A class may include all words or syllables having the same or a similar sound, or a somewhat deviant class, from which we form rhymes and half-rhymes, all words or syllables having just one sound or combination of sounds in common. A class may include all words that can combine with a particular word (combinability or contiguity with that word is the criterion of membership in the class). A class may include all words having the same or similar meaning, or the same or similar syntactic or semantic features (substitutability, synonymy, or similarity are criteria of membership in the class).

The latter kind of class may well include antonyms. Antonyms usually differ from one another in only one semantic feature, sharing all other semantic features. Antonyms are reportedly the most common or frequent responses in experimental word association procedures, supposedly because antonyms involve a change but one that is minimal (Clark 1970). Freud observed the apparent equivalence or intersubstitutability of antonyms or opposites in the construction of dreams. He also noted what one might expect from the minimal difference between antonyms: the antithetical meanings represented by the same primal word. However, he connected this phenomenon to the apparent absence of the negative in dreams and the way in which contradiction is disregarded in dreams (1910, p. 155), rather than to the fact that such antonyms differ with respect to only one semantic feature and share all their other semantic features.

A difference between language and the dream work also suggests a source of some features of primary processes. In language, a linguistic entity may represent or name a class of linguistic entities of which it is not a member (e.g., "verb"). In the manifest dream, the name or description of a manifest dream image—just as in the rebus, the name or description of the pictorial element—can represent a class of linguistic entities just because the name or description of the image or pictorial element is also a member of that class.

Selection of those members of classes of linguistic entities that will be combined to form a translation of dream or rebus is not rule-regulated. Therefore, indeterminacy of meaning is inevitably a problem in the interpretation of dreams (and rebuses) in quite a different way than, for example, it is a problem in interpreting syntactically or semantically ambiguous sentences. But context—the context in which an ambiguous sentence appears, the context of a dream provided by the contents of contiguous conscious mental states reported to the psychoanalyst—determines decisions about meaning in ways that are not so different in the two kinds of interpretation.

The existence or formation of classes of linguistic entities, which may intersect, is one factor that makes condensation possible. That implies that abstraction (of classes) is a precondition of condensation. Given the idea that the primary process, which includes condensation, is commonly held to be a primitive mental process, how can this be the case?

The dream work operates according to an imperative to synthesize—to represent simultaneously a multitude of thoughts by the same manifest dream. From Freud's examples, we see that a manifest dream image or feature may be chosen because its name or description is a member of a class of linguistic entities. Each member of that class is potentially of use in alluding to different thoughts. So, different thoughts may be simultaneously represented by the one manifest dream image or feature.

A particular member of such a class may not allude to any of the latent dream thoughts. It is selected in constructing the dream because of considerations of representability. It names or describes an indifferent manifest dream image but, at the same time, it belongs to one or more classes of linguistic entities whose other members allude to the significant latent thoughts. It functions as an intermediate entity, which points both to the image directly, and indirectly to other linguistic entities in the class of linguistic entities to which it belongs. These other entities, in turn, enter into representations of the latent thoughts.

Displacement, then, as Piaget (1945) has pointed out, is almost an inevitable result of a process of condensation—and neither is necessarily activated by a motive to censor or disguise. For, as I have suggested, one line of Freud's thought has the dream work as simply the way symbolic representations are constructed in sleep.

Metaphor and Metonymy

Some connections between linguistic elements are not based on shared class membership or on combinations governed by rules of language. The connection is based on metaphor (a relation of resemblance of some kind) or metonymy (typically, a relation of part and whole).

Psychoanalysts are continually responding, as they listen to free associations, to resemblances between objects or states of affairs of different kinds or in different realms. As the analysand describes relations between parent and child, relations between different aspects or mental states of the analysand, relations between the analysand and different social and cultural objects and states of affairs (work, play, social life, books, movies, music, world events), and relations between the analysand and the psychoanalyst, the psychoanalyst becomes aware that similar features or patterns recur across the descriptions of these very different kinds of relations.

Implicit comparisons and resemblances between different kinds of objects and states of affairs, existing in different realms, abound in Freud's account of the Irma Dream. Mixed relationships with a patient (who is also an intimate) and with a junior and senior colleague (each of whom is also a friend) are compared. Mixed relationships are linked to the mixed results of treatment. The giving of interpretations (one kind of solution) is compared to the injection of drugs (another kind of solution). A patient's refusal to accept an interpretation is compared to a woman's refusal to yield to a man. Examination of the nose is likened to penetration into female sexual organs. Wives, widows, patients, and dreams all keep their secrets. Foul-smelling wastes are eliminated (diarrhea); offensive psychological symptoms are also eliminated. The connection of dirty injections and foul-smelling toxins with foul-smelling discharges is compared to the connection of words (e.g., interpretations), semen, and food (e.g., strawberries) with "good" eliminations (discovery, birth, and the elimination of a diphtheritic membrane).

The psychoanalyst's response to metaphor, to different interpretations of the same abstract model or relational structure, and to homology is an important part of his expertise.[18] However, we are unable to explicate exactly what about that expertise depends on linguistic competence, and what precisely is the nature of that competence. For such explication, an adequate linguistic theory of metaphor is required.[19] When we have such a theory—and psychoanalysts might very well contribute to its formulation—we shall know a great deal more about how analysands construct and (with psychoanalysts) interpret dreams.[20]

Determinacy of Meaning

I have already mentioned the problem of indeterminacy of meaning in the interpretation of dreams. Freud, on the same topic, commented:

> I had a feeling that the interpretation of this part of the dream was not carried far enough to make it possible to follow the whole of its concealed meaning. . . . There is at least one spot in every dream at which it is unplumbable—a navel, as it were, that is its point of contact with the unknown. (P. 111)

This apparent indeterminacy of meaning is a matter of practical constraints and not of a limitation in principle. It is a matter of what associations are practically obtainable on a particular occasion. For there is a means by which the meaning(s) of a dream can be determined. The allusions of the manifest dream images, although they seem to radiate in all directions, will ultimately converge upon relatively few common points.[21] These nodes of intersection point

18. Chapter 3 contains a consideration of the role of abstract models and homology in the relations between psychoanalysis and other disciplines. Chapter 15 contains a consideration of the role of models and analogy in theory-building and theory-validation.

19. See Chomsky (1965) and Katz (1964).

20. For more about what goes into psychoanalytic interpretation, exemplified by interpretation of the Irma Dream, see Chapter 4.

21. I shall return, in the last chapter of this book, to the theme of convergence, in a discussion there of the way in which such processes of convergence provide evidence and a line of argument supporting the credibility of causal claims in the psychoanalytic case study.

directly to the latent thoughts represented by the manifest dream.[22] The content of a dream is "the fulfilment of a wish." Yet the wish does not exhaust its meaning.

> Certain other themes played a part in the dream, which were not so obviously connected with my exculpation from Irma's illness. . . . But when I came to consider all of these, they could all be collected into a single group of ideas . . . this group of thoughts seemed to have put itself at my disposal, so that I could produce evidence of how highly conscientious I was, of how deeply I was concerned about the health of my relations, my friends and my patients . . . there was an unmistakable connection between this more extensive group of thoughts which underlay the dream and the narrower subject of the dream which gave rise to the wish to be innocent of Irma's illness. (Pp. 118ff.)

Two criteria at least govern the choice of a set of images for use in constructing a manifest dream. Each image should be capable, compared with possible alternatives, of allusion to a maximum number of the latent thoughts to be represented by the manifest dream. The set of images should contain, compared to possible alternative sets, a maximum density of nodes, where processes of allusion arising from each image intersect. The first criterion concerns economy in assigning one perceptible representation to many latent thoughts. The second criterion concerns means for assuring that reflecting upon different dream images under optimal circumstances will result in the recovery of the underlying meaning(s) of the dream.

The method of free association is designed to minimize the influence of processes screening out or suppressing allusions radiating from the manifest dream images. The discovery of nodes of intersection by means of free association is the method of disambiguating a manifest dream. The "unplumbable navel" is a consequence of our practical inability to follow all lines of allusion to theoretically possible points of intersection.

Poetry and music share with the system generating dreams the problem of providing for the recovery of determinate underlying meanings from perceptible symbolic representations. An apparitional or presentational symbol is constructed using materials capable of giving rise to

22. E. Kris (1956a, 1956b) similarly discusses that, in attempts to reconstruct personal histories, a certain apparently remembered event may actually be a node of intersection where multitudes of unremembered details of different events converge on a particular kind of pattern of experience. Such nodes constitute personal myths; these myths may be presented as if they were completely veridical life stories.

a large number of allusions in order to realize with maximum economy the perceptible representation of deep or underlying, and multiple but interrelated, meanings.

Ambiguities constitute a valuable resource in creating the effect or carrying out the function of an art form. But the demonstration of ambiguities in an art form does not compel acceptance of the impossibility of discovering an intrinsic basis for preferring one meaning to another or for relating various meanings to each other (E. Kris 1952). In studying such art forms as poetry and music, as in studying language and the symbolic system generating dreams, one must distinguish between a symbolic system's theoretical potential for generating comprehensible forms, and the practical constraints upon the realization and immediate comprehension of such forms imposed by limitations in the capacities of, or by motivational obstacles existing within, either audience or performer or both.

Censorship and Dream Distortion

Freud postulated that a censor operates in the construction of manifest dreams to disguise unacceptable latent thoughts by distorting their symbolic representation so that its meaning is unrecognizable. The existence of censorship is suggested by a number of phenomena: the analysand's subjective experience of internal opposition to carrying out the procedure of free association; his "inability" to recover memories; his refusal to recreate processes of allusion (or to create processes parallel to those involved in constructing a dream); the repugnance felt by the analysand in the waking state toward the latent wish once it is discovered, and his efforts to get rid of it; and, by analogy, the disguise and distortions of representations of unacceptable wishes in acts of conscious and often ingenious dissimulation on occasions of communicative interactions, in order to avert anticipated, imagined, negatively valued responses from others.

Freud suggested that the dream is a compromise formation of two agencies, one "defensive" and the other "creative" (p. 146). Chomsky makes an analogous distinction between the infinitely generative rules of language and the transformational rules, which are merely interpretive. That is, transformations generate no new propositional content. Like defenses perhaps, such transformations serve as filters. Only acceptable deep syntactic structures are combined by transformational processes to yield the surface structures of perceptible representations.

I want to examine, for heuristic reasons, the functions served by having both a deep-structure-generating component (the creative or productive operations), and a surface-structure-generating component (the transformational nonproductive operations) in language. In particular, an examination of these functions will suggest that the existence of a "creative" agency and a "defensive" agency in the construction of a dream may not simply serve a motive to censor. As in language, which is deeply implicated in dreaming, the existence of two such agencies may be required by a process of symbolization itself.

1. *Economy.* An extraordinary amount of abbreviation or condensation occurs in the process of transforming deep into surface structures. Numbers of pages are often required to make explicit all the meaning-bearing elements and their interrelations in deep structures from which the surface structure underlying even a single complex sentence has been derived. A relatively simple sentence such as "The short, happy boy, who wanted to go to the store, went with his mother" has a structure derived from a combination of the abstract structures underlying such sentences as "The boy was short," "The boy was happy," "The boy wanted to go to the store," "The boy went to the store," "The boy has a mother," "The mother went to the store with him." This simple sentence includes as well indications that the boy is identical in all these underlying propositions, and that "who" refers to the boy and not to his mother, and also includes indications of tense, mood, and gender and number accord.

Similarly Freud declares and his book throughout demonstrates:

> Dreams are brief, meagre and laconic in comparison with the range and wealth of the dream-thoughts. If a dream is written out it may perhaps fill half a page. The analysis setting out the dream-thoughts underlying it may occupy six, eight or a dozen times as much space. (P. 279)

An obvious function served by noncreative transformational processes is economy. As it is, we often have the experience that our thoughts race faster than our capacity to represent them. How much more difficult the representation of thought in language would become if one were required to reflect all propositional content explicitly and directly in the surface structure or arrangement of elements of a sentence.

The dreamer thinks, and represents his thinking in the form of manifest dreams. The manifest dream is not so constructed as to promote ease of communication of meaning. It may nevertheless be regarded as more or less adequately keeping up with and symbolically representing the dreamer's thoughts, so that his thought may be sensually apprehended and contemplated by him.

Any symbolic system provides means for the relatively economic representation of thought. Differences in deep and surface structures, and transformational processes changing one to the other, constitute a minimum—perhaps optimal—device of this sort. Transformational processes may serve a motive to censor, but are not necessarily activated only by that motive in a process of symbolization.

2. *Relative independence of meaning and its representation.* The intrinsic characteristics of a medium or kind of material used to represent thoughts determine what rules or procedures govern its use. These rules or procedures, which are dominated by the intransigence of the properties of the medium itself, are to one extent or another relatively independent of considerations of meaning.[23]

Generating deep structures is primarily governed by considerations of meaning. These deep structures, depending in part upon how they are then changed by transformational processes, can be symbolically represented by more than one medium. Different ways of transforming deep structures into derived surface structures are primarily governed by considerations of perceptible representation in a particular medium.

The range of permissible sounds and sound combinations is limited in any natural language. Deep structures must be transformed into surface structures amenable to permitted compositions of sound. Similarly, the stock of images available to the dreamer is limited. The condition, extent, and nature of this stock of images is certainly determined, but not only determined, by the effects of defense, inhibition, and the constriction of what are possible experiences. In dreaming, transformations of representations of latent thoughts must operate to provide representations that are capable of describing dream images—the materials available.

3. *Pragmatics.* Some transformations, conserving propositional content (meaning) but altering manner of representation, are responsive to characteristics of the situation in which the representation may appear. Considerations of appropriateness may be influential. The manner of representation might affect which elements are highlighted and which suppressed, which are the focus of attention, and which distract attention from others—thereby affecting the attention, interest, and attitude of another in a communicative interaction.

The censorship belongs in this realm of pragmatic factors governing symbolic representation. It is important in thinking about dreaming and dreams to ask about any dream to what degree prag-

23. See Edelson (1971, Chapter 3) for further discussion of the properties of a medium used in constructing symbolic entities.

matic considerations influence its construction over and above the influence exerted by considerations of economy and the nature of the medium. There is no a priori reason to suppose that the tendencies to dissimulation and censorship that are responsive to the exigencies of communicative interaction (including communicative interactions among subsystems of the mind) must also operate in a major way in the sleeping state in every instance of dream-construction. Of course, such tendencies may be activated in anticipating the reporting of the dream in the psychoanalytic situation, and may dominate the final revision of the dream as it is remembered while awakening and in the waking state.

Freud tended increasingly, as time went on, to emphasize the intrinsic consequences of using a particular symbolic system for representing thought. The hypothesis of a primarily defensive agency (censor) interacting with a primarily creative agency (wish) is explicitly altered in a 1930 footnote (p. 146) in which he reminds us that, on the contrary, a dream mav express a wish that belongs to the defensive agency. (This amendment also reminds us to avoid too rigidly separating impulses and defenses as essentially different kinds of psychological entities.)[24] The 1925 footnote I have quoted in the section of this chapter on the dream work, which declares that "dreams are nothing other than a particular *form* of thinking" and that "the *dream-work* . . . creates that form" (pp. 506f.). is written from the perspective I am suggesting Freud increasingly adopted.

What appears to be a consequence of a motive to censor may be primarily a consequence of using such procedures in dream-construction as: (1) the use of one image to represent many thoughts (condensation) and many images to represent one thought (displacement); (2) the selection in constructing a dream of a member of a class of linguistic entities (some of whose members are linked to latent thoughts) primarily out of considerations of its potential for pictorial representation (its capacity to name or describe an image that will represent that class); and (3) the necessity of representing a number of thoughts simultaneously rather than sequentially.

My clinical experience seems to confirm that there need be no difference between the properties of a manifest dream that is easily interpreted by an analysand whose resistance to discovering and accepting the meaning(s) of that dream is relatively low, and the properties of another manifest dream that cannot be interpreted by an analysand whose resistance to discovering and accepting the meaning(s) of this other dream is relatively great. Both of these manifest

24. This point will be taken up in considerable detail in Chapter 9.

dreams are constructed in the same way, and as symbolic representations both offer the same intrinsic difficulties to interpretation. One cannot tell from an inspection of manifest dreams which will be easy and which will be difficult to interpret. Indeed, the same dream—or at least one very much like it—is difficult to interpret at one time and easy to interpret at another.

Similarly, difficulty in understanding a psychotic patient's communications and productions may be a consequence of his use of techniques in symbolizing that are like those of the dream work. (Of course, the contribution of the interpreter's resistance to becoming aware of what such a patient is communicating or thinking cannot be ignored.) The difficulty in understanding may arise not because of any wish to disguise or censor on the part of the patient, but perhaps simply that his level of functioning leaves him no choice but to use such a mode of symbolization (Edelson 1971).

The Dependence of Dreaming upon Language

An important implication of Freud's use of the rebus as a model of the dream[25] is that the latent dream thoughts must be, to begin with, represented linguistically, or must come to be represented linguistically, before a manifest dream can be constructed. "We may suppose that a good part of the intermediate work done during the formation of a dream, which seeks to reduce the dispersed dream-thoughts to the most succinct and unified expression possible, proceeds along the line of finding appropriate verbal transformations for the individual thoughts" (p. 340).

If this is the case, the dream (and perhaps any visual imagery to the extent it is a symbolic entity) cannot be regarded as a more primitive, or as an example of an ontogenetically earlier system of, symbolization than language. Thinking seems to require propositions. Imagery is ineluctably ambiguous. An image cannot represent unequivocally a particular proposition. An image can represent a particular proposition only "under description" (Fodor et al. 1974, pp. 1–21). When and if thinking operates on images, it operates on images "under description." When and if we think in images, we require language to provide the description.

Language as a symbolic system is, according to the rebus model, logically prior and indispensable to the dream as a symbolic system. In that case, dreams should not appear before the development of

25. The importance of models, homologies, and analogy for both constructing theory and establishing the scientific credibility of theory (and, in particular, the scientific credibility of causal claims) is discussed in various chapters throughout this book, but most extensively in the last chapter.

some degree of linguistic competence, and impairments in linguistic competence should result in impairments in generating dreams.

It is especially important to keep in mind here the distinction between generating a dream and the physiological conditions (e.g., REM sleep state, which has been observed in infants) that are necessary but not sufficient for the occurrence of such symbolic activity. There is no a priori reason to reject the possibility that REM sleep state may exist in nature uncorrelated with any symbolic activity (E. Hartmann 1967).

In this connection, Piaget (1945) observes: "In the case of children, we have been unable to find evidence of authentic dreams before the appearance of language" (p. 177).[26] Jakobson (1964), referring tantalizingly to two apparently untranslated articles (in Russian?), comments in a discussion of aphasia: "The inhibition of visual dreams connected with encoding disorders of language has been rightly interpreted as a break-down of that code which provides the transition from verbal to visual signals" (p. 32).

Sapir (1921) makes a related point, although he is describing systems of voluntary communication: "We shall no doubt conclude that all voluntary communication of ideas, aside from normal speech, is either a transfer, direct or indirect, from the typical symbolism of language as spoken or heard or, at least, involves the intermediary of truly linguistic symbolism" (p. 21). Sapir attributes the priority of language not to its auditory elements, but to its organization as a system. "The ease with which speech symbolism can be transferred from one sense to another, from technique to technique, itself indicates that the mere sounds of speech are not the essential fact of language, which lies rather in the classification, in the formal patterning, and in the relating of concepts" (pp. 21f.). Similarly, Edelheit (1969) speculates about the priority of language (he appears to mean speech or not to differentiate sharply between the two) over the visual image (p. 400).

The conjecture that dreaming is dependent on language is consistent with Chomsky's emphasis on the innateness of linguistic competence (noting especially in this connection the rapidity with which the child constructs a language system with only a brief experience with what are at best debased and fragmentary speech utterances), the species specificity of linguistic competence, the early age of language development, and the relative lack of correlation between language development and general intelligence.

26. One would like to know what method of inquiry led to this particular failure.

Psychoanalysis and Other Scientific Disciplines[1]

3

Psychoanalysis: A Discipline on the Boundary

What is there about psychoanalysis that leads so many of its scholars, as well as so many scholars in other disciplines, to take up habitation on the boundaries between it and other disciplines, and to carry out work that crosses these boundaries?

Psychoanalysis as a science takes as its domain symbolic entities—their properties, interrelations, modes of construction, and the meanings they represent. Such entities include neurotic symptoms, the transference neurosis, fantasies, dreams, jokes, and the artifacts of culture. The theory of psychoanalysis is a theory of the mind's symbolic functioning. Psychoanalysis postulates what features the mind must possess to be capable of creating these symbolic representations.

The contribution of psychoanalysis to other disciplines is a conception of the mind that is adequate to the mind's complexity and to its surprising, unique, species-specific achievements in the realm of symbolization. Of course, the findings of another discipline may constrain to some extent what is possible and what impossible for psychoanalysis to postulate about the mind in its symbolic functioning, just as the findings of psychoanalysis suggest to other disciplines what is plausible and what implausible to assume about the human mind in their studies of the physical objects built by humans and the cultural objects and social interactions and institutions created by them.

Many disciplines have a vested interest in a theory of the mind in its symbolic functioning. It will be a part of

1. This chapter is an extensively revised version of Edelson (1977).

the theory of any discipline, belonging to the humanities or the social and natural sciences, that studies man or the works of man. It will be a part of the knowledge applied in any professional or technological endeavor seeking to change or make use of human capacities. It will be of passionate interest to any scholar, scientist, or professional, since it is in the light of such a theory that he reflects upon his own activities.

For psychoanalysis, it is a mixed blessing to be so centrally located, jostled, of such import to so many. There are special vulnerabilities and difficult questions associated with life in a boundary region. What is too little contact with those on the other side of the boundary? What is too much? The prison guard worries as he is tempted, in intimate relations with those he guards, to abandon his own identity and become one of them. He may do so, or instead defensively become impervious or overzealously antagonistic to them. The nurse in a residential treatment center worries, in her continuous interaction with those for whom she cares, "Am I too involved? Too remote?"[2]

Psychoanalysis has always risked being too easily influenced and losing its identity, what is distinctive of itself alone. At the same time, psychoanalysis has always risked defensively becoming so impermeable to influence that it no longer will know what it needs to know, what it can or needs to change.

Many have already decried the impermeability of psychoanalysis. It is for the most part preoccupied with clinical training in institutes with little commitment to or resources for generating new knowledge (Holzman 1976). These institutes are often isolated from the university and so from free exchange and collision with ideas in the intellectual marketplace.

But such impermeability is matched by an excessive porosity—the other side of the coin in a boundary region. The consequences of excessive porosity may be as troublesome as those of impermeability.

It is easier to be indignant about the abuses consequent upon an indiscriminate porosity when it is our neighbors rather than ourselves who commit the abuses. Psychoanalysts, and other scholars as well, are rightfully distressed when a member of another discipline picks something out of context from psychoanalytic theory and calls what he has culled "psychoanalysis." No matter to him that what he uses is outdated or inaccurate, that what he does with it is reductive, and that he perpetuates an oversimplification. (Favorites are ascriptions of decisive influence to a single early traumatic event, or

2. These kinds of intergroup relations are examined in the setting of the psychiatric hospital unit or residential treatment center by Edelson (1964, 1970a, 1970b).

riotously indulgent allusions to sexual "symbols.")[3] The problem of misdepicting psychoanalysis might well be mitigated if the great works *The Interpretation of Dreams, The Psychopathology of Everyday Life, The Three Essays on the Theory of Sexuality,* and *Jokes and Their Relation to the Unconscious* were to replace *Civilization and Its Discontents, Future of an Illusion,* and *Beyond the Pleasure Principle* as favored or privileged works representing psychoanalysis in undergraduate curricula.

Our neighbor's use of psychoanalysis may be inimical, even when he does not debunk or dismiss it. There are those who use psychoanalytic terms as they wish, fitting them to their own purposes, without regard to the possibility that thereby a language may be corrupted or destroyed. There are those who proudly confess that their intention is to find minor, incidental passages in Freud's own writing that can then be used to "sabotage" Freud's text. The reference to psychoanalysis in a so-called psychoanalytic perspective on another topic is often perfunctory—relegated to the periphery in prefatory, passing, or eminently deletable comments. Such a use of psychoanalysis implicitly depreciates it, since the isolated reference to psychoanalysis inevitably contrasts with the sophisticated conceptual apparatus in the borrowing discipline of which our neighbor is a master. So derision and apparent appreciation in the end amount to the same thing.

But, of course, psychoanalysts, as inhabitants of a boundary region, similarly misuse and desecrate other disciplines, and in so doing spoil our own. I believe we ought to set standards for the work we do in relation to other disciplines. Here are four maxims, somewhat obvious and easy to remember.

Beware of Obscurantism

Beware in another discipline, or indeed in any "school" of psychoanalytic thought, of any deliberate obscurity, any boastful evasion of clarity. The psychoanalyst of today does his work in an anti-intellectual milieu. Many in our society are distrustful of science and its applications. They find its notation unfamiliar and forbidding. They can no longer grasp its concepts or its visions. Its consequences (desirable or undesirable) have become beyond imagination and occur at a rate so desperate it obliterates any sense of moving into the future and leaves rather a longing for whatever is static and outside time. Few take pleasure in the rigors and lucidities of conceptual analysis.

3. More about such misuses and misdepictions of psychoanalysis in Chapter 8.

Many idealize instead whatever is primitive, inarticulate, and undifferentiated. This idealization is made popular in movies and on television, in speech that is designed to persuade but is remarkable at the same time for being both a-syntactic and unctuously glib.

Various forms of obscurantism seek to a-*maze* the psychoanalyst, to lure him with a rationalization that is difficult to resist: obscurity of thought and expression alone, it is claimed, can do justice to the inchoate and unknowable—and these are surely attributes of reality the psychoanalyst more than anyone must know.

If a psychoanalyst, forgetting the distinction between properties of his subject matter and those of the methods he uses to study it, finds himself tempted by talk of a truth beyond the reach of science, by words like "Oneness," "Being," "Self," and "Other," characteristically with capital letters, the best antidote is to read forthwith almost any passage from the writings of Bertrand Russell. I recommend especially those passages in which metaphysics has its main value in inciting Russell to analyze, with wit and panache, murk and confusion in the thought and expression of his colleagues.

Russell is an antidote because—like Freud, who sought in his theoretical work ever closer formulations to what he called "the unknown reality" (1900, p. 610)—no matter how fantastic the objects Russell investigated, he was able to depend on what he called his "vivid sense of reality," his "vivid instinct as to what is real" (1956, p. 223). In a passage that suggests the difference between difficulties in comprehending the technical product of a strenuous effort to be precise and difficulties in comprehending the arcane product of a vaunted obscurantism, Russell comments, "It is a rather curious fact . . . that the data which are undeniable to start with are always rather vague and ambiguous," and "that everything you are really sure of, right off is something that you do not know the meaning of, and the moment you get a precise statement you will not be sure whether it is true or false, at least right off."

> Everything is vague to a degree you do not realize till you have to make it precise, and everything precise is so remote from everything that we normally think, that you cannot for a moment suppose that is what we really mean when we say what we think.

He goes on to deplore "the tendency . . . to suppose that when you are trying to philosophize about what you know, you ought to carry your premisses further and further into the region of the inexact and vague, beyond the point where you yourself are, right back to the child or monkey" (1956, pp. 179–181).

Avoid the Facile and Faddish

It is difficult for a psychoanalyst to resist the power and attraction of facile or faddish work in another discipline; the accessibility, the popularity, and the apparent promise of such work are so seductive. However, its content or methods may prove on analysis to be incompatible with the aims, basic assumptions, or subject matter of psychoanalysis. A consideration in this regard of some of the variants of structuralism and semeiotics or of various phenomenological, existential, or dialectical metaphysics is beyond the scope of this presentation. (But see Edelson 1975, pp. 7–13.)

A homely example, however, may suffice to illustrate the point. A psychoanalyst might once have been tempted to reformulate psychoanalytic theory in terms of, and thus have found his work entangled with, paradigms of conditioning in experimental psychology. He might not have noticed that experimental psychologists, with whom he was inclined to ally himself, adhered as a rule to an outmoded radical empiricism and to antitheoretical canons, long since discredited by their own original formulators. In addition, it was rarely if ever made explicit that coercion of some kind is indispensable in most of the experimental situations in which signals become connected with responses (which are mere reactions, not actions). The animal or human subject must be caged, strapped, held, or must hold himself still for the procedure to be carried out. Peirce, in discussing indices and signals, refers to the "brute" connection between an index or signal and the immediate unmediated—I would say "mindless"—reaction to it (Buchler 1955; Hartshorne and Weiss 1932; Wiener 1966). Psychoanalysis can make at best only limited use of such findings. Their generalizability is more restricted than may be apparent to those who generate or those who borrow from them (Edelson 1971).[4]

I shall not go into detail here about the logical fallacies, which may also escape the psychoanalyst, in the definitions of, and in attempts to extend to phenomena outside the controlled experimental situation, such terms as "stimulus," "response," and "reinforcement." These fallacies have been brilliantly pointed out by Chomsky (1959), whose erudition and expertise encompass symbolic logic, the logical foundations of mathematics, the philosophy of science, and psychology, as well as linguistics. It is Chomsky, among others, whose work has also

4. Chapter 5 points out that extrapolating from experimental work with animals in order to answer questions about anxiety, as psychoanalysis conceives anxiety, is of very limited usefulness.

demonstrated the limitations of computer methodology and information theory. A psychoanalyst who might have made irrevocable commitments to computer methods and information theory for scientific work in his own discipline may persist in his enthusiasm, while the reservations of those in other disciplines who test the claims of the methods and theory increase with disconcerting rapidity.

Eschew Intellectual Imperialism

As a guest in the territory of another discipline, the habits of pig or magpie should not be imitated. Current attempts to make psychoanalysis a general psychology may be an expression of intellectual imperialism. Psychoanalysis and psychology do not have the same empirical domains.[5] Psychoanalysis is certainly not *the* science of man, but is one among many such sciences.

An indiscriminate piggishness tries to swallow all the facts and theories of other disciplines, as if all knowledge in these disciplines could be made part of psychoanalysis simply by translating their terms and propositions into the language of psychoanalysis. One thinks in this connection about the attempted assimilation of the work of Piaget, as well as the work of various developmental psychologists, cognitive psychologists, ethologists, sociologists, and anthropologists.[6]

Such translations are likely to be reductive in the sense of "the ideas of another discipline are nothing but psychoanalytic ideas." Translations are not to be confused with the often useful results of logical analyses. A logical analysis, for example, may demonstrate that the propositions of psychoanalysis and those of another discipline are compatible, or stand in some other particular logical relation to one another. Or logical analysis may demonstrate that psychoanalysis and another theory have the same formal structure; they are homologous. That is, each is concerned with a different content, each uses different terms, but both are interpretations of the same abstract system.

It is destructive to what is distinctive about psychoanalytic knowledge to pick up bits and pieces from others and graft these onto the nest psychoanalysis carefully puts together and precariously maintains. But it is truly difficult not to be a magpie if one is unable to

5. Arguments to support this assertion will be found in Chapters 6, 7, and 8.

6. Chapter 9 discusses misdepictions of psychoanalysis arising from enthusiastic borrowings from cognitive psychology, as well as the uncritical adoption of standard paradigms of experimental quantitative research, in designing research to test psychoanalytic conjectures.

define rigorously what is distinctive about psychoanalysis, if one does not know precisely what its exact scope is or how to draw the boundary of its domain.[7]

Master the Discipline That Is Not Your Own

Once you decide the commitment to a body of work in another discipline is worthwhile, study exhaustively, arduously, ardently, with love. Attraction may begin with vague impressions, intuited similarities, and enthusiasms inspired by broad outlines or surface features. Do not be the kind of lover for whom these are enough. Grapple with this new subject. Study its details. Study its technical notation and terms. Search for its underlying structures. Be vigilant in renewing your knowledge of it; keep up with changes in it.

A psychoanalyst approaching another discipline must take care to master it. He should not skim over its surface but remember its details; its technical notations and terms; the special senses in which its concepts are defined and used; the intricate logical connections among, and the differences in the logical status of, its propositions; and its particular methods for providing evidence to support its causal claims.

That means study, usually years of study. That means reading difficult books and papers; catching glimpses of what a worker is up to and what is important about his work; re-reading then, with commitment confirmed, the same books and papers, as well as other books and papers to which each one leads; grasping now what on first reading was passed over; experiencing exhilarating insights into unsuspected relations and exciting implications; re-reading once again to a level of familiarity, even intimacy, where one can use knowledge knowing how concepts have developed and which issues have been resolved and which remain unresolved, and one can be critical without glibness and without loss of love.

Failing this effort, I believe that with respect to another discipline the psychoanalyst should remain silent. For we know only too well, and to our regret, the consequences of falling short of such a degree of mastery and failing to keep up with what has since developed in another discipline.

Nowhere is this more obvious than in the way psychoanalysts have allowed themselves to be influenced by what are now outdated doctrines in the philosophy of science. Throughout the social sciences these doctrines led to a pernicious, constricting scientism, marring psychoanalysis in many ways, some so subtle that even those who set

7. Solving this kind of problem is a major theme of Part 1.

their teeth against this scientism cannot detect the ways in which in their own thinking they have succumbed to it.[8]

The evidence for such consequences lies in discussions that threaten to become as interminable and ultimately futile as the old nature-nurture controversy. Should psychoanalysis dispense with metapsychology in favor of its so-called clinical theory? Is psychoanalysis science or hermeneutics? Is psychoanalysis fundamentally different from other sciences because it has a humanistic rather than mechanistic view of man?[9]

The evidence also lies in the ambivalence with which psychoanalysts regard their own theory of the mind—particularly the use of a system-subsystems model in conceiving of "id," "ego," and "superego." This ambivalence is expressed, for example, in worries about reification and the implausibility of such hypothetical entities. The same ambivalence is expressed when psychoanalysts, acknowledging Freud as a genius, and proud of his great discoveries, are nevertheless likely to acquiesce in talk deprecating him and his theory of the mind. Such talk judges the theory, quite mistakenly, to be a congeries of loose, farfetched, unrelated, and inaccurate or outmoded analogies and speculations, and judges Freud to be regrettably careless, inconsistent, and unsystematic in his formulations, and naive and unsophisticated as a scientist. (His grasp of scientific reasoning and of methodological issues in science was, in fact, in many ways in advance of his own times.[10])

The Status of Psychoanalysis as Science

Is psychoanalysis so different from other sciences that it cannot be considered an empirical science at all? Is psychoanalysis hermen-

8. The reader will note in Part 2 that continued study of the philosophy of science has led to a deep change in my thinking about the problem of achieving evidential support for causal claims in psychoanalysis. I move from a wholehearted espousal of eliminative inductivism in Chapter 12, to a beginning recognition in Chapter 11, to a full recognition in Chapters 13 and 15 that, although the canons of eliminative inductivism are important in attempts to obtain evidential support for *empirical generalizations*, these canons are not sufficient for dealing with a *pattern of causal claims* in a case study.

In the later chapters, I adopt the less traditional "powers" or generative concept of causality, instead of the (until recently) dominant paradigm (the succession concept) of causality. It is the succession concept of causality that is usually associated with eliminative inductivism. I believe that the generative concept is more useful than the succession concept in dealing with the kind of causal story told in a psychoanalytic case study, and in attempts to construct the kind of arguments for scientific credibility most commonly used in a science like psychoanalysis.

9. I deal with the first question in this chapter, the second question in Chapter 11, and the third in Chapter 13.

10. My parenthetical characterization of Freud is supported especially in Chapters 13 and 15.

eutics rather than scientific explanation? Is its knowledge not knowledge of mechanisms at all? Is its knowledge achieved only by means of supposedly nonscientific operations such as intuition, empathy, subjective participation, and unconscious knowing? Since, in Chapter 11, I take up these questions and my answers to them (to all four, my answer is no), I shall confine myself here to addressing briefly some common misconceptions.

1. "Mechanistic" and "machine-like" are not synonyms.[11]

2. Processes of discovery, conceptual invention, or conjecture should not be confused with processes of establishing by argument from empirical evidence that conjectures about a domain, couched within a theoretical framework, can be accepted as scientifically credible. How a scientist arrives at a conjecture, or the theoretical terms he uses to state it, are irrelevant in and of themselves in assessing, one way or another, the credibility of the conjecture. That is, there is no method of arriving at a conjecture that guarantees its truth, just as there is no method of arriving at a conjecture, no matter how disreputable, that justifies the conclusion on that account alone that it must be false.

One may assume that all psychoanalysts have had the following experience in the psychoanalytic situation. Suddenly an idea, a subject, a certain content, a word comes to mind from nowhere. There does not seem to be any way to account for its appearance in the context in which it appears. A while later, totally unexpectedly, the analysand suddenly mentions the same idea, subject, content, or uses the same improbable word. The psychoanalyst may become uneasy. Is one unconscious reading another? This coincidence may lead to the formulation of an interpretation, a guess, a conjecture about the analysand, which the psychoanalyst offers. The analysand replies, "As a matter of fact, I was thinking something like that just now (or a few minutes ago), but I didn't say anything." Must we then judge that this is certainly not a proper way to achieve knowledge? Must we conclude from such an experience that psychoanalytic knowledge is a kind of knowledge different from that obtained by other sciences? Not at all. We certainly must still address the question of the scientific credibility of the conjecture, but we no more need to doubt the conjecture just because the experience is uncanny than any scientist must repudiate an answer to a question simply because he wakes with it from his dream—as a poet might awaken with part of his poem written in his mind or a composer with music sounding in his inner ear.

11. This point is discussed in Chapter 13.

3. There is nothing about the properties of any subject matter that makes it intractable to scientific inquiry. Irrationality may be studied rationally without loss to psychoanalysis as science. This does not mean that any scientific theory can catch in its net everything one might want to say about a particular phenomenon.

There is no justification for requiring that a scientist abandon a subject matter because of limitations imposed by characteristics of available language to talk about it. There is no justification for requiring that a scientist must confine himself to a particular kind of language in expressing his ideas about phenomena in which he is interested.

The use of the single case or case study method cannot be grounds for characterizing psychoanalysis as scientifically defective. A single unrepeated phenomenon can, on occasion, establish the scientific credibility, or the falsity, of a scientific conjecture.

Science is not a matter of subject matter, but of mode of inquiry. A scientist may freely choose any empirical domain for investigation, including the domain constituted by a single mind. This domain includes as entities the wishes, beliefs, knowledge, perceptions, dreams, memories, and capacities of a person, manifested in or inferable from his utterances or the relations among his utterances. The psychoanalyst may seek to discover and to account for the interrelations among the entities in this domain.

What of the problem of generalizing a conclusion reached from the study of a single mind—and, in particular, generalizing it over persons rather than over the mental states or time-slices of that one mind? I make just one comment here about that question.[12] It is instructive to note Chomsky's argument for postulating linguistic universals from the study of a single natural language (1968, pp. 113–114). Different children, exposed to limited and debased utterances, and on the basis of different experiences, all acquire the same or a similar theory or grammar of their language (rules for generating sentences). This theory is underdetermined by the data at their disposal. How do we account for their arriving at this and not some other plausible theory or grammar (plausible, given the data to which they have had access)? We may do so by constructing a device the structure of which is such that, if we suppose that children possess it as part of an innate capacity, the particular grammar and only that grammar will be acquired. But children are not genetically wired to learn one and only one natural language. The constraints we must

12. For further discussion of single case studies and the case study method, see Edelson (1984, especially Chapters 4 and 11) and Part 2 of this book.

attribute to this device, so that one theory of grammar (rather than any alternative) is inevitably acquired by the speakers of a natural language, must then involve linguistic universals. Any hypothesis concerning such universals can be falsified if, for example, it fails to account for the grammar of even one other language.

So, analogously, the intensive study of a single mind, involving very many instances of particular relations among wishes, beliefs, perceptions, memories, and other psychological entities, may lead to "strong empirical hypotheses" about the characteristics of the symbolic function of the human mind, which may be falsified or corroborated by a similarly intensive study of another mind.

4. If psychoanalysis were to convey a more dignified, satisfying conception of man than other psychological theories, that would not make it humanistic rather than scientific. For it would be able to convey this conception only because it was doing justice as scientific theory to the complexity of the phenomena it studied.

5. Science is not the same as technology. Psychoanalysis as science must be carefully and in all useful discourse distinguished from psychoanalysis as therapy. Knowledge is not equivalent to some use or application of it.

There is always a gap between the idealizations of theory and the application of theory in particular circumstances. One may not know, and it may not even be possible to know, everything about a particular situation that must be known to make an effective extrapolation from theory. Nor is it possible to derive in most cases an unequivocal recommendation from knowledge alone. "Ought" cannot be deduced from "is." Application of knowledge involves reference to value-preferences with respect to means and ends, and scientific knowledge alone cannot determine choice among values.

The outcome of a technology cannot decisively confirm or disconfirm the theory applied. Use of soap diminishes the incidence of sepsis. It works, but that does not automatically confirm a "dirt theory" of disease. An adequate theory of the relation between social and personality systems may suggest how a therapeutic community in a residential treatment center should be organized to achieve its objectives but, as Edelson (1964, 1970a, 1970b) makes clear, such theory is not enough. Dilemmas are inescapable. It is impossible to maximize simultaneously equally prized values—for example, equality and liberty. An arrangement that is optimal for one patient may be far from optimal for another with regard to any therapeutic aim at all. There are unalterable limitations in the capacities of personnel and the other resources of the hospital organization. Knowledge does not guarantee that these conditions can be circumvented so that in fact the therapeutic community can be organized as planned or, even it if is, that the intended desired consequence will be forthcoming. Use of psychoanalysis may or may not cure this or that case of illness, but such success or failure in and of itself does not automatically confirm or disconfirm a psychoanalytic theory of the mind.

Possession of scientific knowledge does not automatically confer power to change phenomena. Astronomy and evolutionary theory provide obvious examples.[13]

We cannot even be sure that we know what part scientific knowledge plays in the competence of anyone who applies it, including the psychoanalyst as therapist. Psychoanalysts differ widely in their own testimony about the extent to which their clinical skill depends upon a witting use of knowledge of psychoanalytic theory. In any event, no such testimony can be given privileged status. We know that people have knowledge of which they are not aware and which they cannot explicate, although the possession of such knowledge may be inferred from their use of it. (Linguistic competence is an example.) To what extent a psychoanalyst makes use of psychoanalytic theory or, unwittingly, of linguistic competence in arriving at and making a clinical interpretation is a research question (Edelson 1975). A clinical interpretation may depend in part on knowledge intentionally applied, but the artfulness required not only to come upon or invent the interpretation, but to communicate it in such a way that it can be heard and used by the analysand, has often been remarked.

6. There are two points to be made about metapsychology. The first is that it is not metapsychology, it is not the postulation of certain kinds of entities or processes, that confers upon psychoanalysis the mantle of respectable scientificity. The scientific status of psychoanalysis depends only upon the way psychoanalysis goes about arguing from empirical evidence for the scientific credibility of its conjectures and causal claims.

The second is that metapsychology may not be a substantive theory at all, but a way of talking about and characterizing psychoanalytic theory. The distinction between an object language and a metalanguage is useful here (Copi 1967; Suppes 1957).

If so-called clinical theory is an object language in which the psychoanalyst is able to make statements about phenomena, then a metalanguage is needed to make statements about statements in the object language. We need a metalanguage to talk about our own formulations, including our own clinical interpretations. We want, after all, to assign or ascribe qualities to our formulations; make assertions

13. The inability to control phenomena, to perform experiments, has implications for psychoanalytic research that are explored especially in Chapters 13 and 15.

about their relations (not the relations among objects or properties they describe); categorize them; prescribe their form or content; or evaluate or judge them according to various criteria. All these things must be said in a language different from the object language. If the same terms are used in both languages, then they should be used with the recognition that they are being used at different semantic levels. (The same injunction applies when a theory of linguistics makes theoretical statements in English about the natural language English.)

Psychoanalytic metapsychology, which often seems to be prescribing what psychoanalytic theory should be like, what it should include, what aspects of phenomena it should pay attention to, what kinds of concepts it should employ, may be such a metalanguage. It can only lead to confusion to suppose one is talking about the object world—the domain one is studying—when one uses metapsychology, if metapsychology is indeed the metalanguage of psychoanalysis.

The Relation of Psychoanalysis to Other Theories

Rival Theories

Psychoanalysis may be one alternative among a number of theories claiming to explain the same phenomena. More about that in Part 2 (see also Edelson 1984).

Models, Homologies, and Analogies

Psychoanalytic theory and another theory, although concerned with different subject matter, may have the same structure. The terms of each theory, for example, may be related to each other in the same way. In such a case, we may say that each theory is a different interpretation of the same formal system. One may say that any semantic interpretation of a system of abstract terms, which are related in particular specifiable ways to each other, is a model of that formal system.[14] One may also say that one interpretation of a se-

14. I am using "model" here in a way that is different from the way I used this term in Chapters 1 and 2. Here I use it as a logician uses it.

In those two chapters I referred to the system-subsystems model and the transformational-generative model. There the term "model" designated an abstract system that may receive more than one semantic interpretation. The system-subsystem model may be interpreted using the language of sociology or the language of psychoanalysis. The transformational-generative model may be interpreted using the language of linguistics or the language of psychoanalysis.

Here the term "model" designates one or another such semantic interpretation. Each semantic interpretation is called a model of the abstract or formal system.

mantic system is a model of another semantic interpretation of that same formal system (Suppes 1957).

Metaphors may be considered to be models in this sense. A metaphor is a different interpretation of an underlying abstract system of relations of which we already have an interpretation.

If the structures of two theories exactly correspond, they are isomorphic. The more usual case is partial homology; some statements or sequences of statements in one theory have the same logical form, although not the same content, as some statements or sequences of statements in the other theory.

Metaphors as models are important in the natural sciences. For example, the universe of discourse called mathematics, which is constituted by numbers and their relations, and the universe of discourse constituted by physical spatiotemporal entities in the object world and their relations are isomorphic. Metaphors as models are important in the social sciences. For example, Levi-Strauss has documented that the universe of discourse constituted by mythical entities and their relations and the universe of discourse constituted by cultural and social entities and their relations are at least homologous. Metaphors as models are important in psychoanalysis. The analysand's conception of events in the psychoanalytic situation and conception of events in his past personal history are homologous; this homology provides the basis for the concept of transference.

Edelson (1954) sought to demonstrate that statements about psychic energy in psychoanalysis, although different in content from, are logically homologous to, or have the same formal structure as, statements about energy in physics. Both sets of statements, for example, involve the idea that something remains invariant through a series of transformations. Both sets of statements involve the idea that in any system there are limits, given an available resource, that determine the possibilities for allocation of that resource among foci or subsystems each of which acts in different directions or serves different functions.

Similarly, Talcott Parsons's formulation of primitive terms in his theory of action, and his use of these in defining the structure of systems of action (1937), are logically homologous to structural theory in psychoanalysis.

Parsons believed that this homology depends in part upon contingent facts. His theory of the social system and the psychoanalytic theory of the personality system are not simply different semantic interpretations or models of the same formal system. They are also related to one another because of the way the subject matters with which they deal are related. The theory of the social system specifies factors *controlling* processes of concern to the theory of the personality system, and the latter theory specifies *necessary conditions* for processes of concern to the former theory. In addition, the concept of a system of internalized norms and values, which is part of the content of the concept of super-ego in personality theory, is also important in social system theory, which makes use of the concept of internalized norms and values being shared by members of a collectivity. The internalization of norms and values is a notion that makes sense in talking about personality but not about social systems. That norms and values are shared, and the extent to which they are shared, is a property of social but not of personality systems. The concept of a system of internalized norms and values, which is used in both theories, represents a point of interarticulation of the two theories. Freud anticipated this interarticulation in his own group psychology. His formulation was that group members share the ego-ideal of the leader of a group.

Many analogies in Freud's *The Interpretation of Dreams* imply structural homologies between the construction of dreams and the construction of various cultural or symbolic artifacts, or between the process of making a dream and social processes, such as economic enterprise and political censorship. Freud also uses the analogy between a mental apparatus and an optical system to point to the correspondence between mental contents and virtual optical images, and heuristically in justifying criteria for differentiating subsystems of the mind. Two subsystems are distinct if a function or property assigned to one is incompatible with a function or property assigned to the other. So a mirror, like perception, can only reflect every passing image if it is not at the same time altered to permanently register images, in contrast to a photographic plate, which, like memory, insofar as it is changed in order to register an image permanently, cannot continue to reflect different passing images (1900, pp. 536–539).

A misconception regarding the psychologist K. Lewin's use of topology as a formal system (1936) has led to undeserved neglect of this insightful theorist by psychologists, who tend to suppose that rigor and measurement necessarily go together. Although topology is a branch of mathematics, its application does not involve measurement, but—like the rest of mathematics—it is a formal system of logical relations among abstract entities, and in this resides its utility. The heuristic value of formulating two semantic interpretations of the formal topological system, one involving physical-spatial content and the other psychological content, has gone unappreciated; most psychologists reading him have thought his enterprise vacuous because there was no way to use measurement to assign quantitative values to variables in his psychological theory.

It is impossible to overestimate the value of analogy in detecting relations between theories that might otherwise seem to have nothing to say to each other (Hesse 1966). Such relations of analogy contribute importantly to the invention of theory, and to its revision. We may think of transformation as invariance of content, together with change in the way that content is represented. Analogy, conversely, involves invariance in structure, together with change in content. Analogy invites us to note in what ways patterns or structures are the same in different realms and, also important, in what ways they are not the same. Analogies are fertile, for they raise questions that might otherwise never be asked—about, for example, the unsuspected other ways the realm in which we are interested might be like the one to which we compare it.

Psychoanalytic Interpretation[1]

4

The therapeutic action of psychoanalysis is an effect of interpretation. Psychoanalysis as a method of inquiry relies upon examination of the effects of interpretation. Yet we have no adequate theory either of the psychoanalyst's act of understanding or the psychoanalyst's act of making an interpretation. What do we mean by "the psychoanalyst understands the analysand—the utterances, enactments, dreams, and symptoms of the analysand"? What characterizes the act of interpretation in which the psychoanalyst conveys his understanding to the analysand? What part of that understanding does he convey? In what form? When? Why that part, that form, just then? How is an act of interpretation differentiated from other kinds of interventions? How are its effects identified? Why a particular effect at a particular time? How do we identify that part of its impact that it has because it is true?

Interpretation and Symbolic Function

A vulgar notion of psychoanalysis pictures the analysand reacting to what he out of error or ignorance regards as *indices* or signs of danger, and the psychoanalyst—like a keen-eared, sharp-eyed Holmes—reacting to the analysand's verbalizations and acts as *indices* or signs of the analysand's immediate feelings or unconscious states. That is, these verbalizations and acts are existentially connected to or part of these feelings or states. The psychoanalyst then intervenes to point out to the analysand

1. This chapter makes use of material, which has been revised, from Edelson (1970c, 1977, 1978).

that the analysand is distressed because he is interpreting signs incorrectly, out of mistaken beliefs or ignorance. Or the psychoanalyst intervenes to point out to the analysand unrecognized feelings or unconscious states, signs of the existence of which the psychoanalyst has detected.

An alternate notion of psychoanalysis would have the analysand *making a symbol*, such as his symptom, dream, or transference neurosis, to represent his *conceptions* of his inner reality. Contemplating these symbolic representations, he and the psychoanalyst may come to understand these conceptions.

The idea that psychoanalysis is primarily a process of interpreting indices or signs to mitigate a pathology of signs is similar to the idea that a work of art may best be understood as an index of the feelings and mental states of the artist existing at the time he was making the work of art. The processes of producing art or emitting verbalizations in the psychoanalytic situation are regarded as unwitting expressions, ejaculations, or discharges of concurrently existing feelings and impulses, and therefore indices or signs of these feelings and impulses.

This view of psychoanalysis must disregard Freud's continued use of the term "ideas," when he referred, for example, to that which is repressed, as well as his emphasis that the objects of study in psychoanalysis and the subject matter with which psychoanalytic theory is concerned (including what is conceived as id or instinct) are "psychic representatives," not physiological processes or external situations.

As there is an alternate notion of psychoanalysis, so there is another one of art, which holds that artworks as symbolic entities represent the artist's *conceptions* (his "ideas," in Freud's sense of this term) of attitudes, feelings, inner reality—rather than express (as indices or signs of) the attitudes, feelings, or inner states he had while making the work of art.

The analysand becomes aware as he participates in the psychoanalytic process of himself as human, the *animal symbolicum*, living in a psychic reality of his own creation.

He begins with a conception of himself as a creature or machine-variant merely reacting automatically to his situation, compelled rather than creative. "This is done to me. This happens to me. This is the result of what has been done to me or has happened to me."

His conception changes, as he thinks, "I do this. I make this happen. I conceive—I write, I direct—my own scenario. I populate my own drama with objects as I imagine them to be. I create symbolic representations of these conceptions. The conceptions are mine, and influence my life by means of the representations of them that I make. I am free to contemplate these representations, and so to come to understand my attachments to the sorts of objects with which I people my world, and my wish to play certain scenes over and over, my attraction to certain states of affairs in which I am always ready to play a part. And I am free to conceive differently, and to form new symbols of these conceptions, should I wish to do so."

The analysand overcomes obstacles to the exercise of his abilities to move freely from using one symbolic system to using another. He is increasingly able to translate that which he has represented using the modes of symbolization of one symbolic system, by making use of the modes of symbolization of another symbolic system to represent it. He views himself creating symbolic representations of different kinds, with pleasure rather than distress.

The analysand cannot simply be told by someone else what he knows but does not know he knows. It is in his very effort in the psychoanalytic situation to construct the symbolic entities that will adequately represent his knowledge, that he himself discovers what he knows. In this effort, he achieves higher levels of consciousness. Freud considered this achievement the foundation of psychoanalytic therapy. Wallace Stevens alludes to it in the quiet, tentative, contemplative, yet triumphant, final line of the final and great poem of awakening in his *Collected Poems:* "It was like / A new knowledge of reality" (1961, p. 534).

Describing in this way the changes that come about through the psychoanalytic process imposes a limit on the therapeutic action of psychoanalysis. When misery is, as Freud said, normal rather than neurotic misery—that is, to the extent that the analysand's suffering is caused by his situation rather than created by his own symbolic activity—then and to that extent psychoanalysis may have no specific therapeutic action.

Interpretative Competence

What is the precise nature of the competence the psychoanalyst uses in inventing a psychoanalytic interpretation? A theory of that competence should account for our intuition that such-and-such an utterance is a psychoanalytic interpretation, whereas such-and-such an utterance is not. Developing this theory requires that we identify the features of a psychoanalytic interpretation, for it is just these features for which the theory must account.

Translation

Psychoanalytic interpretation may translate a symbolic representation (a presentational symbol such as a dream, for example) by using modes of symbolization belonging to another symbolic system (discursive natural language, for example) in order to provide a symbolic representation of the same object or state of affairs the nondiscursive symbol presents.

In making such a translation (from one kind of symbolization to another), a psychoanalytic interpretation will reconstruct, often but not necessarily explicitly, the way in which transformational processes have operated in particular sequences upon the contents of deep, basic, or elemental mental states, to produce the mental representation interpreted. Examples of such reconstructions are dream interpretations, as discussed in Chapter 2, or the genetic reconstructions of transformational processes acting through time to form different symptoms or to develop a particular kind of character. The way in which transformations operate through time to bring about such results is discussed in Chapter 9 (with a focus on Freud's explication of the fantasy of a child being beaten) and in Chapter 13 (with a focus on Freud's explanation of changes over time in the Wolf Man's symptoms and character).

Disjunction

The invention of a psychoanalytic interpretation is frequently instigated by linguistic deviance or by abrupt disjunction.[2] By disjunction I mean what I shall call a causal gap in later chapters.[3] A disjunction is an apparent absence of relation or connection among mental contents, among the symbolic representations that are constituents of different mental states. Psychoanalytic interpretations undo disjunction by restoring missing, lost, or deleted relations. In this sense, such an interpretation creates form or structure out of apparently unconnected elements.

The psychoanalyst does not merely seek to make unconscious contents conscious. Relations among contents accessible to consciousness can themselves be unconscious or go unnoticed. The psychoanalyst seeks to join what is kept apart, separated, disconnected, dissociated by whatever means. To overcome a disjunction is to demonstrate a relation. (Something may have been lost when Freud largely abandoned the term "dissociation" after using it in *Studies on Hysteria*.)

A psychoanalyst does not seek to reduce everything to one realm of experience or one kind of mental state—childhood experiences, sexual wishes, transference, or what not. He brings up what is not being talked about. If early childhood were all that was being talked about,

2. Edelson (1975) contains an extended account of the way in which different kinds of linguistic deviance may instigate, or provide an occasion for, an act of interpretation.
3. See especially Chapter 6.

the psychoanalyst would ask, "What is going on in your life at this time?"—and vice versa. If the analysand were to speak only about the psychoanalyst, with nothing said about current experience outside the psychoanalytic situation or about childhood, again the psychoanalyst might inquire about what was missing. It is not one thing or another, one realm or another, one kind of mental state or another, but the relations between them that are important. Interpretations present relations for contemplation.[4]

Ambiguity

An important aspect of interpretative competence is the psychoanalyst's ability—in a state of free-floating attention—to decontextualize linguistic elements, and then to be aware of and responsive to the many kinds of ambiguity (for example, syntactic, semantic, and phonological ambiguities) possessed by linguistic elements out of a given context. In other words, what comes to mind, if the context provided by the analysand is momentarily ignored or not taken at face value, are the many different symbolic representations to which such decontextualized linguistic elements might conceivably belong. What may come to mind is the way in which these linguistic elements, so decontextualized, have recurred in different contexts, and now foregrounded, and perhaps coming into relation with other such foregrounded elements, suggest a pattern not otherwise easily discernible.

Ambiguities are often a means exploited by transformational operations in constructing and changing certain kinds of difficult-to-understand symbolic representations (Edelson 1975). Sensitivity to ambiguities is a means for reconstructing these transformational processes and restoring relation or form where only apparently senseless disjunction has resulted from these processes.

Syntactic, semantic, and phonological ambiguities. Edelson (1972, 1975) has considered the importance of these kinds of ambiguity (e.g., polysemy, syntactic or constructional ambiguities, homonyms) for psychoanalysis. Freud has documented the role of such ambiguities in phenomena of interest to psychoanalysis in his work on dreams, jokes, and parapraxes.

Ambiguities of tone and intention. How can we tell—and how can we express how we tell when we can tell—in what way someone wants us to take his remarks? Are they to be taken as ironic, bitter,

4. That is, *some* interpretations present relations. This discussion of interpretation is not intended to be exhaustive. Since it considers interpretation from a very special perspective, it does not even necessarily focus on aspects of interpretation that may be clinically the most important aspects.

reticent, resigned, quietly affirmative? How can we tell—and how can we express how we tell when we can tell—whether an analysand means what his words seem to mean? How can we tell—and how can we express how we tell when we can tell—what some utterance presumes or implies about an analysand's conception of the psychoanalyst?

Linguistic-logical ambiguities. Syntactic, semantic, and phonological ambiguities and ambiguities of tone and intention do not exhaust the kinds of ambiguities that a psychoanalyst is able to recognize. Knowledge, however unwitting, of linguistic-logical ambiguities as well is part of his interpretative competence.

For example, for a psychoanalyst, any pronoun ("it," for example) is a variable. It is a variable because many words or phrases can be substituted for it. It has a range of significance, a range of possible values. Any word or phrase designating an entity to which "it" might refer is a possible value of "it." The psychoanalyst, suspending the attention that might otherwise be fixed on the context for "it" provided by the analysand, hearing "it" as ambiguous, permits his mind to wander over all the possible values "it" might take.

Similarly, the psychoanalyst is able to abstract from any statement a propositional form, such as "x (no matter who or what x is) is F (no matter what property F is)," or "x is related in some way R to y (no matter who or what x and y are, and no matter what relation R is)." The analysand's statement exemplifies, or is a particular interpretation of, that form.

No matter what content the analysand gives that form as he speaks, the psychoanalyst may shift his focus from content to logical form, and invest the form with other possible contents or interpretations, or indeed regard the form itself as significant. For the psychoanalyst, for example, what may stand out about the structure of each statement in a series of statements is that *three* entities are related to each other, suggesting to the psychoanalyst that he may be hearing about an inner world in which triangles and jealousy predominate. If the structure of each statement in a series of statements is that *two* entities are related to each other, that may suggest to the psychoanalyst that he may be hearing about an inner world inhabited by mother and child alone. Psychoanalysts are sensitive to the pervasiveness of monadic, dyadic, or triadic relations in a stream of associations, no matter what the content, and may be led thereby to locating a source of associations in a fantasy owing its origins to a developmental epoch dominated by monadic, dyadic, or triadic relations. Thus, the attributions "oedipal," "preoedipal," and "narcissistic" might be made in some cases on the basis of the form of associations rather than their content.

For, just as the form of a dream is important in representing its meanings, so the form of the analysand's statements may be important in representing what the statements, not taken at face value, might mean. The psychoanalyst assumes, of course, that the analysand in the psychoanalytic situation, to the extent that he is able to associate freely, is using, in part at least, the mode of symbolization he uses in constructing a dream, although he appears to be using natural language discursively.

Unconnected or apparently unrelated statements may be related because each is an interpretation of the same propositional form. Each assigns a different value perhaps to just one abstract term (*x*, *y*, *F*, or *R*), which appears in that shared propositional form.

Entire sets of statements—about childhood, about the relation between psychoanalyst and analysand—that are apparently unrelated can be related by virtue of the abstract structure of which each set of statements, as in metaphor, is a different interpretation.

In Freud's interpretation of the Irma Dream (1900, Chapter 2), there are a group of associations that can be summarized as interpretations of the following propositional forms.

x rejects (conceals) *z* from *y*.
x accepts *z* from, or reveals *z*, to *y*.

Here are the associations.

A governess conceals dental plates from Freud.
Patients reveal little secrets to physicians.
Irma's intimate friend rejects help from Freud.
Freud's wife rejects help from Freud.
Irma rejects the solution of her illness offered by Freud.
Dr. M. rejects a suggestion of Freud's.
Freud's elder brother rejects a suggestion of Freud's.
Children reveal their bodies to physicians.
Adult females conceal their bodies from physicians.
Freud rejects a solution (the bad liqueur) offered by Otto.
Dreams conceal their secrets from Freud.
Freud conceals information from Freud.[5]
Freud conceals some of his thoughts from his readers.

For *x*, we have the following values: governess, patients, Irma's friend, Freud's wife, Irma, Dr. M., Freud's elder brother, children, adult females, Freud, dreams.

5. "Further than this I could not see. Frankly, I had no desire to penetrate more deeply at this point" (p. 113).

For *y*, we have the following values: Freud, physicians, Otto, readers.

For *z*, we have the following values: dental plates, secrets, help, solution, suggestion, bodies, thoughts.

The relation, rather than the individual entities, in a propositional form may vary. For example, we have the following set of associations, in which the relation *R* of the abstract underlying propositional form varies.

> Freud wants to see . . .
> Freud wants to (does not want to) penetrate . . .
> Freud wants to unveil . . .
> Freud wants to inject . . .

If all the interpretations of *R* have some property or properties in common, the psychoanalyst may be led to postulate that an instinctual aim or wish, which has this property or properties, belongs to the meanings represented by the dream or utterances of the analysand.

Condensation. That Irma, Irma's friend, and governess are all interpretations of *x* may lead the psychoanalyst to postulate that they are identified with each other in the dreamer's mind, perhaps because of some property or properties they share, and that it is this property or these properties that is or are represented by the dream or utterances of the analysand's. If one of the values belonging to a set of such values is the name or description of an image, then that image, in an exemplification of what Freud calls "condensation," is able to represent the entire set of values, or the property or properties the entities in that set share.

Freud's concept of composite-condensation, which is different from identification-condensation, is exemplified by a set of statements describing a dream-image.

> Dr. M. is pale.
> Dr. M. is clean-shaven.
> Dr. M. limps.

Associations include the judgments that the second and third statements are false of Dr. M. but true of Freud's brother. Combining properties in the single image is made possible by the fact that both Dr. M. and Freud's brother are values of *x* in the propositional form

> *x* rejects a suggestion of Freud's.

This complex condensation is discovered from associations in which some descriptions of an image are judged to be false of the person apparently represented by the image, the question "What substitutions would make these descriptions true of some person?" arises, these substitutions are made, and an attempt is made to discover the propositional form that yields true statements when either person is substituted for an abstract variable in that propositional form.

Defense. A defensive operation, such as intellectualization, may work by using a word designating a property instead of using a verb in order to suppress the logical structure of a thought—for example, "x is disobedient" rather than "x disobeys y." The statements

Irma is disobedient.
The patient is recalcitrant.
Freud is reticent.

seem to have the form x is F. But, as we have seen in analyzing the logical structures that underlie associations to the Irma Dream, their actual logical structure involves four constituents—three individual entities and the relation among them—and even that analysis is an oversimplification. The suppression of logical structure tends to interfere with strategies of interpretation that depend on recognizing such structures.

Linguistic-logical ambiguities of use and mention. The psychoanalyst is also able to shift from one level of language to another, namely from use to mention. For example, he may ask, "What occurs to you about the stick in the dream?" and the analysand may respond, "'Stick' is very short," mentioning the word "stick" rather than using it to talk about a stick. The psychoanalyst is prepared to hear this response as a shift away from the use of the word "stick" in the representation of a thought, such as "My stick is very short." In Chapter 2, we have seen that the dream work exploits the same possibilities in making use of linguistic elements.

An analysand in reporting a dream states, "I was watching Alfred." Associations include statements about the man Alfred.

I don't know Alfred very well.
He never has paid much attention to me.

Suddenly there is a shift to associations about the name "Alfred."

"Alfred" can be divided into two—"Al" and "Fred."
A. F. . . . A. H. . . . Ass Hole . . .

"A" and "F" were the initials of the analysand's parents' first names. Every consonant in "Alfred" was the first initial of the name of each of the analysand's four siblings. The first initial of the analysand's own name, "H," was not part of Alfred—the only name of a member of the family not so represented by "Alfred." (The analysand had been left out.) The initial "H" was, however, added almost immediately in an association that was then perhaps not just a deprecatory but a self-deprecatory epithet—"A. H."

Associations to the dream also included statements about the analysand's relation to her family, the strains between mother "A" and father "F" (in the associations the name "Alfred" was divided), the curiosity the analysand felt about their relation ("I was watching . . ."), and the feeling she had of being excluded by them (". . . never paid much attention to me").

Clearly, the expression "Alfred" is ambiguous, since it may represent the man, the name, or the analysand's family. Here, condensation is brought about by mention as well as use of the expression.

Similarly, an analysand's thoughts about a psychoanalyst (whose initials were "T. R.") was represented in a dream about a woman named Esther. The name Esther (Es . . . t[h] . . . [e]r) became "T. R. is an 'es' "—that is, "The psychoanalyst is an ass."

In Closing

To anyone who has read this chapter and Chapter 2, it will come as no surprise that I am inclined to think that an important objective, and perhaps the chief source of the efficacy, of clinical interpretation in psychoanalysis is to make the analysand acquainted with how his mind works—what operations it performs in various kinds of symbolization, and by what kinds of symbolic means it attempts to solve problems arising from acts of symbolization and from intense attachment to the products of these acts.

The reader may think that throughout I have emphasized too much the cognitive aspects of psychoanalytic interpretation. I can see the objection, although I think it ignores the difference between what is required to make an effective clinical interpretation (which includes empathy, tact, a working relationship between psychoanalyst and analysand, and many affective components) and what is required to make a theory of interpretative competence.

I am impressed by the sensitivity of many psychoanalysts to form or structure, and their ability to abstract form or structure from a multitude of variegated particulars. I cannot see that psychoanalytic work is possible without the psychoanalyst's capacity to shift attention from a variety of contents to an invariant underlying form or structure, and from such a structure to its many realizations. We might be astonished at how many psychoanalysts disclaim interest in abstraction, logic, or mathematics, or deprecate the role of cognitive capacities in psychoanalytic clinical work, if Chomsky had not taught us how amazingly complicated capacities such as linguistic competence or knowledge of the rules of language are. Such competence or knowledge is unsuspectedly unique. It is different from other human capacities; it is not simply a manifestation of principles underlying general intelligence. And this competence or knowledge, like that of interpretative competence perhaps, can be possessed and skillfully used without much, if any, consciousness of the mental contents, processes, or principles involved, or of their fundamental nature.

Anxiety[1]

5

Thirty or forty years ago, it would have been inconceivable for a group on the frontier of investigations of anxiety and the anxiety disorders to report their ideas and findings, as has been done in Tuma and Maser (1985), largely without any reference to—without feeling any need even to refute—the ideas and findings of psychoanalysis, and indeed allocating a very limited amount of space to a presentation and discussion of psychodynamic theory and research. Has psychoanalysis then lost not only any intellectual commitment it may once have enjoyed but, now in competition with numerous apparent rivals, its claim as well even to the attention of the present generation of scientists? How has this happened? Have psychoanalytic hypotheses about anxiety disorders been rejected in empirical studies as less credible than rival hypotheses? Such studies are not cited. We witness rather, I believe, the consequences, not only of sociological and cultural changes since World War II, but of a failure of the psychoanalytic community to fulfill an earlier promise. This failure is largely the result, not of an increase in the strength of putative irrational forces opposed to psychoanalysis, or of the unavailability of adequate procedures for measurement, but rather of certain attitudes toward scientific work that are representative of the psychoanalytic community. In this chapter, I spell out what I mean by this assessment, although it is possible, of course, that the diagnosis comes too late to help revive the patient.

I shall develop the following thesis. Differences in the conclusions of psychoanalysis and other disciplines (e.g., neurobiology or neuropsychology) often reflect dif-

1. This chapter was originally published as Edelson (1985a).

ferences in what is meant by the term "anxiety"—differences in domains chosen by each discipline for study.

The thesis can be elaborated in this way. One set of psychological events includes activities subserved or supported by cortical systems. These activities contribute to determining or processing meanings and are carried out in (1) identifying or representing states of affairs, including anticipated states of affairs; (2) evaluatively assessing the meanings or significances of states of affairs, including asessing an anticipated state of affairs as dangerous (here is where psychoanalysis uses the term "anxiety"); and (3) operating upon mental representations (as defenses do) to prevent actualization of a dangerous state of affairs.

A second set of psychological events includes responses to stress that are subserved or supported by subcortical systems. Such responses to stress—involving, for example, arousal, or mobilization of physiological resources preparatory to fight or flight—may or may not occur following the processing of meanings of either anticipated or actual states of affairs. Indeed, as a final common pathway, such a stress response can be precipitated or "triggered off" by a wide variety of internal stimuli, external stimuli, and experimental probes. When an investigator studying such stress responses refers to his model of anxiety, he uses the term "anxiety" differently than does the psychoanalyst; and this difference, of course, raises questions concerning the extent to which animal models of anxiety actually model what is referred to as anxiety by psychoanalysis.

These two approaches seem complementary rather than competitive. My thesis, therefore, leads somewhat naturally to a call for a clarification of just what questions about anxiety and anxiety disorders belong in the province of psychoanalysis and what questions, perhaps suggested by psychoanalytic propositions, belong in the province of some other discipline.

Finally, I discuss the problems of carrying out research on anxiety and anxiety disorders in psychoanalysis, and conclude by making some proposals for dealing with these problems. I emphasize those proposals that are based on the importance for psychoanalysis of conceptual and methodological developments in single-subject research.

Different Meanings of "Anxiety"

Some differences between psychoanalytic and other views about anxiety and anxiety disorders do not seem to involve the kind of competition between rival hypotheses formulated within the same frame of reference with which we are familiar in science. Instead, these differences seem to reflect different definitions of what is to be designated by the term "anxiety."

What is at issue, then, cannot be decided by simple reference to empirical facts. At issue are choices among different domains of study, different entities and different properties of these entities, different ways of cutting the phenomenal world at its joints, different conceptual proposals or inventions—in other words, different concepts of anxiety. I have considered elsewhere questions about the choice of domain, in the context of an examination of the relation between neural science and psychoanalysis (Edelson 1984, pp. 109–120).

A psychoanalyst here will not say, "I disagree with what you say about anxiety disorders" but rather "You are not talking about what I mean about 'anxiety' at all." Redmond, a neurobiologist, has a good sense of how problems of definition may play a role in discussions of anxiety and the anxiety disorders. A "paradigm should be judged . . . by its ability to generate testable hypotheses and to explain a large amount of data parsimoniously" (Redmond 1985, p. 536).

Decisions about the domain to be investigated are ultimately evaluated by (1) the extent to which such decisions enhance a scientist's capacity to generate interesting and testable hypotheses; (2) the explanatory power of these hypotheses; and (3) the scientific credibility such hypotheses ultimately achieve through empirical tests of them. It is with this third criterion that psychoanalysis has had particular difficulty (Edelson 1984).

According to psychoanalysis, anxiety is a subject's appraisal of a state of affairs as dangerous. The state of affairs is anticipated by the subject; it does not actually exist. Thus, it is imagined, not perceived, by the subject. The subject believes (not necessarily as the result of generalizing from causal connections or space-time contiguities encountered in experience) that this imagined state of affairs will become an actual state of affairs if he acts upon some impulse of his own. Further, although the subject is under internal pressure to act on this impulse, it is nevertheless possible for him to delay such action indefinitely and still survive.

Like other affects, anxiety has both a cognitive component (a subject's mental representation of a state of affairs) and an evaluative component (an appraisal of the significance by the subject of this state of affairs as, for example, threatening, thwarting, or gratifying). (An elaboration of this concept of affects may be found in Edelson 1984, pp. 88–90.) The cognitive component is the mental representation of an anticipated (not an existent) state of affairs, which is imagined (not perceived). The evaluative component is an appraisal of the anticipated, imagined state of affairs as dangerous.

Psychoanalysis distinguishes anxiety from fear, for example, by locating the source of danger (i.e., the subject's impulse) in the intrapersonal realm. Therefore, it is impossible for the subject to avoid the danger through physical movements in the service of fight or flight—and that this is impossible is a defining characteristic of anxiety. This formulation is congruent with the theory of affects just mentioned, in which a particular affect is defined, in part, in terms of the type of state of affairs evaluated by the subject.

In the most distinctive psychoanalytic account, it is expression of an aggressive or sexual impulse (or, secondarily, expression of certain affects) the subject has come to believe will produce in actuality a dangerous state of affairs. (These are impulses that can be indefinitely deferred, unlike the impulses to alleviate hunger, pain, or thirst that so often figure in animal models.) That an aggressive or sexual impulse is active is indicated by the presence of a fantasy of a state of affairs in which the impulse is gratified in action. That state of affairs is imagined and appraised as pleasurable. The fantasy is associated with a pressure to realize in actuality what is so imagined. Another state of affairs, which the realization of the impulse in action is expected to produce, is also imagined; anxiety is the appraisal of that state of affairs as dangerous.

Anxiety then, according to psychoanalysis, is always bound up with an ideational content. The activation of a subcortical physiological system ("arousal") or of a behavioral system ("fight or flight") may or may not occur following anxiety. If such activation does occur, it may occur on some occasions and not others or with different degrees of duration or strength in different subjects. In any event, such activation is certainly not identical with anxiety, which is essentially an evaluative assessment (subserved or supported by cortical systems), a "signal" that an imagined state of affairs is both dangerous and highly likely to become an actual state of affairs.

The Path to Anxiety Disorders

Psychoanalytic hypotheses about anxiety, unlike definitions of anxiety, make empirical claims that certain vicissitudes of anxiety are stages on the path to anxiety disorders. Five such stages are described below.

1. Anxiety results in either rational problem solving or defense. Problem solving includes a conscious recognition and assessment of the likelihood of actualization of the anticipated dangerous state of affairs (secondary process thinking, reality testing) and may result, for example, in an utilitarian decision to inhibit impulse-gratifying behavior, in order to avoid pain. (Such an inhibition of impulse-gratifying behavior, or the mobilization of physiological processes leading to or subserving such inhibition, are not, of course, identical with the anxiety that motivates them.)

Because the act of imagining or fantasizing gratification of a sexual or aggressive impulse is associated with a pressure to act upon the impulse, fantasizing may itself come to be appraised as dangerous. This may happen, even though the fantasied state of affairs itself is appraised as pleasurable (the subject has "mixed feelings"). The subject may resort, then, among many possible defenses, to repression; that is, a fantasy of gratification of an impulse is denied access to consciousness in order to deny the impulse access to the motor apparatus.

Defenses, in general, are mental operations. They are directed to and alter processes of forming and maintaining mental representations or the content of such representations. Defenses determine how perceptions of external reality are interpreted (e.g., what motives are attributed to others). They also determine what objects become objects of identification, and the fate of such identifications.

If conscious problem solving or successful repression occurs, anxiety disorders do not develop. Factors in the external world are negative causal factors with respect to the development of anxiety disorders (or positive causal factors with respect to the mitigation of anxiety disorders) only insofar as they promote problem solving (including access to consciousness) over repression and primary process, weaken impulses relative to defenses, or strengthen defenses relative to impulses.

Primary processes accord primacy to the aim of achieving immediate gratification without regard to recognition of or accommodation to external reality. Mental representations that are produced by or are part of a primary process are used in imagining wishes as fulfilled without regard to obstacles or opportunities in external reality. Such mental representations are subject to or are the product of such operations as condensation, displacement, and the substitution of iconic symbols or images for linguistic symbols to represent concepts. Alteration of the balance between defense and impulse by situational factors may not be caused directly by intrinsic features of the external world; usually, rather, the impact of features of the external world is mediated by or depends on the subject's interpretation of them. Such interpretation is often determined by linkages between these features and the subject's unconscious fantasies of wish-fulfillment and danger, linkages that have been forged by primary process operations.

2. If a first act of defense (repression) subsequently fails, if derivatives of unconscious (repressed) contents become increasingly explicit and peremptory, a subject is motivated by anxiety to resort to other defenses. For example, a subject may transform a dangerous internal impulse into a danger from without (an object in external reality is seen as expressing the dangerous impulse or tempting the subject to express it, or as ready to attack or attacking the subject for harboring or expressing the dangerous impulse). (Such transformations are based upon personal-symbolic linkages forged by primary process operations.) Now that the danger is an external one, the subject can resort to fight or flight and may mobilize physiological resources preparatory to such action.

A subject may also subsequently internalize such an attacking object in external reality, identifying with it; here, especially, affects of rage and depression may be generated in response to imagined attacks upon the subject, or imagined interferences with the pleasure seeking of the subject, by the internalized object. This account constitutes one possible explanation of the oft-made observation—that anxiety and depression seem frequently to co-occur.

Factors in the external world are positive causal factors with respect to the development of anxiety disorders only insofar as they undercut secondary process thinking and reality testing, intensify impulses relative to defenses, or weaken defenses relative to impulses. Again, that these causal factors have such effects may not be due to intrinsic features of the external world, but rather to the subject's interpretation of them, and such interpretation, as has been pointed out, is often determined by linkages between these features and the subject's unconscious fantasies of wish-fulfillment and danger, linkages that have been forged by primary process operations.

3. If defensive operations continue to fail, more anxiety is generated.

4. The experience of anxiety, or the manifestations of physiological arousal in preparation for fight or flight, may themselves come to be linked by primary process operations to unconscious fantasies and interpreted as a particular kind of substitute gratification, or as a particular kind of danger. They may be exploited as well (complicating the clinical picture) to obtain gratification or punishment in interactions with others—by patients, for example, suffering from hysteria (an illness that largely goes unrecognized today).

5. Different anxiety disorders are characterized by the occurence of

the dysfunctional consequences of a more or less continuous or intermittently recurring failure of defense; or by the occurrence of the dysfunctional consequences associated with the particular type of defense(s) the subject uses.

Animal Models

Various experimental procedures involving animal subjects purport to provide a so-called animal model of human anxiety or anxiety disorders. Conclusions about human anxiety or anxiety disorders are then drawn from observations of the effects of various kinds of interventions (e.g., the introduction of pharmacologic agents). As interesting and suggestive as such accounts are—see, e.g., Costa's discussion of behavioral models of anxiety (1985), Gray's discussion of a behavioral inhibition system (1985), and Mineka's discussion of the usefulness and limitations of animal models (1985)—in considering generalizations based upon work with them it is appropriate to ask: Does this animal model in fact model what psychoanalysts mean when they speak of anxiety or anxiety disorders?

In general, an animal model will fail to model what psychoanalysis terms "anxiety" if one or more of the following statements is true of it.

1. The responses of the animal that are of interest are those subserved or supported by subcortical rather than cortical systems. That is, what is of interest in the investigator's observations are the manifestations or effects of subcortical rather than cortical activity.

2. The animal responds automatically, physiologically or motorically, to cope with an existent threatening or thwarting state of affairs; an actual frustration of or attack upon the animal occurs. The state of affairs is existent and perceived rather than anticipated or imagined. The activation of a physiological or motoric-behavioral system (as part of the animal's response to that state of affairs) is considered equivalent to anxiety. The investigator is unable to distinguish anxiety from fear, rage, or hopelessness. The investigator tends to regard evaluative assessments of state of affairs (affects) and mobilization of resources for physical action as equivalent.

3. The animal acts to satisfy drives that do not tolerate indefinite deferral of gratification, such as hunger or thirst; in so acting, the animal meets and attempts to overcome an actual obstacle.

4. The obstacle to gratification is arbitrarily imposed by the investigator—and therefore is made up of features of the external situation (external stimuli) that are inescapable and coercive. The connection

between the animal's drive-reducing actions (drinking) and external stimuli that cause the animal pain (electric shocks) is determined by space-time contiguity, or physical causal laws.

5. The animal's utilitarian solution of the problem posed by its recognition that attack upon it is instigated by its own drive-reducing actions (for example, that it is shocked when it drinks) and the activation of the physiological or motoric-behavioral system involved in effecting that solution (avoidance of drinking) are considered to be equivalent to anxiety.

Questions for Psychoanalysis

A number of problems are raised when variables such as unconscious fantasies of wish-fulfillment or danger, whose measurement is expressed in values such as "is present" or "is not present," are used in a theory.

1. How does psychoanalysis obtain measures of variables which refer to the presence or absence of hypothetical or inferred dispositions or propensities? These dispositions may or may not be realized on any particular occasion the subject is observed, and such variables cannot merely indicate the presence or absence of some particular kind of observable action. A particular set of data may provide no evidence of a fantasy or wish fulfillment or a fantasy of danger, or even of a fleeting occurrence of anxiety. It may only provide evidence of the presence of operations of defense.

2. What limits to hypothesis testing are set by the present inability of psychoanalysis to quantify these variables? A crude dichotomous scale (is present/is not present), or at the most an ordinal scale (more/less), are all that is available now, resulting in hypothesis testing that is less precise and rigorous than one might wish.

3. Demonstrating the presence of unconscious fantasies of wish-fulfillment or danger depends upon reference to their hypothetical causal links to observable phenomena other than, and ideally independent of, those manifestations of anxiety and defense that their presence supposedly explains. How does psychoanalysis establish the credibility of such hypothetical causal links, which are often complex if not downright tortuous?

4. What is relevant to observe in establishing the credibility of these hypotheses? For example, the relation between inferred unconscious fantasies and observable states of affairs (i.e., bodily or environmental states of affairs) in the situation of the person-as-a-psychological-system is, according to psychoanalytic theory, a complicated

one. It is not the case that presence or absence of features of a situation are causes, and fantasies are their effects. But what, then, is the relation between fantasy and reality? On the one hand, an inferred unconscious fantasy is the expression of an impulse, which provides a constant, not a momentary, pressure, presumably quasi-independent of situational vicissitudes that might be directly observed. On the other hand, the intensity of an unconscious impulse or wish may vary in response to situational vicissitudes; state of affairs perceived to be opportunities (or incentives or inducements) to gratify such a wish, as well as certain bodily changes, may intensify the wish. How can psychoanalysis deal methodologically with the complications introduced by the phrase "perceived to be"—the complications that follow from postulating the importance of the interpretation a person makes of an observable state of affairs compared to the importance of the intrinsic features of that state of affairs.

So it is, also, with fantasies of danger. Clearly, dread of being overwhelmed by the torment of unsatisfied need is not in any simple way caused by intrinsic features of a situation; nor are fantasies of castration, or of loss of love, or of visitations of shame or guilt, or even of loss of a loved object. But it is just intrinsic features of a situation that investigators are likely to regard as accessible to observation.

5. Finally, how is psychoanalysis to capture empirically the occurrence of anxiety itself? The signal of anxiety, to the extent it acts effectively as a signal, is itself fleeting and often so efficiently followed by defense that it is lost to awareness. Its occurrence must often be inferred rather than observed—even under the special conditions created in the psychoanalytic treatment situation. Here, in this unique situation, the whole process is made to occur in slow motion, as it were, so that—now I mix metaphors—the complete process can sometimes be observed microscopically, with every fine detail in bold relief.

These are all questions for psychoanalytic psychology.

Questions for Other Disciplines

There are also questions about anxiety that are simply outside the province of psychoanalytic psychology. It is to their credit that Michels, Frances, and Shear (1985), in considering such questions, do not espouse a vapid integration of multiple frames of reference and of investigations of vastly different domains. They do make it clear that the psychoanalytic theory of anxiety neither should nor could answer every question of interest about anxiety; and they do start to specify which questions are in the domain of psychoanalytic psychology and which questions it cannot—with its concepts and methods—profitably pursue. We tend to be on our guard lest political or intellectual interest in "integration" prevent recognition that two hypotheses are indeed rival propositions about the same domain, and that both cannot survive. If we also show that some differences are differences in the complementary, but not necessarily competing, questions that investigators of different domains are capable of addressing, we shall have done well.

The First Question

What brings about or causes the linkage between an unconscious fantasy of wish-fulfillment and an unconscious fantasy of danger? Another way to ask this question is: What causes a person to believe, or to respond as if he believes, that gratification of a wish will lead to a dangerous state of affairs? Is this linkage—or this belief—due to contiguity in experience of expression or gratification of the wish and a dangerous state of affairs?

Psychoanalytic psychology, on the whole, answers this last question no. David Rapaport (1959) rejected the notion that a theory of learning can help explain the linkage between wish and danger, if that theory depends on mechanisms of conditioning or reinforcement, which give a place of prominence to the role of situational stimuli acting as indices. But a more complex theory of learning, which, for example, makes use of the distinction between indices (whose meaning is grounded in existential connections), on the one hand, and icons and symbols, on the other hand, or a theory that builds on postulated central and innate mechanisms, might indeed help to explain this linkage (Edelson 1984, pp. 91–92). Since the effects of early childhood experiences can be investigated, and also are accessible to influence, by psychological means, current psychoanalytic psychology itself may have a bias against a theory of learning emphasizing central and innate mechanisms. That would certainly not have been Freud's bias.

The Second Question

How can one differentiate the affective appraisal associated with unconscious fantasies of danger and the physiological and behavioral responses to such appraisal? One may observe a fleeting purposeful anticipatory premonitory "watch out!" buzz, a transient state of tension or readiness, a continuing diffuse unregulated state of arousal, or an overwhelming incapacitating panic. Why this variability? Developmental psychologists and biologists in studies of the maturation of psychological capacities, and biological investigations of such phenomena as inborn differences in autonomic lability, can help answer this kind of question.

The Third Question

What is the fate of the defensive operation? To what extent does it serve adaptation? If it does serve adaptation, to what extent is its survival, in the life of an individual and in the species, independent of the vicissitudes of intrapsychic conflict? Michels et al. (1985) suggest that ethology and evolutionary theory have something to contribute here.

The Fourth Question

What innate or constitutional factors determine the strength and choice of defense; the tenacity of desire; and propensities to fixation, regression, and conflict? Clearly, Freud always regarded such constitutional factors as important, but just as clearly he did not expect the investigation of such factors to be carried out primarily in the psychoanalytic situation or even necessarily by psychoanalysts.

Impediments to Research in Psychoanalysis

I turn now to suggestions for a research strategy that might be adopted by psychoanalytic investigators in testing their hypotheses. The suggestions offered by Michels et al. (1985) seem to me to be good ones, given the impediments to research they have identified. The impediments of which I am most aware, however, are conceptual and attitudinal rather than matters of methods and instruments (Edelson 1984).

One impediment is the lack of a skeptical or critical attitude in the psychoanalytic community toward its own conjectures. (I write as a member of that community.) On the whole, we seem inclined to give priority to the search for confirmations of hypotheses, rather than to the policy of subjecting hypotheses to a risk of falsification. Positive instances of a hypothesis are overvalued, without regard to whether or not all such positive instances are relevant in establishing the scientific credibility of that hypothesis. In other words, psychoanalysts tend to overvalue demonstrations of the explanatory power of a hypothesis over attempts to establish its scientific credibility through empirical tests. David Rapaport in 1959, regretting this community's continued reliance on its wealth of clinical observations, warned psychoanalysts that while "the evidence [for psychoanalytic hypotheses] . . . seems massive and imposing, the lack of clarification as to what constitutes a valid clinical [that is, nonexperimental] research method leaves undetermined the positive evidential weight of the confirming clinical material." Because the required "canon of clinical investigation is lacking, much of the evidence for the theory remains phenomenological and anecdotal, even if its obviousness and bulk tend to lend to it a semblance of objective validity" (Rapaport 1959, p. 111). More than 80 years have elapsed since the appearance of Freud's *Interpretation of Dreams.* Surely that is time enough to move on to undertaking the task Rapaport calls on psychoanalysis to achieve.

A second impediment involves a related conceptual failure. Members of this community frequently neglect the strategy of formulating rival hypotheses within the psychoanalytic frame of reference, then seeking evidence, even in case studies, that will support one hypothesis and eliminate another rival hypothesis (or that at least will provide one hypothesis with much more support than another). Here, J. Platt's explication of strong inference (1964), which details this strategy and shows how advance in science depends upon it, and Ian Hacking's explication of his premise of comparative support (1965), which shows how evidence may provide one hypothesis more support than another (though that evidence is consistent with both hypotheses), are important contributions—apparently unknown to most psychoanalysts.

The final impediment I shall mention is that in establishing the credibility of a hypothesis the psychoanalytic literature, on the whole, does not demonstrate much appreciation for the necessity of presenting an argument, even a weak argument, that one's own hypothesis explains one's finding much better than do plausible alternative hypotheses (plausible, in the light of background knowledge). Freud, who struggled throughout his writings with the question of whether suggestion might not explain the outcome of psychoanalytic treatment, is, on the whole, an exception to this judgment.

Proposals

These conceptual and attitudinal deficiencies can be corrected to some extent even in the clinical case study or in investigations carried out with data obtained from the psychoanalytic situation itself. One does not, for example, have to consider every known plausible alternative explanation of one's findings. Eliminating even one or two alternative explanations is better than making no attempt to eliminate any alternative explanation (Campbell and Stanley 1963). As Cook and Campbell have pointed out in their book of quasi-experimental and naturalistic field studies (1979), one does not have to rely on experimental design methods alone in attempting to eliminate at least some plausible alternative explanations of one's findings; nondesign arguments can also be useful. Luborsky's symptom-context method (Luborsky 1967, 1973; Luborsky, Bachrach, Graff, Pulver, and Christoph 1979; Luborsky and Mintz 1974) though it is naturalistic and nonexperimental, and makes use of data from the psychoanalytic situation, involves a design that is capable of eliminating some alternative explanations of an obtained outcome, and, therefore, of testing hypotheses of interest to psychoanalysis. So, also, the philosopher of science Clark Glymour (1980) has shown that Freud (1909) in the case of the Rat Man study followed the same procedure in testing his hypotheses that Kepler and Newton used in testing theirs.

Psychoanalysis, especially, should not be derided for pursuing intensive studies of single subjects, but instead should be encouraged to turn its attention to and assimilate for its own purposes conceptual and methodological developments in single-subject research (e.g., Campbell and Stanley 1963; Chassan 1979; Hersen and Barlow 1976; Kazdin 1981, 1982; Shapiro 1961, 1963, 1966; Sidman 1960). It should consider the use of (1) individualized ipsative measurements, which can objectify and quantify subjective data; (2) baseline observations, which make systematic comparisons (based on multiple measures of a subject under different conditions) possible; (3) direct and systematic replication, which can deal with problems respectively of internal validity and of external validity or generalization in single-subject research; and (4) probabilistic or statistical reasoning about the relation between hypothesis and evidence. Then (and this is just one possibility) psychoanalysts, following Luborsky's methodological leads, might test psychoanalytic hypotheses about the anxiety disorders by studying and comparing those contexts in the protocol of a single psychoanalysis in which anxiety is reported or is exacerbated and the otherwise similar contexts in which is it not. These proposals are elaborated at much greater length, with additional references, in *Hypothesis and Evidence in Psychoanalysis* (Edelson 1984).

Psychoanalytic Theory[1]

6

Psychoanalysis is an intentional psychology. That is to say, psychoanalytic theory is about such entities as wishes and beliefs, their interrelations, and how they cause the mental states and the acts they cause. As an intentional psychology, psychoanalysis is in particular a science of the imagination, and a science of symbolic functioning. It is a science of the imagination insofar as in its theory psychic reality, attempts at wish-fulfillment, and unconscious fantasy play a central explanatory role. It is a science of symbolic functioning insofar as it studies modes of symbolization that construct mental representations that are constituents of states of imagining; or modes of symbolization that construct mental representations that are evoked by, or are responses to, products of the imagination.[2] Despite the theoretical essays of such psychoanalysts as Heinz Hartmann, who thought psychoanalysis should be a general psychology and who sought through his essays to make it one, it is not a general psychology.

The Domain of Psychoanalysis

No scientific discipline has as its objective the explanation of "everything." A discipline takes as its intended domain a realm of more or less clearly defined phenomena, about which there are specific kinds of questions. The discipline, if it is able to explain any-

1. This chapter is a revised version of Edelson (1986c), and contains material that appeared also in Edelson (1986b, 1986d).

2. Chapters 1 and 2 discuss psychoanalysis as a science of symbolic functioning, and Edelson (1984, pp. 93, 105–108) and Chapters 9 and 10 of this book discuss psychoanalysis as a science of the imagination.

thing or answer any questions at all, will be able to explain these phenomena or answer these questions. If a developing discipline is successful in pursuing its explanatory objectives with respect to its intended domain, its further success will depend in part on its ability to extend its explanatory apparatus, which has "worked" in its intended domain, to other domains.

The entities in the domain of any psychology of mind are mental states or events. These entities have certain properties. For example, they are intentional. Intentional mental states include a psychological activity, carried out by a person, and *directed to, about,* or *of* a mental representation of some object or state of affairs.[3] Some examples of intentional mental states are *intending to do* (including *to say*) *something; thinking* or *believing something;* having a *feeling about a thing, someone,* or *some state of affairs; wishing for* or *wanting something; perceiving something; remembering something;* and *imagining something.*

A mental representation may be linguistic, a linguistically interpreted or described image, or—questionably—nonlinguistic. The objects or state of affairs represented may involve an intentional mental state. Examples: "I feel guilty that I have this wish." "I want to feel that way about him."

The constituents of a mental state include (1) a person's own *mental representation* of an object or state of affairs; and (2) a *psychological activity* (thinking, believing, wishing, perceiving, remembering, imagining, feeling), which is a *mode of relation* between the person and the mental representation.

Some intentional mental states are satisfied by a certain object or state of affairs. A perception, for example, is satisfied by the existence of what is perceived. A belief is satisfied by the existence of a state of affairs that corresponds to what is believed. An intention to do something is satisfied if it is done. A memory is satisfied by the past existence of the state of affairs remembered. A wish is satisfied if a desired state of affairs comes to pass. Sometimes the fact that conditions of satisfaction do not exist are taken by psychoanalysis as what is to be explained. Sometimes such a fact is part of an aetiological explanation in psychoanalysis; the origin or exacerbation of psychopathology is attributed to it.

3. In this account of intentional mental states or events (hereafter, mental states), I more or less follow Searle (1983). "Intention," as used here, should not be confused with "intending." Intending to do something is one kind of intentional mental state. Also, "planning to do something" should be distinguished from an "intention-in-action"—such as the mental state associated with voluntarily lifting one's arm.

The Domain of a General Psychology of Mind

A general psychology of mind focuses on, or gives priority to, explaining psychological activities (thinking, believing, wishing, perceiving, remembering, imagining, feeling), perhaps in particular as achievements. It addresses the questions: What is the mind able to do? How does it do it? What makes it do it poorly or well? Psychological theories of mind explain each psychological activity by characterizing the nature of the psychological capacity that is realized in carrying it out. In other words, the mind is able to do what it does by virtue of the possibly quite different properties and lawful processes characterizing each capacity (Fodor 1983).

A general psychology of mind is primarily concerned with (1) how conscious mental states (including the mental representations that are part of them) are caused by intrinsic features of the environment of the organism (for example, what it is about various capacities of the mind that enable the organism to receive, register, and store the effects of external stimuli in the way it does); (2) how conscious mental states (including the mental representations that are part of them) cause the organism's actions (what it is about the various capacities of the mind that enables the organism to carry out at least some of the various kinds of action—for example, and perhaps notably, speech—that it does carry out); and (3) the causal relations among conscious mental states (for example, how perceptions cause beliefs, and wishes and beliefs in conjunction cause plans to do something or intentions-in-action).

Since the characteristics of a capacity must be inferred from its particular realizations in actual mental states, the psychologist's explication of the capacity is a theory. This theory, together with the facts and empirically observable regularities that constitute the evidence supporting it, become what the neuroscientist is called upon to explain in terms of the physical constituents of the organism. To the work of the neuroscientist the psychologist—not (or only incidentally) the psychoanalyst—contributes an explication of just what the nature of different kinds of capacities are; how these capacities change throughout the development of the human organism; and in what way their characteristics are dependent on or modified by environmental stimuli or conditions. It is the task of the neuroscientist, taking such capacities as facts to be explained, to show how these capacities develop and why they have the characteristics—including the limitations—they do have. The neuroscientist attempts to demonstrate how the features and interrelations of physical constituents and physical subsystems of the organism explain the features and interrelations of psychological capacities.

Features of Persons that Define the Domain of Psychoanalysis

Defining the intended domain of psychoanalysis begins with the following considerations. A person has, at any one time and at one and the same time, *multiple* wishes.[4] Strains result because some of these simultaneous multiple wishes conflict—the states of affairs wished for are incompatible (mutually exclusive)—and because some of these simultaneous multiple wishes compete for the same resources (organismic and environmental) and these, of course, are never unlimited. Certainly, then, at the least it is highly unlikely that the satisfaction of conflicting or competing wishes can all be maximized simultaneously, on some one occasion, within some relevant period of time, or at some one particular place. It would seem that in any person some phenomena, including some mental states, signs of disturbance, or dysfunctional performances, must develop that are simply expressions of (or caused by) such inevitable conflicts and dilemmas and attempts to resolve them.

A person will attempt to resolve the strains resulting from conflicts and dilemmas. Some efforts will be directed toward changing the world—making it more amenable to the achievement of satisfaction. Some efforts will be directed to bringing about renunciations, establishing priorities, or accepting compromises. Compromises may involve partial or deferred satisfaction of each of a number of conflicting or competing wishes; the satisfaction of wishes not in actuality but in imagination; or, in the persistent search for satisfaction, the repeated attribution of the same motives, attitudes, or characteristics to others, or the repeated evocation of the same responses from others, in an effort to create the illusion or support the belief that a wished-for state of affairs exists in actuality that can be enjoyed, or that a feared state of affairs exists in actuality that can be avoided.

In any event, it is unlikely that any attempt at resolution of conflicts or dilemmas can be totally successful—can be effected without some cost in frustration, pain, or misery. If a solution to such a problem involves awareness of the problem, and has been effected by conscious thought as well as deliberate voluntary implementation of plans and decisions following from such thought, we consider whatever misery accrues *normal* misery—the normal lot of mankind.

4. For convenience in this exposition, fears are regarded as *wishes* to avoid particular objects or states of affairs.

In other cases, according to psychoanalysis, the attempted solution to a conflict or dilemma is to get rid of it by rendering it inaccessible to consciousness—and also rendering inaccessible to consciousness its connection to any initial inadequate attempts at solution of it. (Residues of such initial attempts can be found, for example, in character, interests, or preferences.) However, the conflict or dilemma does not go away or does not stay away; it remains or again becomes active. The initial attempts at solution of it, if they persist, are now disconnected from it in consciousness, and can no longer be consciously re-evaluated in the light of subsequent experience, as solutions of it. Further attempts at solution continue, but outside awareness. Therefore, the contents of various relevant but unconscious mental states (wishes, beliefs) are not subject to conscious judgment or to correction by experience. The mental operations carried out on these mental states, most particularly on their contents, in order to effect a solution are not those ordinarily employed when thinking, planning, and decision making are consciously carried out. Such a solution depends on linkages among mental states created by these mental operations—and not on actions that bring about conditions of satisfaction in external reality. The person is not aware of what kind of problem he struggles with; not aware that he considers a certain mental tactic or strategy part of a "solution" to anything; not aware of the mental tactic or strategy itself; not aware of the illusory and unstable nature of this solution; not aware that the continuous effort he expends is required to maintain this solution; and not aware that the pain he suffers, which he considers inexplicable, and which we call *neurotic* misery, is part of what he pays for this solution.

Mental Contents and Causal Gaps

Contents (representations of objects and their features, or representations of states of affairs and their constituents) are causal processes. These causal processes link conscious mental states. They propagate through space-time the causal influence of a mental state or event (a cause) upon another mental state or event (an effect) in another space-time.[5] They connect environmental stimuli in the world and conscious mental states (the direction of the causal process is from world-to-mental-representation), or conscious mental states and actions (the direction of the causal process is from mental-representation-to-world). Observations of the continuity of contents—the persistence of contents from one mental state (mental represen-

5. Salmon (1984) is an especially useful discussion of causality in these terms.

tation) to another, or from world to mental representation, or from mental representation to world—provide evidence for such causal connections.

A mental state is comprehensible (either to a subject or to another) if it is in accord or consistent with a set of wishes and beliefs that are easily accessible to the subject's consciousness. Phenomena in the intended domain of psychoanalysis always involve discontinuities in causal connectedness among conscious mental states; in causal connectedness between environmental stimuli and conscious mental states; or in causal connectedness between conscious mental states and action. There are three possible kinds of causal gaps.

1. There are mental-representation-to-mental-representation causal gaps (neurotic symptoms, and most especially, compulsions, inhibitions, and phobias). Here we have discontinuities in causal connectedness among contents of different mental states (beliefs, wishes, perceptions, memories, thoughts, intentions to act)—some don't seem to follow from, lead to, or indeed have any comprehensible relation at all to any other. "Why am I in a mental state (thought, feeling, impulse) I do not want to be in? Why am I thinking *this,* feeling *this,* experiencing an impulse to do *this?* It makes no sense to me! Why am I unable to bring about a mental state (thought, feeling, intention, memory) I want to be in? Why am I unable to think about *this?* Why am I unable to feel anything about *this?* Why can't I bring myself to do just *this?*" A neurotic symptom is a mental state (a thought, a feeling, an impulse or intention to carry out an action), or an inability to bring about a mental state, that is inexplicable to the person who suffers it, because it seems to have no causal relation to his conscious wishes and beliefs. It occurs whether or not he wants it to occur, and whether or not it makes sense to him.

2. There are world-to-mental-representation causal gaps (dreams and other representational anomalies, possibly jokes).[6] Here we have discontinuities in causal connectedness between actual objects or states of affairs in a person's situation and the person's mental representations of objects and states of affairs. "Why am I seeing something that is not there and could not be there? Why should *that*—it really isn't funny when you stop and think about it—arouse this feeling in me? Why should I have just *this* internal reaction to it?"

3. There are mental-representation-to-world causal gaps (mistakes, failures). Here we have discontinuities in causal connectedness be-

6. There may be advantages to regarding dreams as representations that are an aspect of imagining rather than perceiving (Edelson 1984, p. 93; Wollheim 1979).

tween a person's mental representation of an intended action and the action the person actually carries out. "Why am I unable to do just *that* which I am in the mental state of trying to do? Why am I doing just *that* which interferes with or prevents my doing what I am in the mental state of trying to do? Why did I say *that?*"

These causal gaps are appropriate objects of psychoanalytic explanation insofar as they are not explicable, given present knowledge, as the effect of a defect in the design of the physical organism (genetic endowment); a lesion of the physical organism; or environmental conditions (e.g., inaccessibility of opportunities, facilities, or resources). A stronger requirement would be that the explicability of a phenomenon is inconceivable in terms of such causal factors. The paradigm is a conversion symptom that makes no sense in terms of well-established knowledge about anatomy.

It is the burden of psychoanalysis to demonstrate, using the methods of psychoanalysis, that the causal gaps in its intended domain are caused by (or are an expression of) one or more constituents of an unconscious conflict or dilemma (as in an eruption of an impulse in action, or an intense imagining of a wished-for state of affairs); or are caused by (or are an expression of) a person's attempts to resolve such a conflict or dilemma (a compromise-formation).

Even if the analysand reports one or more compulsions, inhibitions, representational anomalies, or mistakes or failures, it often turns out to be the case that he does not realize, or only comes to discover under special circumstances, that he has avoided noticing or has in some way circumvented experiencing other problems of this sort. A psychoanalysis may encounter defensive strategies designed to avoid symptoms themselves; it may take some time to see through various disguises or compensatory maneuvers to just what the problems are. Rationalization, for example, may conceal certain problems, just as the hypnotized subject, following a suggestion he does not remember, offers reasons for what he is doing. He is not aware of a causal gap. He is not aware that an unconscious idea compels him.

One cannot underestimate the extent to which psychoanalysis was born of Freud's observations of hypnotic phenomena; the fact that causally effective unconscious ideas can be created by an "experimental" intervention and used to bring about behavioral effects is important in supporting the conviction that unconscious entities are real and not just "figures of speech" or theoretical fictions or conventions.[7]

7. Hacking (1983) and Leplin (1984) contain pertinent discussions of scientific realism with respect to theoretical entities.

Mental Contents as Causal Processes Linking Mental States

Psychoanalysis seeks the causes of neurotic misery in the *contents* of unconscious mental states as these undergo various transformations, which are effected by particular kinds of mental operations. These mental operations provide the causal links between the *contents* of unconscious mental states and the form and *contents* of mental states involved in the phenomena that comprise the intended domain of psychoanalysis.

Psychoanalytic generalizations state causal relations among mental states. How are these mental states causally related, according to psychoanalysis? By virtue of their contents. A psychoanalytic explanation refers to the interrelations of mental states in terms of their contents. Psychoanalysis is therefore committed ontologically to the existence and reality of persons; relations between persons and mental representations; and mental representations (the contents of mental states). Psychoanalysis is distinctive in this focus on the content of mental states. It is concerned with questions about the content. It wishes to explain the content. Not *how* is John able to *believe,* but *why* does he believe *that.* Not *how* is he able to form or carry our *intentions,* but *why* does he plan to do—or why does he carry out the intention to do—just *that.*

The Explanatory Task of Psychoanalysis

The psychoanalyst focuses on certain kinds of performances (those exhibiting causal gaps) rather than capacities. He might be said to take capacities for granted.

Psychoanalysis has its raison d'etre in the investigation of phenomena and the treatment of disorders that are expressions in conscious mental life of causal interrelations among unconscious mental states—rather than in the investigation of causal relations between conscious mental states and environmental stimuli, or between conscious mental states and action, except insofar as these too show signs of being influenced by causal interrelations among unconscious mental states.

The Theory of Psychoanalysis

Among the interpretations a psychoanalyst makes to an analysand are explanations of causal gaps in the analysand's conscious mental life. The psychoanalyst's aim is to make intentional mental states the analysand finds incomprehensible comprehensible to him. The psychoanalyst's strategy of explanation is somewhat as follows. (Here, for the sake of clarity, I shall ignore complications that rightly would be considered in a more extended discussion, and shall avoid as much as possible a technical vocabulary.)

1. *Unconscious mental entities.* The psychoanalyst infers the existence of unconscious mental entities—wishes and beliefs.

2. *Unconscious mental contents.* He infers the existence of unconscious mental contents. These unconscious mental contents are fantasies or imagined states of affairs, which are constituents of unconscious wishes and beliefs.

3. *Wished-for states of affairs.* He identifies two major kinds of imagined states of affairs—those the analysand wishes to actualize and those the analysand wishes to avoid.

4. *Sexual wishes.* Those imagined states of affairs the analysand wishes to actualize are states of affairs in which the analysand is experiencing sexual or sensual pleasure.

5. *The problem of aggression.* I am avoiding consideration of aggression, partly because such considerations complicate the story I am telling, and partly because I doubt the coherence of accounts of aggression in psychoanalytic theory.[8] Is aggression primary, secondary to the thwarting of sexual gratification, or as in the case of sadism a component of sexuality itself? Does aggression have a consummatory state analogous to pleasure? Sexual wishes are, in this theory, necessary constituents of conflicts and dilemmas producing causal gaps in conscious mental life. Do aggressive wishes have this status? It is unlikely that sexual *and* aggressive wishes are jointly necessary; and, if sexual *or* aggressive wishes are necessary, then sexual wishes are *not* necessary, because there will be cases in which aggressive wishes will suffice.

Unconscious vindictiveness and fantasies of revenge appear in an unknown number of cases to motivate causal gaps in the domain of psychoanalysis, but in my experience these do not occur "on their own," in the way that sexual wishes appear to do. The vindictiveness is often directed toward a person who has deprived the analysand of sensual gratification by exposing him to the threat or actuality of a painful state of affairs.

Hatred and rage do not appear "primary," as sexual desires do, but often protect the analysand against what is much more difficult to bring to consciousness—longing and desire for a hated person or guilt over wanting to hurt a loved one. Fantasies of revenge may also serve

8. Waelder (1960) points out unanswered questions about aggression in psychoanalytic theory.

to protect the analysand in his imagination from ever again being a victim in the kind of painful state of affairs he wishes to avoid; in his fantasy of revenge he is the one that inflicts the pain rather than suffers it.

We do not know whether hostile wishes that are causally related to causal gaps in the domain of psychoanalysis are (often? always?) part of a defensive strategy against the danger that is threatened should the analysand seek sexual gratification from the depriving or frustrating person or someone identified with that person. Such questions set a program of inquiry for the *core* theory of psychoanalysis. Whatever one's answers to them are, one must deal with the fact that rage appears to be as displaceable as sexual desire; and displaceability is, for psychoanalytic theory, a sine qua non of sexuality.

6. *The problem of wishes to realize ideals and to succeed.* Other imagined states of affairs the analysand wishes to actualize are those in which he is able to fulfill an ideal, and those in which he is successful in meeting the claims of external reality.

The status of these kinds of wishes is problematic. Unlike sexual wishes, these are not necessary constituents of conflicts or dilemmas producing causal gaps in the domain of psychoanalysis. In many cases, these wishes are not even unconscious, although, for example, an analysand may be unconscious of what ideals he does wish to realize in his actions, usually because becoming conscious of them may lead, for example, to shame (they are "childish" ideals).

Sometimes, gratification of these wishes may be means to gratifying sexual wishes, rather than ends in themselves. Sometimes, gratification of these wishes may be part of unconscious strategies for resolving pathogenic conflicts or dilemmas.

Perhaps wishes to realize ideals or to succeed might best be formulated as belonging to the second set of wishes—wishes to avoid states of affairs in which the analysand suffers the pain of loss, failure, guilt, or shame. These questions, like those involving aggression, set a program of inquiry for the *core* theory of psychoanalysis.

7. *Feared states of affairs.* Those imagined states of affairs the analysand wishes to avoid are states of affairs in which the analysand is experiencing pain. In such an imagined state of affairs, the analysand may be experiencing helplessness in the face of loss—loss of love, of a needed other, of a body part, of whatever might be needed to achieve pleasure. The analysand may be experiencing shame; he is being exposed, ridiculed, or belittled. The analysand may be experiencing guilt; he is being scolded or criticized.

8. *Unconscious conflicts and dilemmas.* The psychoanalyst infers the existence of unconscious beliefs that states of affairs the analy-

sand wishes to actualize will inevitably be followed by states of affairs he wishes to avoid, or will make impossible other states of affairs the analysand wishes to actualize. In other words, the psychoanalyst infers the existence of unconscious conflicts and dilemmas.

9. *Anxiety.* The psychoanalyst infers that the analysand is unconsciously tempted to actualize a state of affairs that he believes will have one or more of these dread consequences, and that his unconscious realization that he is so tempted generates anxiety.

10. *Anxiety as a motive for resolving conflicts and dilemmas.* The psychoanalyst infers that the anxiety motivates the analysand to develop and carry out strategies for resolving the conflict or dilemma in such a way that anxiety may be avoided.

11. *The nature of unconscious strategies for resolving conflicts and dilemmas.* The psychoanalyst infers the nature of the strategy—what mental (symbolic) operations the analysand has used upon what materials (ultimately, mental representations) to achieve this resolution.

Strategies have not been adequately described and classified. Some strategies are used to resolve almost any conflict or dilemma; they are, so to speak, permanently in place. Others are called up in response only to one particular conflict or dilemma. The usual list of defense mechanisms by no means exhausts the possible strategies. In addition, the defense mechanisms are often confused with mental operations that come into play in any unconscious mental processes; employment of such a mental operation may not necessarily serve defense. Condensation may be one such mental operation.[9]

What strategy will be selected? This is another question that needs further thought and suggests paths for further investigation. One possibility is, of course, that the analysand's capacities will predispose him to employ some rather than others. There is also a sort of least action principle postulated under the name "overdetermination" or "multiple determination." While this has been interpreted to mean that there are multiple causes of each causal gap in the domain of psychoanalysis, it may also be interpreted as a criterion for choice of a strategy. That strategy will be chosen that eventuates in the gratification of the greatest number of wishes. Unless one has, of course, even in the individual case, a finite list of wishes to be gratified, this may turn out to be nothing more than a heuristic: don't be satisfied with finding that a symptom gratifies in part at least one wish; look for others as well. The heuristic has considerable value in the treatment situation. On the other hand, the danger is that such a

9. Chapter 2 contains an extended discussion of this point.

principle will turn out to be an ad hoc device for evading failure. If the symptom has not yet disappeared, it is because it still gratifies as yet some undiscovered wish; so, the theory is saved and the psychoanalysis must continue.

12. *Causal gaps as effects of unconscious attempts to resolve conflicts and dilemmas.* The psychoanalyst shows by tracing the effects of the various mental operations employed that a causal gap in the analysand's mental life results from a more or less successful (depending upon the extent to which anxiety is avoided) attempt—or part of an attempt—at such a resolution, which realizes a state of affairs that to some extent and in some way yields pleasure, and yields it in such a way that the analysand is able to some extent and in some way to escape the consequence he wishes to avoid.

An apparent paradox results when actualization of a danger itself is imagined to gratify a sexual wish (yielding masochistic pleasure, for example); alleviates guilt or ends an anxious wait for punishment; becomes a condition for sexual gratification; or is a ploy for achieving it ("I have paid for it so now I can have it").

An analysand may perceive any situation as one involving or at least promising pleasure or pain, for example, by attributing to others in his situation motives, intentions, or attitudes toward himself. To do so is one strategy for avoiding anxiety (maintaining eternal vigilance, always prepared to detect and avoid what one fears or grasp what one wants). Interpreting any situation in such a way that it yields some pleasure and enables the analysand at the same time to take steps to evade an imagined danger is one way of effecting a resolution of an unconscious conflict or dilemma. To interpret reality in this way as part of a strategy for avoiding anxiety or effecting a resolution of an unconscious conflict or dilemma introduces, of course, a degree of stereotypy in the analysand's relations to external reality, a stereotypy that in itself may be dysfunctional (maladaptive).[10]

Here, of course, is the basis for the importance of transference phenomena both in psychoanalytic therapy and research.

13. *The status of the origin of wishes and beliefs in childhood.* The psychoanalyst may also infer the *origin* of such unconscious wishes or beliefs in the analysand's infantile experiences—with his body and with parents and siblings. But, except that current unconscious wishes for pleasure are postulated to be infantile (that is, the kind of wish a child has), this inference is not crucial to the psychoanalyst's explanation of a causal gap in the analysand's conscious mental life.

10. Edelson (1984) discusses these points at greater length.

The cause of that gap lies in the presence of efforts to resolve some unconscious conflict or dilemma here-and-now. Psychoanalysis is not essentially a historical discipline.

Here I may be taking issue with Grunbaum's critique (1984) of the foundations of psychoanalysis, for it rests heavily on taking as fundamental the claim that the etiology of neurosis lies in childhood. It is true that Freud and others make this claim, and as a statement of remote rather than proximal etiology, it is, of course, part of psychoanalytic theory. This claim is not, however, logically prior; nor do I believe, for reasons I have given in the text of this chapter, that it has priority in any research program seeking to establish the scientific credibility of psychoanalytic theory.[11]

14. *The status of adaptation.* The psychoanalyst also may show the analysand what *consequences* the resolution he has attempted has for adaptation, but again this is not part of his explanation of the causal gap in the analysand's conscious mental life.

15. *The genetic and adaptive points of view.* The implication of this account is that the so-called genetic and adaptive points of view of metapsychology (Rapaport and Gill 1959) do not have the same status as the topographic, structural, dynamic, and economic points of view. The latter points of view are ways of saying that the cause of a causal gap in the domain of psychoanalysis has topographic aspects (it involves unconscious mental entities, contents, and processes); structural aspects (it involves different kinds of aims or wishes); dynamic aspects (it involves unconscious conflicts or dilemmas); and economic aspects (it involves the relative strength of competing or conflicting wishes, and involves also the mental operations that effect displacements of the affect or intensity associated with sexual wishes).

On the other hand, the genetic point of view proposes that the chain of events through time that have brought about the existence of the current cause of the causal gaps in the analysand's conscious mental life also be investigated. Each such event will itself have topographic, structural, dynamic, and economic aspects. What is to be investigated here, for example, are how the analysand's current unconscious wishes and beliefs came into being over time, how they became unconscious, and how he came to use just the strategies he now uses to resolve unconscious conflicts and dilemmas.

The adaptive point of view, in addition, proposes that the effects of causal gaps in the analysand's conscious mental life on the analysand's relation to external reality also be investigated. Here, a question

11. Chapter 13 contains an extended discussion of the assertions in this paragraph.

about the quality of object-relations looms large: is it an effect rather than a cause (although, of course, itself having further effects)?

So, additional questions of interest are: How did the current cause of a causal gap in the analysand's current mental life come into being through time (over the course of the analysand's development)? What are the effects of causal gaps in the analysand's conscious mental life upon his relation to his situation? These questions do become part of the program of psychoanalytic inquiry, even if one rejects, as I do, contra Hartmann, that the genetic approach, as a program for *psychoanalytic* inquiry, must involve investigation of all aspects of psychological development, or that interest in adaptation, as a program for *psychoanalytic* inquiry, must involve investigation of all aspects of an analysand's relation to reality.

However, developing answers to these two kinds of questions, even when the scope of such questions is restricted, which can be shown to be scientifically credible, using only the data obtained in the psychoanalytic situation itself, is indeed problematic.[12] Therefore, Hartmann, among others, was inconsistent in emphasizing the importance for psychoanalytic theory of formulating answers to the two questions of genesis and adaptation, while at the same time insisting that the major tests of psychoanalytic theory must make use of data obtained in the psychoanalytic situation.

16. *The theoretical apparatus of the psychoanalyst.* The theoretical apparatus used by the psychoanalyst in his explanations of causal gaps in the analysand's conscious mental life is, in fact, the core of psychoanalytic theory. The psychoanalyst does not take at face value the analysand's accounts of the nature of his conflicts and dilemmas or the analysand's explanations of causal gaps in his conscious mental life.

Psychoanalytic theory is distinctive in postulating an *ideogenic* etiology for these causal gaps.[13] The psychoanalyst uses a theory that postulates the existence of unconscious mental entities (wishes and beliefs); unconscious mental contents (fantasies or imagined states of affairs that are the objects of these wishes and beliefs); unconscious conflicts and dilemmas (relations among these wishes and beliefs); and unconscious strategies for resolving unconscious conflicts and dilemmas, making use of mental operations that, for example, operate on one mental content and transform it into another. The set of mental operations includes but is not equal to the set of defense mechanisms. The theory further postulates that infantile sexual

12. As Edelson (1984) and Grunbaum (1984), among others, have shown.
13. Wollheim (1971) points out the importance of this postulate.

wishes are necessary constituents of any unconscious conflict or dilemma producing a causal gap in an analysand's conscious mental life. In summary: unconscious conflicts and dilemmas activate unconscious strategies, which involve symbolic operations on mental representations. The use of these operations produces causal gaps in conscious mental life.

Elaborations of psychoanalytic theory in the interest of extending psychoanalytic theory to other domains should wait until this core theory has been shown to be scientifically credible, using data obtained in the psychoanalytic situation—since if a theory cannot be shown to offer better explanations than its rivals in its intended domain it is not likely to be a credible alternative in any other domain. Hartmann for one appeared confident that this task has already been accomplished, but in this belief he was mistaken.[14]

Heinz Hartmann's Influence on Psychoanalysis as Science

Effects of Attempts to Make Psychoanalysis a General Psychology

I shall conclude this chapter with some comments about the work of Heinz Hartmann, because it exemplifies the effort to make psychoanalysis a general psychology, and demonstrates some of the pernicious effects of pursuing that objective. I will not attempt here to explicate or assess what and how much that was valuable he added to psychoanalytic theory as he went about this task.[15]

In short, adopting such an objective obscures just which domain of phenomena psychoanalysis is most likely to be able to explain. Furthermore, psychoanalysis may pursue this objective by imperialistically incorporating the knowledge of other disciplines—often merely translating their terminology into that of psychoanalysis—but the price of this "enrichment" is that psychoanalytic theory becomes increasingly unwieldy and difficult to test. The difficulty is evident if one believes, as Hartmann did, that the data used in carrying out such tests will, for the most part, be obtained in the psychoanalytic situation. Surely, justifying the propositions of a general psychology is too heavy a burden for the slender reed of one particular method of investigation or mode of inquiry to bear.

Psychoanalysis as a theory of motivation should be distinguished from a theory about one or more psychological capacities—perception, memory, cognition, or language. Hartmann's

14. See, e.g., Hartmann's comments on the scientific aspects of psychoanalysis (1964, pp. 297–317, and especially p. 307) and Edelson's (1984) critique of them.
15. Schafer (1970) has already described and evaluated these additions.

thesis of the autonomy of "ego" apparatuses, among other theses, stated in the terminology of psychoanalysis, carries the unfortunate implication that psychoanalytic theory should include subtheories about each such capacity. (Why "ego" apparatuses, since such capacities serve other kinds of aims than those subsumed under the term "ego?")

Psychoanalysis as a theory of motivation must also be distinguished from a theory of performance, which is concerned with what use a subject makes of a capacity and what causes him to make just that use of it, when the principal causal factors of interest are states of the physical organism, features of the environment, and perhaps the conscious, "rational" intentions of the subject. These factors instigate, constrain, or enhance performances making use of intact capacities. Psychoanalysis does not include but presupposes knowledge of such factors. It does not attempt but rather presupposes a theory of rational action.

The wish to make a general psychology of psychoanalysis or, at the least, to extend psychoanalysis prematurely to other than its intended domain, has distracted it from the work of developing increasingly precise and testable answers to the distinctive questions it raises about the phenomena in its intended domain. That is to say, it has interfered with the work that psychoanalysis must do if it is to develop as a science.

The Work Facing Psychoanalysis as a Science

What is that work? A science must establish that certain empirical regularities exist in its domain. A conjecture or hypothesis that an empirical regularity exists must be supported by evidence. The data must be consistent with, follow from, or be entailed by the hypothesis. The data must favor the hypothesis over rival hypotheses. Factors in the situation of observation that might have led to obtaining the data supporting the hypothesis, even if the hypothesis were false, must be ruled out.

Then, an attempt must be made to explain this regularity and other such regularities as well. Usually, only a few explanations are plausible candidates; and one among these must be shown to be the better explanation of those being considered.

A better explanation is one that describes the mechanisms by virtue of which the effects of interest are produced. It explains the most facts (e.g., the greatest number of different empirical regularities found in the domain). It successfully predicts facts that are previously unobserved and unexpected (improbable, in the light of background knowledge). It may enable the scientist to set up conditions under which entirely new and otherwise nonobservable phenomena can be observed. It successfully functions explanatorily in the greatest number of different contexts—it explains different kinds of phenomena in the same domain, the interrelations among which are otherwise unsuspected, and it can be extended successfully to other domains.

Such an explanation may involve a model (a metaphor) that, when it is checked against data, often yields anomalies. The anomalies in turn lead to corrections of the model. Metaphors are fertile, if they suggest the kinds of solutions required to deal with these anomalies.

Such an explanation usually postulates theoretical (not directly observable) entities. Increasingly, evidence is obtained that argues against the idea that these theoretical entities are simply fictions or conventions. Different methods under different circumstances yield the same inferences about their properties. The structures of such entities come to be described in ever greater detail. Knowledge about their properties and structure enables the scientist to use them to affect causally other kinds of entities, or to measure other kinds of entities, in other domains.[16]

Upon What Does the Scientific Status of Psychoanalysis Depend?

What is the basis for making or rejecting the claim that psychoanalysis is a science: that it postulates the existence of certain kinds of entities or processes, about which it formulates conjectures; or that it uses a sound methodology to establish the scientific credibility of its conjectures?

Hartmann's theoretical essays, which merely allude rather nonspecifically and imprecisely to the existence of various kinds of data, might very well suggest to a reader, because of their preoccupation with metapsychological issues, that the scientificity of psychoanalysis is based on the "scientific" nature of the entities or processes it postulates (e.g., kinds of "energies," conservation principles), rather than on the methodology it employs to establish the credibility of its hypotheses about empirical regularities and the adequacy of its theoretical explanations of these regularities.

Even if Hartmann were attempting to demonstrate the explanatory power of psychoanalytic theory, which in fact his essays do little to demonstrate, explanatory power alone is not enough to confer scien-

16. Hacking (1983), Leplin (1984), and Salmon (1984) discuss theoretical entities in, and other characteristics of, scientific explanation. Chapter 13 gives an account of Freud as a scientist doing this kind of explanatory work in the Wolf Man case. Chapter 15 argues that psychoanalysis is capable in principle of doing this kind of scientific work.

tific credibility, as the above description of scientific work should show. An explanation must have other properties than its consistency with hundreds of observations to establish its credibility, and Hartmann's relative neglect of methodological issues might lead a reader to suspect that he was unaware of this or mistakenly considered that coping with such problems has a low priority in the development of psychoanalysis as a science.

Although Hartmann's essays have been accused of being too abstract, they unfortunately often can be read as mere translations of observable clinical phenomena into an arcane and not especially fertile terminology. He argues that changes in theory have brought about changes in technique, but he does not address questions about establishing the efficacy of these compared to the efficacy of previous or alternative techniques. For example, he does not address the possibility that the emphasis on "ego psychology" has had dysfunctional effects on psychoanalytic practice, leading perhaps to overcaution (even to the point of passivity when the patient is having difficulties participating in the treatment); postmature interventions (no less deleterious in their effects than premature interventions); superficial and perhaps, therefore, less radically effective interpretations; an unwarranted extension of the domain of technique;[17] and relative neglect of especially steamier aspects of the transference.

What evidence has been obtained that bears decisively upon such questions? Not much. But the relative efficacy of psychoanalytic interventions is far from a trivial issue with respect to the credibility of psychoanalytic theory (although Hartmann pays little attention to it), since a theory that does not extend the range of one's power to alter known, and even create new, phenomena is not likely to be preferred over other theories.

The psychoanalyst makes inferences about an analysand based on what the analysand says conjoined to psychoanalytic theory. If these inferences turn out to be false, their falsity would suggest some aspect of the theory must be incorrect. Therefore, the question of establishing ways to decide that an interpretation is (approximately) true is critical for establishing the scientific credibility of the theory.[18] Hartmann is not helpful here either.

In fact, his emphasis on the theoretical essay at the expense of an emphasis on the case study (where theory interacts with data) is disastrous for the development of psychoanalysis as a science. A case study is not, as it often seems to be in the literature since Freud,

17. There is an excellent discussion of this problem in Lipton's discussion of Freud's technique in the Rat Man case (1977).

18. See Meehl's discussion of subjective inference in psychoanalysis (1983).

merely a vehicle for illustrating theoretical conjectures; for piling up instances "confirming" theoretical conjectures; for ingeniously inventing or generating theoretical conjectures; or for applying theoretical conjectures—held to be scientifically credible for what are probably insufficient reasons—to a unique or singular case or instance. The case study is most importantly an argument for the scientific credibility of a hypothesis about an empirical regularity, or an argument that a theoretical explanation of such regularities is better than rival explanations. A great deal depends on improving the quality of such arguments in case studies in psychoanalysis, and to this task Hartmann contributed little if anything.[19]

Summing Up

There is much to like and learn from Hartmann's work, not the least of which was his even, considerate, respectful tone in dealing with controversies or ideas with which he did not agree.

He made some exceptionally valuable distinctions (e.g., between cathexis of contents and cathexis of functions); although I believe in his discussions of the "structural point of view" he muddled the distinction between sets of different kinds of aims and the different "apparatuses" or capacities that might serve any one or all of these aims.

His attention to metapsychology can be construed to be a measure of his devotion to the objective of making psychoanalytic theory a truly causal explanation of phenomena in its domain.

He was a man of great learning, with a deep understanding of bodies of knowledge other than psychoanalysis (see, e.g., his fine scholarly essays on rational action, moral values, and adaptation). He may seem to some to have overemphasized the importance of transactions across the boundary between the natural sciences and psychoanalysis; he actually did much more to indicate the importance of transactions across the boundary between psychoanalysis and the social sciences. Although the example of the scope of his knowledge might have done something to offset the increasingly defensive parochialism of some segments of the psychoanalytic community with respect to knowledge of other sciences, his work does not seem to have done much to create a meaningful rapprochement between psychoanalysis and the social and natural sciences.

He probably increased the respect and tactfulness with which psychoanalysts came to interpret their analysands' psychopathological traits and actions, and did something to mitigate the exaggerated

19. Chapters 11 and 13 show how the achievement of such a task might begin.

respect accorded "ego" aims and achievements and the odium attached to and prejudice against "id" aims that seem occasionally to creep into psychoanalytic discourse, by pointing out that symptoms and characterological pathology may serve adaptation, and that at times and for some purposes automatisms may in fact be more adaptive than reflective action.

Nevertheless, I conclude that the example set by his theoretical essays may have had a pernicious if not devastating influence on the development of psychoanalysis as a science. His writings about metapsychology, almost completely divorced from observations, and his relative neglect of methodological issues, may, unfortunately, have suggested to his readers that the scientificity of psychoanalysis is based on ontological rather than methodological grounds (on the scientific respectability of the kinds of entities and processes it postulates—such as the different kinds of "energies" with which Hartmann was so concerned—rather than on the quality of the arguments used in claiming that data obtained in the psychoanalytic situation support or confer scientific credibility on psychoanalytic theory). This was not Freud's position (if he held it early in his career, he abandoned it with the Project).[20] Hartmann's focus on the theoretical essays sans a convincing (i.e., precise and detailed rather than allusive) reference to data might also seem to some to justify, even if he never meant it to do so, submitting therapeutic practice to the test of compatibility with theory, rather than attempting to answer questions about therapeutic process, efficacy, and outcome through rigorous empirical studies.

Hartmann's attempt to turn psychoanalysis into a general psychology only contributed to making the theory more unwieldy and untestable, besides obscuring the logical relations among psychoanalysis and other sciences. In fact, the knowledge of some of these sciences is presupposed by psychoanalysis and some of them have the task of answering questions set by psychoanalysis that it cannot answer.[21]

The style of Hartmann's essays suggests a man of great modesty. Would that he had been more so in his objectives for psychoanalysis as a science.

20. Grunbaum (1984, pp. 2–9) discusses the difference between ontological and methodological grounds for the scientificity of psychoanalysis as seen by Freud.

21. An assertion I have elaborated in considerable detail in Chapters 3, 5, and 7.

7 Psychoanalysis and Neuroscience[1]

Some of us will, of course, reflect with high hopes upon the possible implications that exciting achievements in neuroscience have for the convergence of neuroscience and psychoanalysis. But why should we anticipate such a convergence? Do we foresee a solution to the mind-body problem? Is the theoretical reducibility of psychoanalysis by neuroscience within our reach? Do the contributions made by neuroscience to the explanation and treatment of psychopathology increasingly replace and render obsolete those contributions made by psychoanalysis?

In this chapter, I shall advance arguments for five positions, which here, in introduction, I merely assert.

1. The mind-body problem, as it is usually stated, is a metaphysical problem. ("Metaphysics" here refers to philosophical preoccupations with "what is the ultimate nature of Being, what really exists" sorts of questions, not with the use of philosophy for conceptual analysis, explication, and clarification.) Insofar as the mind-body problem is stated as a metaphysical problem (in this sense of the term "metaphysical"), no empirical evidence can contribute to its solution.

2. If we want answers to our questions about the relation between mind and body that are dependent upon and responsive to empirical knowledge, then the mind-body problem is best conceived as a question about the relation between two scientific disciplines. More specifically, it is best conceived as a question about the relation between the theories of a psychology of mind and the theories of neuroscience.

What relationship, consistent with a metaphysical

1. This chapter is a somewhat revised version of Edelson (1986d).

belief in materialism, is possible between a psychology of mind and neuroscience?

a) A psychology of mind might be *replaced* by neuroscience. If that day comes, the language of neuroscience will suffice for all purposes to which the language of a psychology of mind is now put.

b) A psychology of mind might be *theoretically reduced* by neuroscience. If that day comes, we shall be able to demonstrate that the concepts of a psychology of mind correspond to concepts of neuroscience, and that the explanatory generalizations of a psychology of mind can be logically derived from explanatory generalizations of neuroscience.

c) A psychology of mind might be an *autonomous* science. Holding to a metaphysical belief in materialism, it concedes that *every* mental state is embodied or has some particular neurophysiological correlate(s). But, as an autonomous science, it can neither be replaced nor theoretically reduced by neuroscience. (The great difference between what is conceded to materialism here and what is entailed by theoretical reductionism is not generally understood or fully appreciated.)

3. Special (i.e., nonphysical) sciences such as psychology, sociology, anthropology, and economics are autonomous. Each asks its own kind of questions of nature. Each sets itself distinctive problems to solve. The concepts of a special science—which reflect the way it classifies phenomena, the way it cuts nature at its joints—are not necessarily coextensive with, nor do they necessarily overlap, the concepts of any other science. The theories of a special science will probably not be ultimately logically derivable from the theories of physics and chemistry. In particular, the theories of psychoanalysis probably cannot be logically derived from the theories of neuroscience, or ultimately from the theories of physics and chemistry. In other words, psychoanalysis is not reducible by neuroscience. (In point of fact, it has never been demonstrated that the theories of neuroscience are themselves logically derivable from the theories of physics and chemistry, although it seems generally taken for granted that this will turn out to be the case.)

4. Psychoanalysis and neuroscience do not, and perhaps cannot, explain the same kinds or aspects of psychopathology. If either offers a scientifically credible and sufficient explanation for a kind or aspect of psychopathology, the other probably has no adequate explanation, and perhaps cannot have any adequate explanation, for it.

5. There is a body of theory in a psychology of mind that is not psychoanalytic but that is of some importance to both psychoanalysis and neuroscience. Such theory is concerned with the nature

of psychological capacities. Some part of a theory of psychological capacity does not require any reference to what is represented in mental representations for the discovery of patterns and regularities in nature or the formulation of explanatory generalizations. That is, it does not require any reference to the contents of mental states. That part of a theory of psychological capacity in a psychology of mind may be amenable at least to a partial reducibility by neuroscience.

To converge is to turn and approach each other, to come nearer together, to develop or come to possess like features, to meet at the same point. But neuroscience and psychoanalysis at present appear to be competitors for societal resources and intellectual commitments and, as such, to some extent, to be inclined to denigrate or ignore each other. We might hope, however, that their convergence will take the form of their becoming increasingly alike in the extent to which the conclusions each reaches independently in its own domain are methodologically reliable and, therefore, scientifically credible. We might hope that they will come nearer together, in that—though their tasks be different, though the problems they have set themselves to solve be different, though their concepts and techniques be different—someday they shall reach comparable high levels of scientific achievement. Then, for that reason alone, each shall command the respect, interest, and appreciation of the other.[2]

The Mind-Body Problem

A discussion of the relation between two scientific disciplines—neuroscience and psychoanalysis—may shed more light on the ancient mind-body problem than has many a metaphysical discussion in philosophy about the ultimate stuff the world is made of.[3]

Monism says there is only one ultimate stuff.

Neutral monism says this stuff is neutral, but can be described equally well in either a physical or a mental language. What this neutral stuff might be no one can say.

Idealism says the one ultimate stuff is mental. Most scientists do not find this a congenial stance—although some psychoanalysts of a hermeneutic persuasion, in their preoccupation with the metaphor of

2. I shall not discuss here the problems psychoanalysis faces as it attempts to reach this goal. See Edelson (1984), and Part 2 of this book.

3. In this discussion of the mind-body problem, I have drawn especially upon Borst (1970); Bunge (1980); Dennett (1978); Fodor (1968, 1975, 1981a, 1981b, 1983); Hebb (1980); Nagel (1979); Rosenthal (1971); and Searle (1983).

patients as linguistic texts to be deciphered or interpreted, seem in their discourse quite willing to forget the body altogether.

Materialism says the one ultimate stuff is physical.

Eliminative materialists, holding that a psychology of mind will eventually be *replaced* by neuroscience, go off the deep end and say, "Get rid of all mental terms! Such terms are manifestations of prescientific superstition and folkways." They really do seem to believe that people will eventually learn to say "my C-fibers are firing" instead of ignorantly complaining of pain.

This rather bizarre position is, of course, linked conceptually to *behaviorism.* Behaviorism discards mental terms in favor of descriptions of manifest behavior. It limits psychology to studying functional relations between environmental stimuli (physical events) and behavioral responses (physical events). *Radical behaviorism,* which is not only sans mind but sans nervous system as well, fell into disrepute as it became clear that states, events, or processes inside the organism played a causal role in producing behavior.

Logical behaviorism, an attempt to save the behaviorist program, retains mental terms but defines them as behavioral dispositions—that is, as propensities of varying strength to behave in certain ways (physical events) under certain stimulus conditions (physical events). Mental terms then have the same logical status that disposition terms such as elasticity and brittleness have in physics. However, a disposition is only one kind of cause. Event-event causation is, on the whole, more important in causal theories. Many examples of apparent event-event causation in mental life cannot be encompassed without considerable strain, if at all, by logical behaviorism (Fodor 1981a, pp. 115f.).

In any event, neither eliminative materialism nor behaviorism conforms to usual conceptions of what science is or how it proceeds.

1. A scientist may select *any* aspects of phenomena (i.e., choose any *concepts*), concerned only that these enable him to detect and describe *patterns* or *regularities* in his domain.

2. A scientist seeks to formulate law-like *generalizations* explaining these patterns or regularities.

3. A scientist must provide an *argument* justifying the *scientific credibility* of his explanatory generalizations. The cogency of the argument depends critically upon the use it makes of *empirical evidence.*

If these statements characterize science adequately, then what motivates these eliminative materialists and behaviorists to promulgate their irrelevant counterproductive conceptual and methodological strictures? Metaphysical allegiances. Though metaphysical beliefs about the mind-body relation masquerade as empirical hypotheses about the world, they are actually value-preferences for a particular kind of world.

Identity-thesis materialism says that mental terms and physical terms both name the same physical states, events, or processes, just as water and H_2O or the Evening Star and the Morning Star are different names for the same thing. It does seem somewhat incoherent, however, to regard a mental state and a physical state, apparently two kinds of things, as really one thing, when one has spatiotemporal extension and the other does not. The usual criterion for identity is that every property one entity has, the other also has, and that neither entity has any property the other does not also have. But the experience of having pain, for example, and the firing of C-fibers do not seem to satisfy this criterion; their properties seem very different indeed.

Theoretical materialism says that the relation between what is mental and what is physical is like the relation between temperature and the mean kinetic velocity of molecules. Temperature, for example, or a mental state, event, or process, is a property of a system, which is caused by, or is an expression at another level of observation of, the properties or interrelations of the constituents (e.g., molecules or neurones) of that system. Theoretical materialism has the virtue of neither flinching from nor denying the reality of mental facts. It seems, on the whole, therefore, the most coherent and attractive form of materialism. As we shall see, however, if theoretical materialism is the right position to take, then it follows that belief in the possibility of theoretical reduction is the right position to take.

Dualism says there are two ultimate stuffs, mental and physical.

Parallelism says there is no causal relation between them. By some miracle, mental processes and physical processes just run along synchronously.

Epiphenomenalism says the body causes the mind, as if the mental were some kind of effluvium floating in some imaginary space above the physical, but having no effect upon it. An occasional teacher is still fond of drawing in a blackboard diagram a causal arrow from one circle, the body, to another circle, the mind.

Animism says the mind controls the body, but the body has no effect on the mind. This idea conjures up images of spirits or demons—ghosts in the machine—telling our bodies what to do, but remaining aloofly undisturbed by them.

Causal interactionism says physical processes cause mental processes and in turn mental processes cause physical processes (see, e.g., Popper and Eccles 1977). Despite the obvious apparent interdependencies, and no matter how many permutations of mutual influence or how many kinds of causal pathways (including feedback loops) are postulated, the stumbling block for any form of dualism has always been how to conceptualize something material being caused by or causing something immaterial without suspending the laws of conservation of energy and mass. Causal links are presumed to be structures or processes propagating causal influence through space and time (Salmon 1984). What could the causal links between the mental and the physical possibly be? In addition, no doubt, an aesthetic preference for the austere grandeur of Oneness over the extravagant gaudiness of Twoness—expressed in science as a preference for simplicity (a difficult and elusive concept)—plays a part in the decline of dualism.

More recently, an apparent variant of dualism has taken the form of drawing perhaps overfacile analogies between computer hardware and the brain, on the one hand, and computer software and the mind, on the other. The brain and computer are more different than alike.[4] Just for starters: neurones, unlike the elements of a computer, are continuously active, must be active to be excited, and can fire spontaneously. Connections between neurones, unlike those of computer elements, are plastic or variable, not fixed. There is no reason to suppose that all mental functions (not only thinking but feeling and imagining, for example) are the same, namely, computations.

It is true that hardware of different designs (i.e., with different wiring) can run the same program. But a neuroscientist is not likely to accept the quasi-dualistic conclusion that there are functions (rather than "stuff") constituting something we call mind, which can be carried out by different physical realizations (not only computers, but robots and Martians as well), and, therefore, mind is different from and independent of the body (although the body is one possible incarnation of it). The neuroscientist is likely to regard as an uninteresting truism that any number of possible black boxes designed in different ways can have the same output. The neuroscientist wants to discover the way that the *actual* brain is designed, such that it has the properties it has—not the possible ways various machines held to be similar to brains might be designed.

There are three points to be made about all this metaphysical argument.

1. No conceivable evidence, no correlations between mental events and physical events, can decide between monism or dualism. For such correlations are equally consistent with either view. That

4. See, for example, Bunge (1980).

this is so ought to lead us to the conclusion that the assertions of metaphysics are not empirical hypotheses.

2. The assertions of metaphysics belong to ideology. Different ideologies spring from competing or conflicting value systems, and lead to different conceptual commitments, and to different priorities, strategies, and programs in research.

Materialists value *ontological simplicity* ("ontology" referring to the study of the question, what sorts of things exist in the actual world?). Rejecting dualism, they heed Occam's injunction not to multiply kinds of entities. They austerely imagine the world as made of one and only one kind of stuff.

Materialists also value *scientific simplicity.* As theoretical reductionists, they are likely to believe that no matter what subject matter a science has, the regularities it discovers eventually will be derived from the theories of physics and chemistry.

Dreams of a unity of science based on ontological or scientific simplicity express values not facts. The dream of a unity of science based on ontological simplicity: everything in the world that can be studied scientifically is a physical thing and, if it is made up of constituents, these too are physical things. The dream of a unity of science based on scientific simplicity: no matter what subject matter a science has, the regularities it discovers eventually will be derived from the theories of physics and chemistry, by means of bridging statements connecting concepts in that science with concepts in physics or chemistry.

I shall argue that theoretical reduction of psychoanalysis by neuroscience is highly unlikely, if not indeed impossible. (I am using "theoretical reduction" in a strict technical sense. "Theoretical reduction" is not just the idea that all mental states are embodied or have neurophysiological correlates.)

But this does not mean I am not a materialist. I shall also show that an alternative view, which holds that psychoanalysis is an autonomous discipline, not reducible by neuroscience, is nevertheless consistent with a thoroughgoing materialism. But not materialism in the strong sense—commitment to both ontological and scientific simplicity. Rather, this alternative view is consistent with a commitment to an everything-is-ultimately-physical ontological simplicity only. The belief that every particular mental state is caused by, or is an expression of, some particular physical state—materialism in the weak sense—is a tenable common ground upon which neuroscientists and psychoanalysts might meet.

No one so far as I know, however, has ever demonstrated that commitment to either of these conceptions of the unity of science is a necessary condition for scientific work. The fact that biology has not yet been reduced by physics and chemistry—although the expectation that it eventually will be is a useful heuristic—has not inhibited scientific progress in biology. I see no compelling reason why the investigations of psychoanalysis should be constrained by allegiance either to ontological or scientific simplicity.

3. However, dreams of ontological and theoretical simplicity aside, the unity of science does and should matter to psychoanalysis, when that unity is conceived as a matter of shared methodology rather than ontology or theoretical reduction. The unity of science is based on the fact that all sciences, if they are sciences at all, share the same method of acquiring knowledge of the world. (By method here, I do not mean techniques for obtaining information.)

The objective of every science is to discover patterns or regularities in the interrelations among abstracted aspects of phenomena, to locate events in some causal nexus, or to show that events and processes occurring in the world are instances of generalizations. Every science recognizes and attempts to meet standards for obtaining, and assessing the quality of, evidence for empirical claims. Every science recognizes and attempts to meet standards for judging the adequacy of proposed explanations. Every science recognizes and attempts to follow canons for deciding that evidence favors one hypothesis over its rival.

Science is, above all, skeptical of our beliefs. Every science seeks to satisfy our desire for knowledge by subjecting our superstitions and our wish-fulfillments to a social process of indefatigable criticism. Therefore, every science institutionalizes ways of subjecting cognitive claims to rigorous criticism, to repeated severe tests, to controlled inquiry. (Controlled inquiry, by experiment, by complicated and subtle analytical tools, or by nondesign arguments about the features of a nonexperimental situation, attempts to eliminate certain factors as responsible for an outcome and to determine the differential effects of others in bringing it about).[5]

If this is the vision of the unity of science to which you commit yourself, you will strive above all to improve the extent to which investigations in your discipline exemplify scientific method. I do not exempt psychoanalysis here.

Regrettably, there is a tendency instead to regard psychoanalysis as struggling with significant issues, but as inevitably therefore more diffuse, imprecise, and methodologically unreliable than neuroscience. Its role in their relationship might be to raise important

5. This description of scientific method is to some extent a paraphrase of that given by Nagel (1979, pp. 9ff.).

questions, which its methods are inadequate to address, for neuroscience to answer (Reiser 1984, p. 26). The implication seems to be that psychoanalysis, then, is ultimately dependent upon the methodologically reliable findings of neuroscience to evaluate the explanatory fertility of its concepts, and to test and revise its hypotheses. Regarding psychoanalysis in this way runs the risk of depriving it of incentive to strive toward greater methodological reliability in its conclusions.[6]

I should prefer to think rather that either discipline could raise questions for the other discipline to answer, and would do so, if and when these questions were peripheral to it—peripheral with respect to the problems that are its major concern. In any event, unless the two disciplines become able to formulate conclusions at comparable levels of methodological reliability, talk about theoretical reduction of one by the other is bootless.

Theoretical Reduction

To achieve the *theoretical reduction* of psychoanalysis by neuroscience (in the technical sense of "theoretical reduction"), three objectives must be met.

1. Each discipline must achieve explanatory generalizations that are of comparable scope and that have passed comparably rigorous tests of their scientific credibility. Otherwise, what can we relate to what? We are not relating body and mind. That is metaphysics. We are relating explanatory generalizations in one discipline to comparable generalizations in another.

2. Bridging laws must relate concepts in psychoanalysis to concepts in neuroscience. These laws are theoretical laws, taxonomic or causal, and not statements of identity.[7] Those who hold that these bridging laws are merely statements of identity claim that mental terms and physical terms both name the same physical states, events, or processes. It is presumably this kind of thinking that leads some neuroscientists to say such things as, "mind *is* brain," or "pain *is* the firing of C-fibers," or even "psychology *is* neuroscience." However, bridging laws are not merely definitions, true by virtue of the meanings of the words themselves; they require empirical investigation to

6. For extended discussion of what might be expected of psychoanalysis as a science, see Edelson (1984) and Part 2 of this book.

7. In describing two kinds of bridging statements, I make use of discussions by Fodor (1981b, pp. 127 ff.) and Nagel (1979, pp. 95ff.) of the logic of reduction of one science by another.

establish their truth. An example of a bridging law (which is a taxonomic rather than a causal law, and therefore the kind of bridging law that tends to encourage the identity-thesis) is: if anything whatever is water, it consists of molecules of H_2O, and if anything whatever consists of molecules of H_2O, it is water.

However, if bridging statements were all statements of identity, one theory would be as good as the other in explaining the facts of interest. One says no more or less than the other. The identity-thesis leads to the untenable conclusion that the explanatory generalizations of psychoanalysis do not say anything different from what is said by the explanatory generalizations of neuroscience, and that one might just as well replace psychoanalysis with neuroscience.

If bridging laws are theoretical laws, then the relation between mental and physical is like the relation between temperature and mean kinetic velocity of molecules. A mental state is a property of an organism, in the same sense that temperature is a property of an object. Temperature is caused by—or is an expression at another level of—the properties or interrelations of the internal constituents of an object (i.e., its molecules). So, changes in the temperature of an object are *explained* as caused by changes in the mean kinetic velocity of molecules. In the same way, a bridging law relating mental and physical states would have explanatory power. It might assert, for example, that a kind of mental state or process can be explained by a kind of neurophysiological state or process.

In formulating such a bridging law. the theoretical reductionist must not, of course, imagine some physical stuff causing some mental stuff, and thereby run the risk of suspending the laws of conservation of energy and mass. His view is rather that as matter arranges itself in increasingly complex systems, new properties emerge. These properties do not belong to the subsystems of these complex systems or to their ultimate constituents.

Life itself is, of course, such a property. Mental states, events, and processes are other such properties. Such properties as these only emerge as properties of enormously complex biophysical systems—the ultimate constituents of which are physical and nothing but physical, no matter how complex their organization or interrelations.

If a person is thinking about a particular state of affairs, the state of the constituents of that person's brain, events or processes in that brain, or causal interactions among subsystems of that brain, cause or explain the property is-thinking-about-that-particular-state-of-affairs (possessed by this biophysical system). Similarly, from the point of view of the theoretical reductionist, when we are tempted to speak of interactions between a person's mind and body, or of interactions between a person's thoughts and feelings, then what is occurring, and what, of course, we should be studying, are causal interactions between subsystems of that person's brain, or causal interactions between a subsystem of that brain and some other physiological system of that body.

However, mental states, events, or processes, which are truly novel emergent properties, are not to be viewed as *nothing but* these ultimate physical states, events, or processes. The mental terms that name these properties are not naming the same thing physical terms name, nor can a mental term and a physical term have the same meaning. Mental states, events, or processes, as properties of an organism at one level of description, are ultimately caused by the properties, interrelations, or organization of internal constituents of the organism—that is, by physical states, events, or processes at another level of description.

3. The third objective to be achieved if psychoanalysis is to be reduced by neuroscience: explanatory generalizations in the reduced science (in this case, psychoanalysis), which relate one *kind* of mental state, event, or process to another, must be paralleled by explanatory generalizations in the reducing science (here, neuroscience), which relate one *kind* of physical state, process, or event to another.

Let us take an example of such a parallel in physical science (see fig. 1). We might say, basing our assertion upon a *macroscopic*-level explanatory generalization, that a rise in temperature at an earlier time

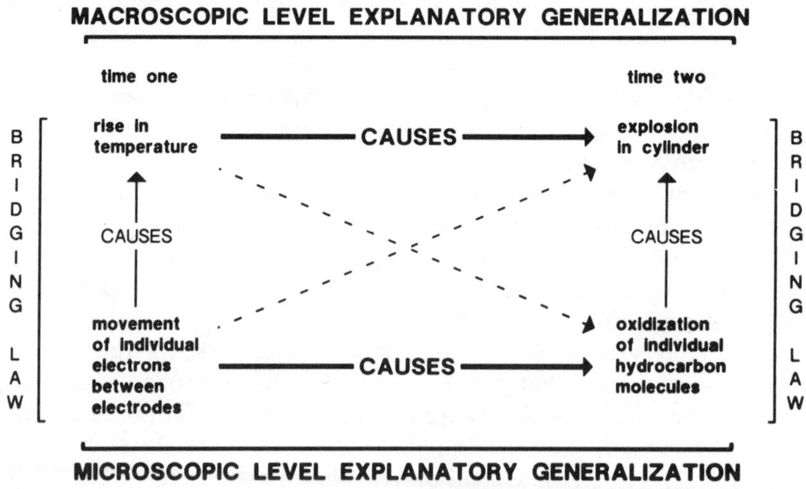

MACROSCOPIC LEVEL EXPLANATORY GENERALIZATION
time one
time two
BRIDGING LAW
rise in temperature
CAUSES
explosion in cylinder
BRIDGING LAW
CAUSES
CAUSES
movement of individual electrons between electrodes
CAUSES
oxidization of individual hydrocarbon molecules
MICROSCOPIC LEVEL EXPLANATORY GENERALIZATION

Fig. 1. Theoretical reduction: example from physical science. (Adapted from Searle 1983, p. 269.)

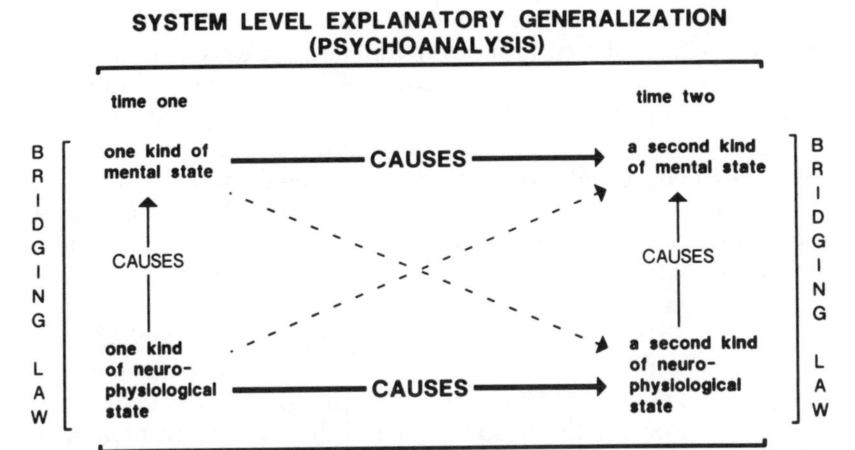

Fig. 2. Theoretical reduction: analogy.

causes an explosion in a cylinder at a later time. We might also say, basing our assertion upon a parallel *microscopic*-level explanatory generalization, that the movement of individual electrons between electrodes at that earlier time causes oxidization of individual hydrocarbon molecules at that later time. By a bridging law, the rise in temperature (macroscopic level) at the earlier time is caused by the movement of individual electrons between electrodes (microscopic level) at that same earlier time. By another bridging law, explosion in the cylinder (macroscopic level) at the later time is caused by oxidization of individual hydrocarbon molecules (microscopic level) at that same later time.

We may also say then, without fear of incoherence, although we are moving causally from one level to another, that movement of individual electrons between electrodes (microscopic level) at the earlier time causes an explosion in the cylinder (macroscopic level) at the later time. And that rise in temperature (macroscopic level) at the earlier time causes oxidization of individual hydrocarbon molecules (microscopic level) at the later time.

By analogy (see fig. 2; also Searle 1983, p. 270), for the theoretical reduction of psychoanalysis by neuroscience to be possible, we must imagine that mental states occur at a level analogous to the macroscopic level in the previous example. Let us call that level a system-level description. Neurophysiological states occur at a level analogous to the microscopic level in the previous example. Let us call that level an internal-constituent-level description. A bridging law might explain one kind of mental state (system level) as caused by one kind of neurophysiological state (internal-constituent level). Another bridging law might explain a second kind of mental state (system level) as caused by a second kind of neurophysiological state (internal-constituent level). An explanatory generalization in psychoanalysis (system level) causally relates the two kinds of mental states. Another parallel explanatory generalization in neuroscience (internal-constituent level) causally relates the two kinds of neurophysiological states.

So, without fear of incoherence or commitment to dualism, we might then be able to say that an earlier occurrence of an instance of the first kind of neurophysiological state (internal-constituent level) causes a later occurrence of an instance of the second kind of mental state (system level), and that an earlier occurrence of an instance of the first kind of mental state (system level) causes a later occurrence of an instance of the second kind of neurophysiological state (internal-constituent level).

If it were possible to build such bridges, connecting methodologically reliable conclusions of neuroscience with methodologically reliable conclusions of psychoanalysis, then one discipline would indeed influence what phenomena are regarded as problematic and selected for study—and how phenomena are described—by the other discipline. The discoveries of one discipline would indeed set constraints on the possible explanations available to the other. Such influence would travel a two-way street, acting as stimulant and corrective in both directions, for evidence obtained at one level would then have no necessary superiority over evidence obtained at another (Hebb 1980, pp. 42f.).

Under these circumstances, neuroscience and psychoanalysis might come to address the same problems, focus upon and attempt to explain the same phenomena, ask and attempt to answer the same questions. It might even be that using different techniques and concepts, and obtaining very different kinds of information, they nevertheless might arrive at the same answers.

Theoretical Reduction of Psychoanalysis by Neuroscience

But the following argument suggests that a *complete* theoretical reduction of a psychology of mind by neuroscience may be in fact practically if not logically impossible.

1. Psychoanalysis—like sociology and economics—is a special sci-

ence. That is, by definition, it is not a physical theory. I take this point as self-evident and shall not discuss it further.

2. Psychoanalysis is what Fodor (1981b) calls a representational theory of the mind.

3. Psychoanalysis is also a computational theory of the mind—although perhaps not quite in the sense Fodor (1981b) uses that phrase in cognitive science.

4. Special sciences are autonomous. That is, in general, special sciences cannot be reduced by physics and chemistry. Specifically, psychoanalysis cannot be reduced by neuroscience.

Psychoanalysis is a representational theory of the mind. Psychoanalysis is a theory of mental states. In brief, a mental state includes (a) a person; (b) the person's own internal symbolic-semantic representation of an object or state of affairs; and (c) a psychological activity (manifesting a general psychological capacity), which is a relation between the person and his mental representation. Examples of such relations are: perceives; remembers; believes; thinks; feels; wishes; voluntarily intends to. Mental representations have form and content.

"John wishes that his father were dead" and "John believes that his father will soon die" are examples of mental states. A mental state consists at the least of a person (e.g., "John"), a mental representation of an object or a state of affairs (e.g., "John's father is dead"), and a relation between that person and that mental representation (e.g., "wishes"). The type of mental representation and the type of relation together determine the type of mental state a mental state is.

When we speak of the contents of mental states, we refer to objects and states of affairs as these are internally represented by persons who are related by psychological activities to their mental representations. These contents are objects and states of affairs as represented or described by a person, and not as they might be represented or described by another who observes that person.

It is because we see the contents of mental representations (sometimes transformed by mental operations) recurring in different mental states, that we see that a mental state is the cause of another mental state. Similarly, because we see objects or states of affairs in the environment also appearing as the contents of mental representations, we see that a mental state is the effect of an environmental event or the cause of an action. A psychoanalytic explanation focuses especially on the causal interrelations of mental states—rather than the cause-effect relations of mental states and environmental states or actions (which may more properly belong to the province of a general psychology of mind)—and typically accounts for the causal interrelations of mental states in terms of their contents. What is causal in psychoanalysis are mental states and in particular the contents of mental states.

A distinct type of relation ("perceives," "remembers," "believes," "thinks," "wishes," "feels,") is characteristically embedded in a particular kind of causal nexus. That is, what type of relation a relation is depends upon what are its typical causes and effects. For example, a perception typically causes a memory and a set of memories typically causes a belief.

A distinct type of mental representation is determined by both the formal and the semantic features of its contents. Just what mental state a person is in is determined in part by the mental representation to which that person has a certain kind of relation. To substitute for one description of a mentally represented object or state of affairs another description, different either semantically or syntactically from the first, of the *same* object or state of affairs, is to change the mental representation and, therefore, the mental state itself. (This suggests one source of some methodological conundrums faced by psychoanalysis as a science.)

A mental process consists of operations on or transformations of mental representations.

Psychoanalysis is a computational theory of the mind. Mental events are operations on the form or symbolic-semantic content of mental representations. These operations change form and transform contents. Examples of mental operations of particular interest to psychoanalysis (as distinct from those that might be of interest to a cognitive scientist): (a) displacement, or shifts in emphasis or attention from one content to another; (b) condensation, or the combination, fusion, or transfer of contents (e.g., the transfer of a property from one object to another); (c) splitting, or the separation of contents from one another, each becoming part of different mental representations to which a person has different relations; (d) the decomposition of linguistic entities (e.g., words into syllables), with the products of such decomposition each becoming the object of other mental operations; and (e) formal transformation, changes in the way contents are represented. Examples of formal transformation include the transformation of linguistic representations into images, and "syntactical" rearrangements of the constituents of a state of affairs. Examples of syntactical rearrangements include rearrangement of the order of linguistic entities in a linguistic representation, as in a switch of subject of a verb and object of a verb; rearrangement of the spatial relations of elements in an image; and the reassignment of linguistic entities to different syntactic categories. Even a brief consideration of the mental operations that, in addition, determine the vicissitudes of the person's emotional response to his own mental representations is beyond the scope of this chapter. See Chapter 9 for a discussion of this problem.

Mental processes of particular interest to psychoanalysis are sequences of causally connected mental states, or clusters of simultaneous mental states, or both, whose mental representations are linked by such mental operations.

Special sciences are autonomous. What is the argument that leads to the conclusion that special sciences cannot be completely reduced by physics and chemistry, and that in particular psychoanalysis cannot be completely reduced by neuroscience?

a) *The rejection of theoretical reduction is compatible with materialism.* The first step in the argument is to note again that this conclusion is not incompatible with materialism. It is possible to concede that, for every mental state an organism is in, that organism is necessarily also in some physical state. Mental states are always embodied. But that concession is much weaker than the one demanded by theoretical reductionism. For theoretical reduction to be possible, each *kind* of mental state (each theoretically relevant *type* of mental state) that enters into causal relations with another kind of mental state must also be lawfully related (*regularly,* in *every* organism, and in every case that has been or *ever will be* observed) not just to any old particular neurophysiological state but rather to a *kind* of neurophysiological state (a theoretically relevant *type* of neurophysiological state) that appears in explanatory generalizations of neuroscience.

b) *Different sciences investigate different domains and each asks very different kinds of questions about its own domain.* The second step in the argument is to point out that each special science chooses what kinds of entities it shall investigate, and upon what properties and relations of these entities it shall focus, in the expectation that it is the study of just these properties and relations that shall turn out to yield regularities, causal patterns, and explanatory generalizations. What makes a special science different from other sciences? The particular kind of entities—and the particular properties of, and relations between and among, these entities—it chooses to study. These choices determine what kinds of questions it will address to nature. The kinds of questions in turn determine what kinds of explanation it will consider relevant or appropriate—just those that are answers to the kinds of questions it asks. What constitutes an appropriate explanation in one science does not necessarily constitute an appropriate explanation in another. Two different sciences are different sciences just because they have chosen different kinds of facts to explain, and so even when they seem to be studying the same concrete phenomenon, they ask very different questions about it.

A fact of interest to a special science may be an abstraction from—a different aspect of—the same concrete phenomena of interest to other sciences. So, if an organism is a concrete phenomenon, facts about that organism are statements that may be couched in different languages, involving concepts that are distinctive to different special sciences. Facts may be about neurones, psychological capacities, wishes and beliefs, or social roles. Neurones, psychological capacities, wishes and beliefs, or social roles are all abstracted aspects of the same concrete organism. So one science will choose to study the neurophysiological systems of human organisms; another, the social roles occupied by persons and how these influence a person's participation in a social collectivity; and, still another, the needs motivating the actions of persons.[8]

Facts about the body and the mind, as abstracted aspects of a concrete organism, are likely to be very different in different special sciences. Consider a sociology such as that conceived by Talcott Parsons. His conceptual framework involves a means-end schema and functional explanation. Encoded culture is the "gene" that transmits patterns from one generation to another, and thereby controls the values shared by members of a society and what collective ends are consonant with these values. The society's way of organizing to achieve its collective ends is to assign roles to individuals, and to subject these individuals to norms and sanctions consonant with its value system that are attached to these role assignments. Role assignments control which of the needs out of all those that constitute a personality system—and how these particular needs—will be satisfied. In this sense, the society controls the personality system; and, in turn, the personality system is viewed as controlling the body. It makes use of the body in carrying out actions required by roles and for the satisfaction of needs. The direction of control in such a concep-

8. It is not generally appreciated that facts are abstractions from concrete phenomena. Although "realists" tend to deny the reality of abstract entities, a scientist may in fact have a good deal of interest in, and accept the real existence of, abstract entities as well as physical particles. Numbers are such abstract entities. So also are words, which are abstractions from concrete acoustical or orthographic tokens of them; a word may be considered a class of its variously realized physical tokens, and is certainly then not identical with any particular member of that class.

A similar point should be made here. The same concrete entities can be described by a theory in either a thing-language (which emphasizes its physicalness) or an event-language. The result of these different descriptions are notational variants *not* substantive differences. That is, the choice of notation makes no difference with respect to the empirical consequences of the theory.

tual framework is from culture to social system to personality system to body.

The body's capacities, on the other hand, "set limiting conditions"—which are to be distinguished from "controls"—on what needs of a personality system can be satisfied. The personality system's needs set limiting conditions on what roles it is motivated to occupy, and therefore on the degree of success a social system has in achieving collective ends. The achievements of a social system set limiting conditions on what the nature of the culture is that is transmitted and how well it is transmitted.

To express these ideas in a somewhat different way: Whether or not a variable is an independent, causal, or stimulus variable, on the one hand, or a dependent, effect, or response variable, on the other hand, will depend on the total conceptual framework in which it is embedded. Therefore, answers to questions about what causes what in relations between body and mind depend on which special science is asking the questions.

A special science is free to invent or choose concepts and to state its facts in ways that lead to the discovery of lawful relations among these concepts that account for these facts. The differences in what kinds of fact are of interest do not simply reflect differences in level of observation. They reflect different ways of cutting nature at its joints, and result in addressing different questions to nature and developing different strategies for answering these questions.

From these differences it follows that neither neuroscience nor psychoanalysis can capture with its conceptual apparatus the regularities or patterns and the explanatory generalizations captured by the other. A neuroscientist's explanation of a mental state in terms of the properties and interrelations of constituents of a neurophysiological system cannot answer the questions a psychoanalyst is asking about that mental state. The former is likely to be trying to explain just what conditions make it possible for an organism to be in that kind of state (e.g., dreaming). The latter is likely to be trying to explain the contents of that mental state (e.g., why is this person dreaming *that?*).

Assume a psychological activity of a person (such as wishing, believing, perceiving, remembering, thinking, feeling, or intending an action), directed to, or of or about, that person's mental representation of an object or state of affairs. The neuroscientist is likely to be trying to explain how an organism can carry out such a psychological activity (e.g., remembering), that is, how an organism realizes some psychological capacity (e.g., memory). He asks: "What conditions and mechanisms make it possible for an organism to remember?

What constraints are there on remembering? How, by what mechanisms, can an organism gain access to a memory—any memory?" The psychoanalyst is likely to be trying to explain something about the mental representation—that is, something about the contents of a mental state. He asks: "Why is this person remembering *that?*" and by "why," he means: What *other* mental states, such as wishes or beliefs, with what *other* contents, cause this person to remember *that?*

c) *Concepts appearing in explanatory generalizations of psychoanalysis are not necessarily coextensive with concepts appearing in explanatory generalizations of neuroscience.* The third step in the argument that psychoanalysis probably cannot be reduced by neuroscience, then, is to give five reasons that suggest at the least that *not all* concepts entering into the explanatory generalizations of a psychology of mind can be related by bridging laws to concepts entering into the explanatory generalizations of neuroscience.

First reason.[9] (See fig. 3.) Suppose that an explanatory generalization of psychoanalysis might take a form something like this. Given: a *class* of mental states (a kind or type of mental state, which is to be distinguished from any of its particular tokens or realizations). It is characteristic of members of this class that some psychological activity—no matter what it is, believing, wishing, voluntarily intending an action, or whatever—is directed to a person's mental representation of his anus, rectum, bowel, or his gastrointestinal tract in general, or to any of the activities and products associated with these. In other words, what type of mental state a mental state is, in this explanatory generalization, is determined, not by the kind of psychological activity that is a constituent of that mental state, but by the kind of contents that are part of that mental state.

Also given: a *class* of mental states, in which some psychological activity—no matter what it is, believing, wishing, voluntarily intending an action, or whatever—is directed to a person's mental representation of any economic institution, or a person's mental representation of any activity involving anything that happens to fill the role money fills in our world. Take it that the two *kinds* of mental states, as so conceived, are related by an explanatory generalization in psychoanalysis.

The "money" kind of mental state is a concept entering into an explanatory generalization of psychoanalysis. But what about the

9. The main line of this argument about the irreducibility of the special sciences—although not as applied here to psychoanalysis and neuroscience—can be found in Fodor (1981b, pp. 127ff.).

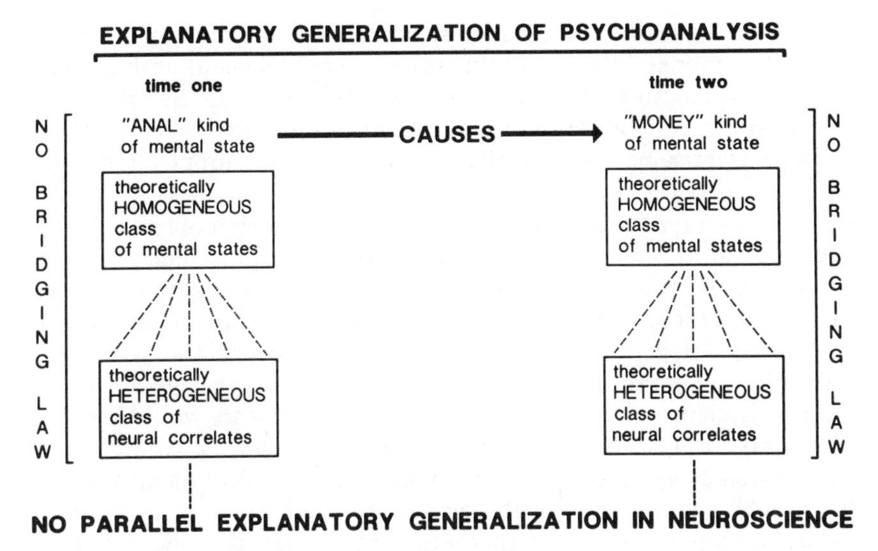

Fig. 3. Bridging laws *not* possible. (Adapted from Fodor 1981b, p. 139.)

neurophysiological correlates of such mental states? The class of "money" mental states will include mental states in which very different psychological activities (remembering, feeling, perceiving, thinking, wishing, voluntarily intending to act, believing), which are supported by very different kinds of neurophysiological subsystems, are directed to mental representations of very different objects and states of affairs.

All the representations will be of states of affairs involving monetary activities or transactions. But from the point of view of the general concept "money kind of mental state," which enters into explanatory generalizations of psychoanalysis, it does not matter *what* relations (believing, wishing, voluntarily intending an action) *any* person has ever had or *ever will have* to these representations. It does not matter what counts or *ever will count* as a "monetary activity or transaction." It does not matter what objects or states of affairs are represented—*buying* or *paying* for a wide variety of goods or services (rent, food, sex), in different ways (paying by check, with cash, by credit card, by barter), in very different kinds of environments. It does not matter what aims are emphasized: *producing* or *acquiring, saving* or *secreting,* or *spending* money (or *any* one of the symbolic or functional equivalents of money).

But, if one is a materialist, every one of these differences in psychological activity or in object or state of affairs represented must correspond, of course, to a difference in neurophysiological substrate.

Otherwise, it makes no sense to speak of them as different. It is true that the "money" class of mental states is homogeneous with respect to the explanatory generalizations of psychoanalysis. But the neurophysiological correlates of these mental states are *heterogeneous.* They comprise no theoretically relevant class of neurophysiological states.

However, remember what is required for reduction of one science by another. If reduction of psychoanalysis by neuroscience is to be possible, all "money" mental states would have to be caused by, or be an expression of, one *kind* of neurophysiological state. That *kind* of neurophysiological state would have to appear in explanatory generalizations of neuroscience. In at least one explanatory generalization of neuroscience, that *kind* of neurophysiological state would have to be related to a second similarly theoretically-relevant *kind* of neurophysiological state. That second *kind* of neurophysiological state would be one that causes all "gastrointestinal" mental states. But, alas, for the enterprise of theoretical reduction, the neurophysiological correlates of these mental states too are heterogeneous. They, too, comprise no theoretically relevant kind of neurophysiological state.

I am assuming here that causal relations among entities and subsystems of entities in the domain of neuroscience can be investigated and formulated without reference to what exactly is represented in mental representations, that is, without reference to the particular contents of mental states (although not necessarily without reference to the *capacity* to form mental representations). The successful achievements of neuroscience to date bear me out in this. (It is just as well. I do not think that a neuroscience whose explanatory generalizations were at the mercy of the contents of new books or of changes in cultural institutions, for example, would have good prospects as science.)

Consequence: there is no explanatory generalization in neuroscience corresponding to the one in psychoanalysis.

A number of my colleagues have had difficulty with the notion of "heterogeneous" neurophysiological correlates. Of course, I mean a set of neurophysiological correlates without any theoretically interesting shared properties. That is, these neurophysiological correlates will not comprise a class of neurophysiological states falling under a concept in neuroscience, the use of which will lead to the discovery of patterns or regularities in neuroscience, or which will appear in any explanatory generalization of neuroscience.

"But why," asks my colleague, "is it not possible that some day we will find what all these neurophysiological states have in common?"

"You mean," say I, "aside from the fact that they are correlated with a class of mental states identified by psychoanalysis, what *other* interesting theoretical properties do they have, such that they will lead to the discovery of patterns or regularities in neuroscience itself, or appear in explanatory generalizations of neuroscience accounting for phenomena in the domain of neuroscience?[10]

"Perhaps you would like to consider again the motley crew of neurophysiological states that would have to fall under one concept in neuroscience, in the light of what you *already know* about the brain and the nervous system and the other systems of the body to which they are related. What kind of concept, having theoretical significance for *neuroscience,* can you imagine?

"Also, do you not think that, at the least, it is possible that one of the particular neurophysiological states embodying some one member of the first kind of mental state might cause a neurophysiological state that does *not* embody any of the members of the second kind of mental state?

"Please give these questions some thought before dismissing what I am saying out of hand."

If, of course, the difficulty arises because my colleague believes that only neurophysiological explanations correspond to what he is willing to consider a real or fundamental explanation in science, then further discussion between us on this point is futile.

Second reason. The equipotentiality of different regions or subsystems of neurophysiological systems (that is, their ability to perform the same function) suggests that, on different occasions or under different circumstances, different states of a neurophysiological system may underlie or cause the same particular mental state. This same possibility is suggested by the fact that sundials, clocks, and various other kinds of physical devices all "tell time" (perform the same function), and also by the fact that the same program can be run by very differently wired hardware (Dennett 1978).

On at least one occasion, Freud expresses himself as a functionalist, in this sense, and adopts the attitude of the functionalist

10. Notice that the maneuver of constructing two concepts out of the two sets of heterogeneous neurophysiological states—(this state *or* this state *or* this state *or* this state *or* . . .) CAUSES (that state *or* that state *or* that state *or* . . .)—will not work.

> I think, for example, that it is a law that the irradiation of green plants by sunlight causes carbohydrate synthesis, and I think that it is a law that friction causes heat, but I do not think that it is a law that (either the irradiation of green plants by sunlight or friction) causes (either carbohydrate synthesis or heat). Correspondingly, I doubt that "is either carbohydrate synthesis or heat" is plausibly taken to be [the kind of concept that appears in scientific laws]. (Fodor 1981b, p. 140)

toward the body, rather than the attitude of the dualist he is ordinarily taken to be. He makes the statement quoted below in what Strachey characterizes as "perhaps his most successful non-technical account of the theory of psycho-analysis, written in his liveliest and lightest style" (Freud 1926, p. 181). In that work, Freud writes:

> It will soon be clear what the mental apparatus is; but I must beg you not to ask what material it is constructed of. That is not a subject of psychological [i.e., psychoanalytic] interest. Psychology [i.e., psychoanalysis] can be as indifferent to it as, for instance, optics can be to the question of whether the walls of a telescope are made of metal or cardboard. (Freud 1926, p. 194)

It is only because here he takes the attitude of a functionalist that he is able to write in the same work, with grand disregard of other things he has said about the relation of mind and body throughout his writings:

> I am bound to admit that, so long as schools such as we desire for the training of analysts are not yet in existence, people who have had a preliminary education in medicine are the best material for future analysts. We have a right to demand, however, that . . . they should resist the temptation to flirt with endocrinology and the autonomic nervous system, when what is needed is an apprehension of psychological facts with the help of a framework of psychological concepts. (P. 257)

Third reason. Psychoanalysis finds it necessary to make certain distinctions with respect to emotional states, if it is to discover patterns or regularities in its domain and to formulate law-like generalizations explaining these. Jealousy is distinguished from envy, for example, largely on the basis of differences in the contents of the mental representations—the kinds of states of affairs represented—involved in each of these emotional states. Similarly, anxiety is distinguished in this way from fear, and different kinds of anxiety are distinguished from each other.

While a materialist (in the weak sense) agrees that for every difference in mental state, there must be a difference in neurophysiological state, it does not follow that these differences at the level of neurophysiology make a difference when it comes to conceptualizing a kind of neurophysiological state that will enter into the explanatory generalizations of neuroscience. For that purpose, jealousy and envy both may be relevantly physically encoded as "prepare to fight" and anxiety and fear both as "prepare to flee"—while "autonomic arousal," for example, may suffice as the single criterial attribute determining membership in a theoretically relevant class of neurophysiological states. Of course, if the psychoanalyst were to confine himself to classifying mental states so that they could be coordinated by bridging laws with such classifications of neurophysiological states, he would have to abandon what are for him crucial distinctions, and he would in so doing lose considerable explanatory power in his domain of interest.

Fourth reason. Which of many possible neurophysiological states actually, in a particular person and on a particular occasion, underlies, causes, or is expressed by a particular kind of mental state is very likely to depend upon the context of the mental state—the network of other mental states in which the mental state on a particular occasion is embedded and the background of capacities, skills, and habits to which it is related.[11] How could such variations become embodied in the explanatory generalizations of neuroscience? It is not even clear how a psychology of mind can deal in its theories with this kind of complexity and degree of variation.

Fifth reason. Sentences involving relations between persons and mental representations do not conserve their truth values when an expression designating one mental representation is substituted for an expression designating another mental representation, even when the two mental representations refer to what in fact is the same object or state of affairs. It may be true to say of John that he wishes to kill his father, but it may not be true to say of him that he wishes to kill the man who is even now unbeknownst to John planning to divorce John's mother, even though "his father" and "the man who is even now unbeknownst to John planning to divorce John's mother" are expressions referring to the same person. For John does not know his father is planning to divorce his mother, and so he will not represent him under such a description and such a description will not be part of his mental state. Concepts of neuroscience that depend on the intrinsic nature of the existent objects or states of affairs impinging upon an organism are unlikely to be coextensive with concepts of psychoanalysis, which have to do with a person's relation to his own mental representations of objects or states of affairs.

If theoretical reduction is an unlikely, if not an impossible task, as this argument at the least suggests it is, then the relation between neuroscience and psychoanalysis is that which holds between autonomous disciplines.

11. See Searle (1983, pp. 141ff.) for a consideration of network and background as used here.

The Domain of Psychoanalysis

Psychoanalysis is in difficulty. The credibility of its explanations and the efficacy of its treatment are in doubt. Its theoretical ambitions, for example, its ambitions to become a general psychology of mind, are inflated, even grandiose. Indeed, its theory has become so swollen, such a hodgepodge of accretions, that it is almost impossible to formulate so that it can be tested. But that does not appear to matter to it, for it sometimes seems to be suicidally indifferent to the task of demonstrating that its conclusions are methodologically reliable and therefore scientifically credible.

It is tempting to try to tie what appears to many to be a failing or a failed enterprise to the thriving neurosciences. So it has been said that the neurosciences are or will be in a position to test the most general propositions of psychoanalysis, for its theoretical terms refer to *nothing* if they do not refer to neural events (Rubinstein 1965, 1967, 1976). This temptation should be resisted, for dependence of psychoanalysis upon neuroscience becomes a way to evade scientific responsibilities.

Autonomous disciplines describe phenomena at different levels. What are these different levels? In studying the activities of an organism, we may choose to describe the facts in at least three different ways. (I shall ignore in this discussion description at a fourth level—that of the social system.)

First level: neurophysiological mechanisms. We may choose to describe the facts in the language of physics and chemistry, and focus on explaining patterns or regularities in terms of neurophysiological mechanisms.

Second level: nonrepresentational psychology of mind. We may choose to describe the facts in the language of psychological functions or capacities. What is the mind able to do? What "architecture," design, or "wiring" might account for what the mind is able to do? We want to know what constrains and controls psychological activities such as perception, memory, cognition, emotion, or voluntary intentions to act—without regard to particular contents of mental states.

In other words, here, the contents of the mind—mental representations of objects or states of affairs—have no explanatory role. In fact, we limit our interest to just what it is about what the mind is able to do that is *not* affected by particular wishes and beliefs of the organism. We ask: "What are those belief-and-wish-*impenetrable* features of various capacities of the organism enabling it, or limiting its ability, to register, process, and store the effects of external stimuli in just the way it does; or enabling it, or limiting its ability, to carry out actions (e.g., and notably, speech) in just the way it does?"[12] For the most part, mental states are of interest to a nonrepresentational psychology of mind as manifestations of psychological capacities—and therefore as clues to features of these capacities, and evidence for hypotheses about them.

To the work of the neuroscientist a nonrepresentational psychology of mind—*not,* or only incidentally, a representational psychology of mind or psychoanalysis as a theory that is part of a representational psychology of mind—contributes an explication of just what the characteristics, including the limitations, of different kinds of capacities are; how these capacities change throughout the development of the human organism; and in what way their characteristics are dependent on or modified by environmental stimuli or conditions. Since the belief-and-wish-impenetrable features of a capacity must be inferred from performances, the explication of the capacity is a theory. It, together with the facts and empirically observable regularities that constitute the evidence supporting it, become what the neuroscientist is called upon to explain in terms of the physical constituents of the organism.

Like other scientific theories, theories about psychological capacities are tested. They are evaluated, for example, in terms of the degree to which evidence supports them against their rivals, and the degree to which they generate successful predictions about new facts and interrelations. As a theory in a nonrepresentational psychology of mind characterizing a capacity becomes scientifically credible, the capacity it characterizes itself becomes something to be explained by neuroscience. (I am describing logical relations here, not actual sequences in time.)

It is not generally appreciated that neuroscience is dependent on this kind of work in a psychology of mind to provide it with what is to be explained. Without an adequate formulation of the belief-and-wish-impenetrable features of the human organism's capacity for language, for example, neuroscience cannot even begin to describe adequately, much less to explain, the effects of lesions resulting in aphasias. It is the task of the neuroscientist, taking such capacities as facts to be explained, to show how these capacities develop and why they have the characteristics—including the limitations—they do have. He attempts to explain what the mind is able to do—to demonstrate how the features and interrelations of physical constituents and physical subsystems of the organism explain the belief-and-wish-

12. See Fodor (1983) and Pylyshyn (1985).

impenetrable features and the causal interrelations of psychological capacities as these capacities are characterized by a nonrepresentational psychology of mind.

The identification of this level of description suggests a new way of looking at the possibility of theoretical reduction. For it may be that only with respect to those theories (Chomsky's linguistic theory is an example of one such theory) characterizing the organism's different capacities at a level of description making no reference to mental representations—to the contents of mental states—that a part of a psychology of mind may be reducible by neuroscience (Fodor 1983).

Third level: representational psychology of mind. We may choose to describe patterns and regularities among contents of the mind. Mental contents are mental representations of objects and states of affairs, to which psychological activities (such as wishing, believing, perceiving, remembering, thinking, feeling, and voluntarily intending to act) are directed. Explanatory generalizations at this level involve (a) not objects and states of affairs as they might be described by someone other than the person but objects and states of affairs as represented by the person himself; (b) rule-governed transformations of contents (as in cognitive processes of inference); and (c) principles such as rationality—a person's action is explained by what he wishes together with what he believes to be a means of bringing about the state of affairs he wishes to bring about. A psychology of mind at this level investigates how the contents of mental states such as wishes and beliefs that are accessible to consciousness influence psychological capacities—that is, how mental representations influence what a mind is able to do. In other words, at this third level, a general psychology of mind is interested in belief-and-wish-*penetrable* features of various psychological capacities.

A general psychology of mind, nonrepresentational, or representational but nonpsychoanalytic, focuses on, or gives priority to, explaining the psychological performances, and perhaps in particular the achievements, of persons (what the mind is able to do, poorly or well). Some theories of a general psychology of mind will explain these performances or achievements by characterizing the nature of psychological capacities. The mind is able to do what it does by virtue of the properties of these capacities. The characteristics of particular kinds of performances are determined not only by the properties of particular kinds of capacities, but by the way in which experience (i.e., learning) and physical or social features of external situations influence how, the extent to which, and the purposes for which capacities are expressed or realized on occasions in performances.

So far, with reference to this third level, I have discussed a part of a psychology of mind that operates at the level of mental representations but makes no use of the concepts or generalizations of psychoanalysis. When it is concerned with mental states, it is primarily concerned with (a) how conscious mental states are caused by intrinsic features of the environment of a person; (b) how mental states accessible to consciousness are causally interrelated; and (c) how conscious mental states cause a person's actions. In accounting for irrational actions, this psychology appeals to such factors as error and ignorance—that is, a person's beliefs about means of gratifying wishes may be based on incorrect or inadequate knowledge.

Psychoanalysis is but a part of a psychology of mind. I have argued elsewhere about why it is necessary to define a more circumscribed domain for psychoanalysis than is usual (1984, and Chapter 6 of this book). There are those who, on the contrary, regard psychoanalysis as a general psychology of mind. In my view, when Reiser (1984), for example, explores the relation between psychoanalytic and neuroscientific knowledge, he writes not only about psychoanalysis but about a general psychology of mind, or he writes about psychoanalysis conceived as a general psychology of mind. Consider in this connection the strong interest he expresses in his work in the nature of psychological capacities such as memory.

We know that Freud was of two minds in his ambitions for psychoanalysis, but at least on some occasions he expressed himself in favor of defining a limited domain for it. In replying to a grandiose theoretical paper:

> Freud said . . . that he felt like someone who had hugged the coast all his life and who now watched others sailing out into the open ocean. He wished them well but could not take part in their ventures: "I am an old hand in the coastal run and I will keep faith with my blue inlets." (Waelder 1960, p. 56)

Freud in the same vein writes:

> Psycho-analysis has never claimed to provide a complete theory of human mentality in general, but only expected that what it offered should be applied to supplement and correct the knowledge acquired by other means. (Freud 1914a, p. 50).

And again, surprising his editor:

> It must not be forgotten . . . that psycho-analysis alone cannot offer a complete picture of the world. If we accept the distinction which I have

> recently proposed of dividing the mental apparatus into an ego, turned towards the external world and equipped with consciousness, and an unconscious id, dominated by its instinctual needs, then *psychoanalysis is to be described as a psychology of the id (and of its effects upon the ego).* In each field of knowledge, therefore, it can make only *contributions,* which require to be completed from the psychology of the ego. If these contributions often contain the essence of the facts, this only corresponds to the important part which, it may be claimed, is played in our lives by the mental unconscious that has so long remained unknown. (Freud 1924, p. 209, italics added to phrase in second sentence)

Psychoanalysis is distinctive not only in confining itself to the representational level of description and explanatory generalization, and in focusing at that level on the contents of mental states. It is also distinctive in the kinds of problems that attract its attention, and in its explanatory strategy for dealing with these problems. (These problems and explanatory strategies have been discussed in detail in Chapter 6. For the sake of continuity of argument, I shall repeat a bit but not most of that relevant material here.) Psychoanalysis, indeed, might be said, in its concern with mental contents and their interrelations, to take psychological capacities for granted.

Psychoanalysis asks questions about mental contents. It wants to explain mental contents. Not *how* someone is able to *believe,* but *why* does he believe just *that*? Not *how* is he able to form or carry out *intentions,* but *why* does he intend doing, or why does he carry out the intention to do, just *that*? It wants to explain in what way certain mental states, beliefs and wishes, influence psychological capacities—in other words, in what way unconscious beliefs and wishes influence other mental states, thoughts, perceptions, memories, intentions to act.

Psychoanalysis focuses on the interrelations among mental states. How are mental states causally interrelated, according to psychoanalysis? By virtue of their contents.

Contents are not only what are to be explained but are the causal processes of interest to psychoanalysis. Mental representations propagate causal influence through space and time. Let *x* and *y* be distinct states of affairs. John wants *x*. John believes that *x* only if *y*. John intends to bring about *y*. The contents *x* transform to *x only if y* and the contents *x only if y* transform to *y*, as wish produces belief and belief leads to intention.

Psychoanalytic explanation makes use of schemas such as: John wants his father to protect him; Joe reminds John of his father; John wants Joe to protect him; John believes that Joe will protect him only if he (John) shows Joe how helpless he (John) is; John intends to show Joe how helpless he (John) is. It is clear that (a) the content of John's wishes and beliefs are crucial for explaining the content of John's intention; and (b) referring to John's neurophysiological states, even if it were an explanation of John's intention, does not give us the explanation we want. In fact, we lose explanatory power by moving from mental contents to neurophysiological states.

But—a departure from other representational theories of mind—what is a mental process in psychoanalysis? Typically, not a cognitive process of inference, governed by a principle of rationality. Rather, strange operations on, or transformations of, mental representations—like the dream work and defense mechanisms.

Objects and states of affairs—existent or mentally represented—are the causal processes connecting conscious mental states; connecting environmental stimuli and conscious mental states (the direction of the causal process is from world to mental representation); or connecting conscious mental states and actions (the direction of the causal process is from mental representation to world). Observations of the continuity of contents—the persistence of contents from one mental state to another, from world to mental representation, or from mental representation to world—provide evidence for such causal connections.

Phenomena of interest to psychoanalysis involve causal gaps—discontinuities in causal connectedness among conscious mental states; in causal connectedness between environmental stimuli and conscious mental states; or in causal connectedness between conscious mental states and actions. The various kinds of causal gaps have been described in Chapter 6 of this book.

What is distinctive about these causal gaps is that they cannot be explained as the result of (a) impairment in neurophysiological systems or defects in capacity; (b) intrinsic features of situations or social contexts to which a person is exposed; or (c) departures from rationality due to error or ignorance. That is, psychoanalysis limits the set of phenomena to which its explanations are applicable to just those phenomena that cannot be explained as determined by the nature or impairments of psychological capacity or by features of the environment or social context. It assumes that, if one were to be limited to the facts and theories of a general psychology of mind or neuroscience, these causal gaps would be inexplicable. Occurrences of such inexplicable causal gaps in the presence of intact capacity and, therefore, unimpaired neurophysiological systems, on the one hand, and in the presence of situations or social contexts that cannot satisfactorily account for them, on the other, comprise the subject matter of psychoanalysis.

It is the burden of psychoanalysis to demonstrate, using the methods of psychoanalysis, that such causal gaps are caused by (or are an expression of) one or more constituents of an unconscious conflict or dilemma (as in an eruption of an impulse in action, or an intense imagining of a wished-for state of affairs); or are caused by (or are an expression of) a person's attempts to resolve such a conflict or dilemma (a compromise-formation).

The Diagnosis and Treatment of Psychopathology

If neuroscience and psychoanalysis differ in level of description and domain, what then is their relation when it comes to the diagnosis and treatment of psychological disorder?

A person wishes, fears, believes.[13] His wishes and beliefs are causally efficacious in relation to each other, in relation to other kinds of mental states, and in relation to actions. Most important for understanding the kind of psychopathology of interest to psychoanalysis, a person has, at any one time and at one and the same time, *multiple* wishes. The consequences of this important feature of persons has been discussed in Chapter 6, and what follows depends on that discussion. One consequence is that in any person some symptoms or signs of disturbance, some dysfunctional performances, must develop that are simply expressions of (or caused by) what are, in the light of this important feature of persons, inevitable conflicts and dilemmas, and his attempts to resolve them.

In particular, the causes of neurotic misery (as distinct from the causes of normal misery) are to be found in the *contents* of unconscious mental states, as these undergo various transformations, which are effected by particular kinds of mental operations. These mental operations provide the causal links between the contents of unconscious mental states and the form and contents of mental states involved in a neurotic disorder.

There is no more reason to suppose that a defect at the neurophysiological level (or an environmental deficit) must exist and exert a causal influence in the production of neurotic misery than to suppose that there is such a defect existing and exerting a causal influence in

13. For convenience in this exposition, fears are regarded as *wishes* to avoid particular objects or states of affairs.

the production of normal human misery. That human beings devise inadequate solutions to conflicts and dilemmas is inevitable, given the very existence and nature of such conflicts and dilemmas, which at best are not amenable to completely satisfactory solutions, and at worst are constituents of the pathogenic intrapsychic conditions under which neurotic solutions are attempted, and given ordinary limitations in the capacities of persons and in the conditions under which solutions to conflicts and dilemmas must be attempted. These solutions have undesirable and disastrous side effects—crippling psychological dysfunction and very real suffering.

There always will be psychological disorders, then, for which such a method as psychoanalysis or psychodynamic psychotherapy is the indicated treatment, even though some disorders now treated by such methods may well ultimately come to be seen as more appropriately treated by physico-chemical interventions.

For a disorder to be included in the proper domain of psychoanalysis, it must be true that, given present knowledge, there is nothing in the design of the physical organism (here, we think of specific genetic endowment) and no defect in or lesion of the physical organism that is sufficient to cause (has the power to produce) the disorder. This means more than "ruling out" physical causes of it. It means that from what we know about the physical organism, there is no specific genetic endowment or lesion that could produce just this disorder. A conversion symptom such as a paralysis, whose distribution corresponds to no anatomical facts, and indeed contradicts what is known about anatomy, but does correspond to the patient's conception or mental representation of anatomy, is a paradigmatic example.

Furthermore, for such a disorder to be included in the domain of psychoanalysis, there can be no pattern of actual situational circumstances, either in the past or the present, involving, for example, the unavailability of opportunities, facilities, resources, inducements, or rewards, or the presence of danger, threat, deprivation, or punishment, that is sufficient in itself to cause or maintain it.

Another way to express this idea is that when there is "something wrong" at the psychological level, and there is no neurophysiological or situational condition, event, or process fitting the description "something is wrong" that is sufficient to produce the "something wrong" at the psychological level, we have a psychological disorder that is an appropriate subject of psychoanalytic explanation.

The neuroscientist may mistakenly believe that a disembodied mind has crept into this discourse here. So he wants to resist the idea that there could be "something wrong" at the psychological level *without* there being "anything wrong" at the neurophysiological level. But in so doing he commits himself, even if unwittingly, to the implausible belief that the neurophysiological system is so constructed that, well-designed and undamaged, it guarantees an organism cannot fail to solve the problems it faces—guarantees, in other words, successful adaptation.

Of course, when I say no physical or situational factor has the power to cause a particular kind of psychological disorder, I do not say that no physical or situational factor is causally relevant to that psychological disorder. Causal relevance means that the presence or absence of a factor (something like the presence or absence of oxygen in getting a fire started) makes a difference in whether an event occurs, or in the form it takes if it does occur, but does not in itself have the power to cause it. The particular kind of psychological disorder that develops, for example, may depend on genetic endowment—which, determining the patient's capacities, influences the kinds of strategies he is likely to employ in attempting to resolve conflicts and dilemmas—as well as on examples set by others, and identifications with others, in the patient's situation.

The implication is twofold.

1. To identify whether or not a disorder belongs to the domain of psychoanalysis presupposes a good deal of knowledge on the part of the psychoanalyst: (a) about the structure and processes of physical organisms in general, not only the status of a particular patient's organism (*and this is reason enough for the psychoanalyst to interest himself in the doings of neuroscientists*); as well as (b) about the causal powers of situational factors in general, not just those that have impinged on a particular patient (*and this is reason enough for the psychoanalyst to interest himself in the doings of social scientists*). This does not imply, obviously, that psychoanalysis as a scientific discipline is responsible for or capable of generating such knowledge.

2. What belongs to the domain of psychoanalysis may change as knowledge of physical organisms and the causal powers of situations changes. That some disorders once thought to belong in the province of psychoanalysis come to be regarded as belonging in the province of neuroscience is not a matter for regret. Any scientific development that clarifies in just what domain a phenomenon belongs is always to be welcomed. One may assume, nevertheless, that there are and always will be clear, unequivocal cases of those kinds of disorders whose features are such as to defy explanation in terms of physical or situational factors, given what is already securely known about physical organisms and the causal texture of situations. These are just

those disorders that can be explained as expressions of causal interrelations among mental states. However, because there will always be cases in which not enough is yet known about physical organisms or the causal texture of situations for one to be sure that no physical or situational causal explanation of a disorder is possible, there will always be borderline cases and the boundary of the domain of psychoanalysis will always be somewhat fuzzy or fluid.

The psychoanalyst believes that, while it is quite conceivable that over time the set of phenomena of interest to him will diminish, an irreducible set of phenomena will remain in which "something wrong or strange" in mental life will have no sufficient explanation in neuroscience or a general psychology of mind. (That is, they will have no sufficient explanation involving, for example, "something wrong or strange" in the neurophysiological system, in psychological capacity, or in the social context or situation.) So, for example, as previously noted, a conversion symptom may correspond to no anatomical facts, and indeed contradict what facts are known, though it does correspond to the patient's mental representations of anatomy; it cannot be explained by postulating a lesion in the nervous system. This belief of the psychoanalyst's in the persistence of his domain is quite on a par with the neuroscientist's belief that all mental facts and regularities among mental phenomena will ultimately be derivable from the laws of physics and chemistry. Each subscribes to his own faith in the ultimate destiny of his discipline.

Of course, a patient may suffer from more than one illness at a time, and there is no reason to doubt that the effects of a neurophysiological defect and the effects of an unconscious conflict or dilemma can coexist. However, attempts to correct actual or supposed neurophysiological defects by the introduction of intrinsically efficacious physical agents are not likely to prevent or eradicate unconscious conflicts or dilemmas, any more than the verbal-symbolic interventions of a psychoanalyst are likely to prevent or eradicate neurophysiological defects.

The mitigation of effects of neurophysiological defects that interfere with a patient's accessibility to verbal-symbolic interventions, of course, may render that patient more accessible to such interventions. Furthermore, the mitigation of effects of neurophysiological defects that constrain a patient's abilities to solve problems presented by conflicts and dilemmas may make it easier for that patient to abandon neurotic solutions and less likely that, once having abandoned them, he will find himself forced to return to them.

Similarly, where mental representations of the effects of neurophysiological defects have themselves entered into the contents of conflicts and dilemmas resulting in neurotic disorder, the mitigation of either one—neurotic disorder or effects of neurophysiological defect—may mitigate, or at least change the form of, the other. So, also, if neurotic disorder deleteriously influences a patient's willingness to cooperate, for example, in the appropriate treatment of his neurophysiological disorder, mitigation of the former by verbal-symbolic means may increase his acceptance of such treatment.

It is with respect to the treatment of different kinds of coexisting disorders such as these, then, that neuroscience and psychoanalysis may find a common ground upon which to collaborate.

Conclusion

In place of the vision of a unity of science that depends on deriving the theories of every special science from the theories of physics and chemistry, let us put instead the vision of autonomous disciplines, each with its appropriate ontological commitments (it studies the entities it has chosen to study), each committed to its own level of description, its own distinctive concepts, its own questions to answer about the domain it has selected to investigate, its own explanatory strategy, and its own criteria as to what will count as an answer or an explanation. Let us hope for an eventual rapprochement between neuroscience and psychoanalysis that is based not on the hegemony of one discipline over another but is based rather at the least upon both having reached, in their equally exciting quests for knowledge, comparable levels of scientific rigor and methodological reliability. But then let us regard it as proper that, toward that day, there should be for now those in both disciplines who, without regard to their neighbor, assiduously tend each one his own garden.

Applied Psychoanalysis[1]

8

Applied psychoanalysis is the application of psychoanalytic knowledge to explanatory, methodological, or technological problems arising in disciplines or human endeavors other than psychoanalysis. (Strictly speaking, psychoanalysis as treatment is also applied psychoanalysis, for it involves the application of psychoanalytic knowledge to technological problems. However, the theory of the therapeutic action of psychoanalysis does not ordinarily fall under the rubric "applied psychoanalysis," although thinking about it as applied psychoanalysis might in fact prove instructive.)

What is Psychoanalytic Knowledge?

I have argued in Chapters 6 and 7 for regarding psychoanalysis as having a limited domain, rather than for regarding it as a general psychology. Rescuing the core of psychoanalytic knowledge from its mass (and mess) of accretions is a matter of some urgency, if one wants to test it. Those who agree will have an idea of what counts as an application of *psychoanalytic* knowledge that is very different from that held by those who regard psychoanalysis as a general psychology, or who would define its domain, and therefore its core knowledge, differently than I do.

The core of psychoanalytic knowledge consists of those observations in its distinctive domain that are most surprising (given other knowledge) and at the same time most securely established (repeatedly avail-

1. An abbreviated version of this chapter was presented as part of a panel on applied psychoanalysis at the midwinter meetings of the American Psychoanalytic Association, December 19, 1986.

able to the same observer and to different observers, under certain conditions such as the psychoanalytic situation provides). It consists of those concepts psychoanalysis applies to entities or events in its domain (the distinctive way psychoanalysis classifies or orders these entities or events). It consists of those theoretical claims about the relationships among such concepts that are most well-entrenched (because of the number of tests they have survived and the amount of probative evidence that favors them over rival theoretical claims). In any event, well-entrenched or not, these are the theoretical claims that must be true at least of the intended domain of psychoanalysis, for if these claims do not hold in that domain attempts to extend psychoanalytic theory to other domains are pointless.

My own view is that the core of psychoanalytic knowledge is knowledge about the psychological causes of mental contents. Mental contents are objects, states of affairs, or events, not as they "are," but as they are symbolically represented by a person in his own mind. Such mental activities as wishing, believing, perceiving, remembering, thinking, imagining, dreaming, feeling, or intending to act are *of, about,* or *directed toward* mentally represented objects, states of affairs, or events. Mental representations, which have form and contents, are entities in the domain of psychoanalysis.

Psychoanalytic knowledge is knowledge of the causes of those mental contents that are accessible to consciousness, when these do not seem to the person conscious of them to have any causal links—or to be explicable in terms of their relation—to other mental contents accessible to consciousness, or to environmental objects, states of affairs, or events to which he has been exposed and about which he therefore may be said to have information or knowledge. (The more difficult case, of course, is when a person claims he knows the causal links, but his account of them is not accepted by others—who must, of course, justify their skepticism. So, a person may rationalize some actions without realizing what others observe—that, for example, these actions were caused by suggestions given to him in a state of hypnosis.) Psychoanalysis attempts to explain such causal gaps.

The core of psychoanalytic knowledge is not one unified theory. It consists rather of at least five overlapping components, which are not deducible from one another:

a) a theory of sexuality and, in particular, of psychosexual phases;

b) a theory of dreams;

c) a theory of parapraxes;

d) a theory of the pathogenesis and internal structure of neurotic symptoms; and

e) a theory of the therapeutic action of psychoanalysis, which is concerned with the way in which aspects of the psychoanalytic situation—the use of the couch, the frequency of the sessions, the stance of the psychoanalyst, the method of free association, and different kinds of interpretative interventions—produce transference phenomena, and promote the development and resolution of a transference neurosis.

Each of these theories is about its own particular kind of phenomena. The theories, which have certain concepts in common, nevertheless are not deducible from each other or from some small set of axioms. They are used rather to recruit different kinds of evidence converging upon the same conclusion, or to make different contributions to a pattern of observations and to a complex set of hypotheses explaining this pattern.[2]

Psychoanalytic theory does not appear to consist of general laws. Rather, it seems to consist of a set of concepts, a proposal that entities or events and their relations should be regarded in the light of these concepts, and a claim that in each case studied these concepts will be related in some way. Just exactly how in each particular case they will be related—what form their relation will take, how strong the relation will be—usually cannot, at least at present, be predicted from the theory. A research program for psychoanalysis is set by such questions as: What in a particular case is the form or strength of a relation between certain concepts? What determines changes from case to case in the form or strength of the relation between those concepts?[3]

The core of psychoanalytic knowledge includes the following concepts and theoretical claims.

1. Phenomena in the domain of psychoanalysis—that is, those associated with causal gaps—are *ideogenic*; they are caused by wishes and beliefs. In particular, they are caused by *unconscious* wishes and beliefs. (The *contents* of these wishes and beliefs, or the *relations* among these wishes and beliefs, or the *significance* of the contents of these wishes and beliefs may be unconscious.)

2. These wishes and beliefs are constituents of *conflicts* or *dilemmas.*

3. These conflicts or dilemmas originate in the mental life of *childhood.*

4. *Sexual wishes* are constituents of such conflicts or dilemmas.

2. Chapters 13 and 15 develop these points.

3. See Herbst (1970) and Rozeboom (1961) for an explication of such a research program—a way of proceeding somewhat systematically down the path of discovery in science.

5. These sexual wishes lead to *wish-fulfillments* in imagination rather than to reality-accommodated thought or instrumental action. (*Fantasies* exemplify wish-fulfillment in imagination.) It is such wish-fulfillments especially that become the focus of conflicts or dilemmas. They cause persistent dispositions (for example, a disposition to reproduce in actuality, over and over, and usually under certain conditions, manifestations of a particular fantasy). Thus fantasies promulgate causal influence through time, determining to some extent at least internal and external responses to objects, states of affairs, and events. As models, their influence extends to *nonsexual* as well as sexual areas of life, and indeed therefore to psychological functions themselves.

6. The core of psychoanalytic knowledge includes the discovery that the mind works in two very different ways. This discovery is usually discussed in terms of differences between primary process and secondary process. Psychoanalytic knowledge is knowledge of causes of mental contents that derive their causal powers from the way in which the mind works when it works according to *primary process.*

Another way of stating this is that psychoanalytic knowledge is knowledge of the relation of wishes to imagination, and of the vicissitudes and consequences of this relation, rather than knowledge of the relation of wishes and beliefs to instrumental action, and of the vicissitudes and consequences of that relation.[4] In the first relation, the mind operates in such a way that primacy is accorded to wish-fulfillment in imagination. In the second relation, the mind operates in such a way that primacy is accorded to the gratification of wishes in instrumental action. Instrumental or rational action is action governed by a conjunction of wish and reality-accommodated belief. Discussions of adaptation sometimes assume mistakenly that instrumental action is necessarily adaptive and that wish-fulfillment is not.

It follows that my own judgments of any application of psychoanalytic knowledge will depend upon the extent to which this core knowledge is used. In particular, I would regard any application of psychoanalytic knowledge that made no implicit or explicit use of the discovery of the two ways in which the mind works, that scanted the causal role of unconscious conflicts and dilemmas revolving around sexual wishes, or that failed to emphasize wish-fulfillment in imagination (the causal role of fantasy systems) rather than the pursuit of goals through instrumental action as deficient in its depiction of psychoanalytic knowledge.

4. For an especially lucid and insightful explication of this view of psychoanalysis, see Wollheim (1969, 1971, 1974a, 1974b, 1974c, 1979, 1982, 1984).

Misdepictions of psychoanalytic knowledge are common:

1. *That, according to psychoanalysis, dreams are caused by unconscious infantile sexual wishes.*

In fact, anything that disturbs sleep may instigate the production of a dream. Examples of such instigators are intrusive external stimuli, painful somatic stimuli, wishes of any kind that have been stirred up, or thoughts about unresolved problems. Of course, immanent changes in the state of sleep itself will cause changes in what is capable of disturbing it.

Disturbances of sleep are analogous to the striking of a match to produce fire. But just as the striking of a match will not produce fire unless oxygen is present, so a disturbance of sleep will not produce a dream unless some kind of wish (not necessarily sexual) is present. If a disturbance of sleep is not caused by a wish, the dream-instigator causes a wish to be recruited—calls up a wish to help in the construction of a dream. If someone goes and gets fuel to start a fire, we do not ordinarily call the fuel, which could sit there forever without bringing about any fire, the cause of the fire.

An unconscious infantile sexual wish—in fact, any kind of wish—is not sufficient to cause a dream, doe not have the power to cause a dream, will not produce a dream, unless it has been excited or activated in some way, either during the previous day (so that it is, as a disturber of sleep, an instigator of dream-production) or during sleep itself in response to a disturbance of sleep. Some kind of wish is, however, necessary, if not under all circumstances sufficient, to cause a dream, because a dream is an attempt at wish-fulfillment. (Perhaps in adults, only an *unconscious sexual* wish, long and perpetually frustrated, has the power to serve such an attempt.) A wish-fulfillment pleasures and soothes the sleeper, and prolongs his sleep, in the way nursemaids are reported to have pleasured and soothed and put to sleep a sleepless infant by masturbating him. If any wish has the power to cause a dream, it is the wish to sleep.

As any other human endeavor may fail, so the attempt at wish-fulfillment may fail; then, presumably, we have a dream that is not a wish-fulfillment or that is aborted. The whole matter of the causation of dreams is clarified if we regard the sleeper as trying to solve a problem—to remain asleep despite the presence of disturbances of sleep. A particular strategy—the construction of a wish-fulfillment—is usually effective, although of course under some circumstances attempts to carry it out will partially or completely fail.

2. *That, according to psychoanalysis, the dream work is motivated by intentions to censor and disguise a wish-fulfillment.*

The dream work—condensation, displacement, considerations of representability, secondary revision, and the use of iconic symbols—

is the way the mind works in acts of imagination and in particular in constructing will-fulfillments. These operations are necessary if what is produced in imagination is indeed to be a wish-fulfillment (as opposed to a veridical image, for example)—or, what is more usual, a fulfillment of multiple wishes simultaneously (the representation by *effective imagining* of a number of wishes and thoughts simultaneously).[5] These operations contrast with those the mind uses in making cognitive inferences leading to the gratification of wishes in instrumental action.

This is not to say that a wish to censor or disguise may not play a part in determining the way a dream is constructed, remembered, or reported. But even in the absence of such a wish, the procedures encompassed by the dream work will be used to construct a dream, for that is the way mental contents are represented when the mind works in this way and toward such an end.

After all, as a psychoanalysis proceeds, and during periods when the analysand's resistance is low, it is easier for both analysand and psychoanalyst to understand a dream; yet the dream work remains the means by which the dream is constructed. The ubiquitous use of operations like the dream work in art and in joking, for example, calls into question that these operations serve only a defensive function. That this is one way the mind works in forming and operating on symbolic representations to achieve a variety of ends seems more likely.

3. *That, according to psychoanalysis, parapraxes result from unconscious conflicts.*

Parapraxes may result from unconscious conflicts but, more generally, they are the result of attempts to resolve conflicting intentions, whether unconscious or not, by means of operations belonging to primary rather than secondary process.

4. *That, according to psychoanalysis, free association leads without error to the causes of phenomena of interest and therefore is a method that certifies causes.*

Like all methods of investigation, free association is not theory-free. It involves certain presuppositions that, given the peril of circularity, cannot be tested by its use (Edelson 1984). Free association is not done to "prove" that something is the cause of something, or because it leads unerringly to the right answers, but to make accessible mental contents that might otherwise remain inaccessible. It is an attempt to suspend, usually temporarily, and for brief periods of time, critical judgments of any kind, and thus the domination of one

5. "Effective imagining" is Wollheim's phrase.

way the mind works (secondary process) over another way the mind works (primary process). Unconscious mental strivings are freed by the method to influence conscious mental contents more directly, namely, in the direction of wish-fulfillment rather than reality-accommodated thought or instrumental action.

More importantly, the analysand who attempts to free associate continues, from time to time and in particular contexts, to be reluctant or unable to reveal various thoughts and feelings. Such instances of reluctance and of apparent incapacities of expression—in other words, the existence of causal gaps in the stream of discourse—provide the psychoanalyst with the very observations he intends to explain as well as with evidence for his theoretical conjectures about them.[6]

The misdepiction of psychoanalysis that is perhaps most ubiquitous and mischievous in applied psychoanalysis is the one that follows.

5. *That, according to psychoanalysis, "bad" experiences in childhood (for example, losses, "bad" parenting, seductions)—in general, disturbed object-relations or traumatic events in childhood—are intrinsically efficacious causes of the phenomena of interest to psychoanalysis and in and of themselves sufficient to produce the kinds of disturbances in functioning or performance, or the kinds of psychological disorders, later in life that belong in the domain of psychoanalysis, without regard to what the child makes of them, how he interprets them, or what use he makes of them (including the use of them as material in constructing fantasies for his own purposes).*

That the child is father to the man, that as the twig is bent so grows the tree, that, for example, the death of a parent or sibling, divorce, or abuse in childhood affect developmental outcomes are not distinctively psychoanalytic ideas. Their use in applied psychoanalysis, in my opinion, is not an application of *psychoanalytic* knowledge.

Causation in psychoanalysis typically (and certainly in its theory of pathogenesis) involves a complemental series. Pathogenic events, on the one hand, and what the child does with these in constructing, responding to, and revising or disposing of sexual wish-fulfillments, on the other hand, have a different weight or make more or less of a contribution to the final result in each case.

I emphasize here the explanatory role played in psychoanalysis by the vicissitudes of psychosexual phases, which result in fixations (dispositions to fantasy). They play a causal role more important than memories of traumatic events or disturbed object-relations, just be-

6. See, for example, the discussion of free association by A. Kris (1982).

cause psychoanalysis is more concerned with the vicissitudes of the relation of wish to imagination than with the vicissitudes of the relation of wishes and reality-accommodated beliefs to instrumental action.

Responses to traumatic events or disturbed object-relations involve the development and adoption of a more or less permanent mode of adaptation to them, which is not necessarily born of conflict, or maintained by conflict, and which may or may not be dysfunctional. But dispositions to regress in the face of obstacles to gratification-by-means-of-instrumental-action, and dispositions to conflict (that is, to respond to regressive wish-fulfillments with shame, guilt, or anxiety), play an important causal role, according to psychoanalysis, in vicissitudes of the relation of wish to imagination that lead, for example, to neurotic symptoms.

What causes these dispositions is a question for research. The role traumatic events and disturbed object-relations themselves play in determining such dispositions will vary from case to case. Every psychoanalyst has data suggesting the extent to which dispositions to produce particular kinds of fantasies, and internal responses to these fantasies, in themselves can bring about, for example, psychological dysfunction in thought and feeling, which in turn may bring about traumatic events and disturbed object-relations. In these circumstances, traumatic events and disturbed object-relations are effects, or effects as well as causes, rather than simply causes of dispositions.

Uses of Psychoanalytic Knowledge

Disciplines may be grouped according to the kind of systems, objects, or events they study: (a) cultural systems or artifacts; (b) social systems or interactions; (c) personality systems or persons; (d) physical systems, physical objects, organisms. These disciplines face three kinds of problems: explanatory; methodological; and technological.

Explanatory Goals

Psychoanalytic knowledge may be usefully applied whenever a discipline faces a deviation from its own laws, something it cannot explain given its own knowledge, and it suspects that its model of the world needs adjusting just where that model includes a model of the human mind. For example, to the extent economic theory makes use of a model of the mind that assumes perfect rationality, one assumes there will be economic phenomena it cannot completely explain. To the extent sociological theory makes use of a model of the mind that assumes it is entirely governed by the vicissitudes of socialization and by internalized norms and values, one assumes there will be sociological phenomena it cannot completely explain.

The participation of individuals in social systems depends in part on motivational states, but concepts that make sense applied to personality systems are, strictly speaking, not in the domain of a science of social systems. A science of social systems, however, may draw on psychoanalytic knowledge to refine its formulations about deviant participation in a social system, suspecting that it is just here that intrapersonal factors may play a role over and above, or in addition to, those factors concerned with properties of the social system (or of that other system—the cultural system—which like the personality system is on the boundary of the social system).

Similarly, a science of culture or a humanistic discipline that has explanatory goals, such as history, literary or art criticism, and biography, may make use of psychoanalytic knowledge to understand:

a) puzzling contents of a work of art, which mar the work or are not understandable in terms of its other features, the conventions according to which it was constructed, or the material stringencies that determined choices in constructing it;

b) the way people respond to works of art and literature, and the use they make of such works, perhaps especially those responses and uses, for example, that are idiosyncratic and could not have been predicted simply from a knowledge of the work itself, the kind of work it is, or the author's or the artist's intentions;

c) the nonrational uses people make of history, both personal and social history, and the nonrational responses people have to particular contents in a history;

d) puzzling features of contents of history, such as stories that persist as part of history despite the knowledge they are false.

Theories of history or art and literature are unlikely to be able to completely explain such phenomena without a theory of the mind that embodies knowledge of the relation of wish to imagination, and therefore knowledge of the way the mind works when it works according to primary process, which is the way it works in the realm of imagination, and of the causal role of wish-fulfillments, created by imagination, and exemplified by fantasies as well as dreams. Wish-fulfilling fantasies as enduring mental dispositions cause mental states, influence interpretations of experience and the extent to which and the way in which psychological capacities are realized in performance, and so contribute to the causation of action.

Methodological Goals

Psychoanalytic knowledge could conceivably also be useful in helping an investigator in a discipline identify sources of systematic bias in himself, especially in a case study research where he participates in the system or interacts with the subject he studies as he obtains information about the system or subject. In the social sciences, the intrapersonal processes of the investigator, his values and biases, his conscious and unconscious wishes, fears, anxieties, and defenses against impulses and dysphoric affects, especially as these are mobilized in transference reactions, influence what information he is able to obtain from the human subjects with whom he interacts and how he interprets the information he does obtain. Unavoidably, he affects his subjects and they affect him. These factors might account for his having obtained the data favoring his own conjectures over rival conjectures, even if his own conjectures were false. The possible influence of such factors, if ignored or uncontrolled, casts doubt on the conclusion that his data necessarily support his conjectures.

Psychoanalytic knowledge might help investigators to recognize intrapersonal factors that make for choosing one problem to investigate and neglecting another; failing to see or take seriously that there are conjectures that are rivals of one's own and what some of these might be; or failing to detect reasons for obtaining data favorable to one's own conjectures even when these conjectures are false. It is one of the paradoxes of science that some psychoanalytic investigators, especially in the way they use (misuse?) the case study method, make so little use of psychoanalytic knowledge in thinking critically about their own research and conclusions.

In addition, psychoanalytic knowledge can provide a theoretical foundation for the development in various kinds of research of tools, devices, or instruments used to obtain and interpret or classify observations or information, for example, projective tests and interviews.

Technological Goals

Psychoanalytic knowledge can be used, not to define worthwhile ends, but, given some end, to suggest the means of achieving it. Here, its use is analogous to the use of physics in engineering. Implications can be drawn from psychoanalytic knowledge about ways to achieve goals by the professions, for example, of education, law, and medicine and psychiatry. I write a big CAUTION here, because psychoanalysis is a science of the imagination—and here we are in the realm of practical, rational, or instrumental action.

Methodological Canons of Applied Psychoanalysis

In conclusion, I shall state six methodological canons of applied psychoanalysis. I shall simply state them without much in the way of argument or example. They are intended to be controversial, to provoke discussion, and to inspire further attempts to develop such canons.

1. Make explicit just what psychoanalytic knowledge is being applied. This presupposes a prior sifting through and organizing a vast corpus, deciding what is core and distinctive in it, and selecting what is relevant to the purpose at hand.

a) Do not misdepict psychoanalytic knowledge. Study carefully and repeatedly the observations, concepts, and theories of psychoanalysis. (One or two readings of *The Interpretation of Dreams* does not qualify as preparation for the application of psychoanalytic knowledge about dreams to problems in other disciplines.)

b) Note changes in psychoanalytic concepts and theories over time, and assess whether or not these changes are motivated by new observations.

c) Do not assume that general knowledge is psychoanalytic knowledge simply because it has something to do, for example, with childhood, dreams, or sexuality. A psychoanalyst quite properly makes use of a great deal of general knowledge about human beings, social systems, and culture in carrying out a psychoanalysis; that does not make such knowledge distinctively psychoanalytic.

d) Indicate where rival conjectures, relevant to your application, exist in psychoanalysis, and justify your acceptance of one rival psychoanalytic conjecture over another by citation of evidence, if possible, or at least by showing that it has proved superior with respect to explanatory power or fertility.

e) Indicate the evidential status of psychoanalytic conjectures used in your application—and the degree to which these conjectures are central or peripheral. Judge which theoretical claims are well-entrenched and which are merely opinions (no matter how generative they are, how powerful the explanations they make possible are, or how many observations are consistent with them). Justify your judgments.

Applied psychoanalysis has an important obligation, and conformity to this first canon would go far toward discharging that obligation. Applied psychoanalysis should be a source of steady pressure on psychoanalysis as a science to make its causal claims explicit, and to devote itself to establishing through empirical tests the scientific credibility of these claims. It is regrettable that, instead, those carrying out applied psychoanalysis frequently come to conclusions in other disciplines or human endeavors based on their *presupposition* (unaccompanied by any attempt to warrant that presupposition) that the psychoanalytic conjectures used in coming to these conclusions are true. The unhappy outcome is then that such an application becomes for psychoanalysis another demonstration of its credibility, fertility, or explanatory power; the application fails to challenge psychoanalysis to do scientific work.

2. Use discretion. Do not attempt to explain a complex set of phenomena in another discipline as due solely or simply to factors of special interest to psychoanalysis. In other words, avoid reductionistic explanations, which are simplistic if not arrogant. An application of psychoanalytic knowledge should be tactful and modest—that is, informed by awareness and including explicit mention of the limitations and proper domain of that knowledge.

a) One way to avoid a reductionistic use of psychoanalytic knowledge is to limit what is to be explained. At best, what is to be explained psychoanalytically will be some single puzzling fact about an entity or system rather than everything about that entity—a single fact about a man's life, for example, rather than his entire life; a single word, line, pattern, or episode in a poem or play rather than the whole poem or play; a single feature of a painting rather than the whole painting; a single aspect of an historical event rather than the entire event. Why does Rebecca West reject Rosmer's proposal after doing so much to bring it about? Why does she become so agitated on hearing apparently innocuous information about her father? Why did Van Gogh cut off his *ear?*

I don't see how—given the limited domain of psychoanalysis—there can be such a thing as a completely psychoanalytic biography, a completely psychoanalytic understanding of a work of art, a completely psychoanalytic explanation of the existence of a social institution.

b) Another way to avoid a reductionistic use of psychoanalytic knowledge is to specify just what contribution, and what weight to give that contribution, a factor of interest to psychoanalysis makes to an explanation.

c) A third way to avoid a reductionistic use of psychoanalytic knowledge is, in choosing what to study, to consider where some possibility exists of obtaining the kind of data to which psychoanalytic ideas and methods of investigation are most applicable. For example, it is more likely to be fruitful to use psychoanalysis to explain something about a man's life if a great deal of personal information about that life (such as might be found in personal documents or from interviews of those who knew him well) is available, rather than if the man lived very long ago and little reliable information is available about him. It is more likely to be fruitful to investigate psychoanalytically the *responses* of available subjects to a work of art (painting, music, or literary) than to attempt to explain psychoanalytically some features of that work of art, or to infer something about the mental states of its inaccessible progenitor.

d) A fourth way to avoid a reductionistic use of psychoanalytic knowledge is to choose an especially appropriate phenomenon, that is, one for the understanding of which psychoanalytic conjectures are especially likely to be relevant: a bad piece of art rather than a good piece of art; a failure rather than an achievement; an inexplicable redundancy or irrelevancy; an apparently unmotivated, inexplicable, or dysfunctional departure from an artistic or social convention rather than conformity to such a convention. Why does Lauren Bacall in one scene of *The Big Sleep* absentmindedly move to scratch her thigh while talking to Humphrey Bogart? Why does Cocteau in *Beauty and the Beast* have the Beast turn into a prince when Beauty's brother's friend (her suitor) is killed, rather than—as in the fairy tale and therefore as expected—on her bestowal of a kiss?

3. Conduct an application of psychoanalytic knowledge, if possible, in such a way that it constitutes a challenge to or test of psychoanalytic knowledge. That would mean:

a) there is some way of specifying whether the application is successful (in other words, it is possible that it may fail); and

b) there is some way of arguing that a failure in applied psychoanalysis, such as the failure of a prediction in another discipline based on psychoanalytic conjectures, if it occurs, raises a question about the credibility of the psychoanalytic conjectures used. That is, the application has been carried out in such a way that its failure cannot be attributed to ad hoc or nonlawful conditions under which it was carried out, or to knowledge in that other discipline that was in question.

Even in the humanities, for example, it seems to me that there might be some attempt to judge the success of a "reading" of a work of art by making a prediction that if the "reading" is correct, some members of its audience at least, having certain characteristics, should have a particular response to it.

An instance of applied psychoanalysis so conducted might help to establish the scientific credibility of psychoanalytic conjectures, because here the factor of suggestion—so troublesome for research in the psychoanalytic situation itself—might be eliminable.

4. Solid, thorough knowledge of the discipline in which psycho-

analytic knowledge is being applied—and respect for its concepts, methods, and hypotheses—are indispensable. Do not mix levels of description. The temptation to treat other kinds of systems as "psyches" is to be resisted. For example, do not describe a social system as having an id, ego, and superego.

5. In giving a psychoanalytic explanation, offer rival explanations and present the evidence for preferring one explanation over another. Consider in presenting evidence not only what information is available but the quality of that information (how reliable or dependable it is).

There is an excellent discussion of this point in Runyan (1982). Of particular value is Chapter 3, "Why Did Van Gogh Cut Off His Ear? The Problem of Alternative Explanations in Psychobiography" (pp. 38–50). Runyan gives thirteen alternative explanations of why Van Gogh cut off his ear, and assesses the reliability of the available evidence favoring or counting against each one. In this book, which should be consulted by anyone interested in the case study method, Runyan has gone far in formulating methodological canons for life histories and psychobiography.

6. Avoid grandiosity. Do not play God in making recommendations as to what action to take or what policy to adopt. There are a number of pitfalls in applying psychoanalytic knowledge in such technological enterprises as education, law, and medicine.

a) There is no way to derive a prescription as to what is a desirable end from knowledge alone. Psychoanalytic knowledge cannot tell us, for example, what a mature adult should be like, since maturity as a concept is heavily value-laden, nor can it tell us, without reference to desirable ends about which there may be value-determined differences of opinion, what the one best way to raise a child is.

b) One cannot assume automatically that because a technology is successful that the knowledge on which it is based is true. One cannot assume automatically that because a technology is unsuccessful that the knowledge on which it is based is false. Psychoanalytic therapy may be successful, but not necessarily because psychoanalytic conjectures are true. Psychoanalytic conjectures about some form of psychopathology or other may be true, even though such psychopathology cannot be successfully treated by psychoanalysis.

c) The efficacy of a procedure or line of action, or the impact of a decision, in law or education for example, cannot be warranted by the supposed scientific credibility of the psychoanalytic conjectures on which it is based. Claims about the efficacy or impact of a procedure, line of action, or policy must themselves be empirically tested.

The Importance of Applied Psychoanalysis to Psychoanalysis Proper

Why are psychoanalysts themselves so interested in applied psychoanalysis? We can guess from observing the major reason for Freud's interest.

Freud made causal inferences in the clinical situation about the role of unconscious (and therefore hypothetical) fantasy systems and the role of unconscious (and therefore hypothetical) mental operations designated by such terms as "primary process," "dream work," and "defence mechanisms," to explain such phenomena as parapraxes, dreams, and neurotic symptoms. One warrant for the credibility of these causal inferences is provided by evidence for the *existence* of such fantasy systems and mental operations in realms other than the clinical situation.

It would be difficult to argue that the evidence obtained in these other realms is produced by such extraneous factors as the psychoanalyst's suggestion or a desire to please the psychoanalyst or to fulfill his expectations, which might be supposed to be producing data favoring his inferences in the clinical situation.

Just as Freud referred to hypnosis and word-association experiments as providing independent evidence for the existence of unconscious causally efficacious mental states, so he used studies of art, literature, autobiography (as in the Schreber case), and jokes to provide independent evidence for the existence of the kinds of fantasy systems and mental operations that play a causal role in his clinical explanations. At a time when psychoanalysis is being challenged to make explicit rational grounds for believing its clinical inferences, applied psychoanalysis continues to have this same serious job to do for psychoanalysis proper.

Imagination, Fantasy, and Defense[1]

9

The Issues

Investigators who seek evidence for the psychoanalytic theory of defense in studies of memory and forgetting miss the mark. That people tend to forget unpleasant experiences, a hypothesis apparently much in favor as a target of research in this area, is not even remotely a psychoanalytic hypothesis or a proposition that belongs to the psychoanalytic theory of defense.

An experimental psychologist claims that he is unable to find any evidence for such psychoanalytic notions as repression. It turns out that he studies those cognitive processes that determine whether access to memories is achieved or lost, or that he studies cognitive processing (registering and/or storing) of perceptual input. Hearing this, a psychoanalyst can only suppose that the experimental investigator looks where what he looks for is not. The failure to find the evidence is related to basic misconceptions of the core theory of psychoanalysis, and in particular of its theory of defense.

Defense

The Target of Defense

The target of defense is the content of a fantasy that is a sexual wish-fulfillment.

Information about the human capacity of imagina-

1. Parts of this chapter were presented at the Conference on Repression, Dissociation, and the Warding-off of Conflict-Related Cognitions and Affects, sponsored by the MacArthur Foundation Program in Conscious and Unconscious Mental Processes and held at Yale University, June 23–25, 1986. This chapter also contains reactions to other papers and discussions at that conference, which I had no opportunity then to express.

tion, not memory, and about sexuality, not cognition, is especially relevant for research in this area. But much if not all research in psychology in this area (I refer here primarily to quantitative or experimental research) focuses on the nature of memory as a human capacity and on dysfunctions of remembering, or more broadly on cognitive capacities and operations, rather than on the nature of imagination as a human capacity and on dysfunctions related to the causal powers of fantasy.[2]

If we start from Breuer and Freud's *Studies on Hysteria* (1893–1895), and Freud's work in particular, where his attempt is to explain neurotic symptoms, we may emphasize, as he did there, unconscious mental states, conflict, and defense. But if we start from *The Interpretation of Dreams* (1900) and the very important later paper "Formulations on the Two Principles of Mental Functioning" (1911b), where Freud's attempt is to explain dreams as well as neurotic symptoms, and a wide range of other mental phenomena as well, we may emphasize, as he did there, wish-fulfillment and the difference between primary and secondary process.

Taking conflict as a point of departure tends to lead to the assimilation of the phenomena in which psychoanalysis is primarily interested—neurotic symptoms, parapraxes, and dreams—to the principle of rationality. Once we grasp what the subject's unconscious desires and beliefs are, then we shall see that mental operations on mental states have been governed by a principle of rationality after all. Whatever irrationality the phenomena psychoanalysis investigates manifest results from the fact that—because the subject's beliefs were formed in childhood, or since unconscious mental states have been rendered inaccessible to the corrections or influences of experience—the subject holds mistaken or false beliefs (the subject is in error) or has insufficient knowledge (the subject is ignorant). If I am not mistaken, this perspective on the phenomena of interest to psychoanalysis permeates ego psychology and the way it influences what is viewed as the therapeutic action of psychoanalysis. It is especially attractive to those who wish to integrate psychoanalysis and the cognitive sciences. What follows in this chapter will suggest the difference it makes to take wish-fulfillment, fantasy, and primary process instead as a point of departure.

The Motive for Defense

The motive for defense is ultimately the quest for pleasure, when that quest takes place under conditions of threat. The motive for de-

2. See Wollheim (1969, 1974a, 1974c, 1979, 1982, 1984) for an account of the role of imagination in psychoanalytic thought.

fense is not just a matter of a quest for security or an attempt to avoid pain. A defense is an attempt to solve a problem: how to achieve gratification under conditions of threat (Schafer 1968b).

Information about wish-fulfillments, not unpleasant experiences, and about problem solving, not forgetting, is especially relevant for quantitative/experimental research purporting to test psychoanalytic theory.

The Means of Defense

The means of defense are primary processes. That is, defenses are typically symbolic operations on mental representations of a certain kind—fantasies, which are both constructed and, in the interests of defense, revised by means of primary processes.

Information about primary process, not rational cognition (or error or ignorance as deviations from rational cognition), and about problem solving in the world of imagination, not adaptive learning in the actual world, is especially relevant for quantitative/experimental research purporting to test psychoanalytic theory.

Indications of Defense

A manifest indication of defense is resistance, not necessarily forgetting. Forgetting may be, in some but not all cases, an outcome of a process of defense; even when it is an outcome, forgetting does not constitute the entire process.

A defense is a strategy for achieving gratification under conditions of threat. It may involve, *in part,* the expulsion of the *contents* of mental states or the *significance* of such contents from consciousness.

The subject resists efforts to make him aware of such contents or their significance, and this "fighting off" constitutes evidence for the existence and causal role of a defensive strategy of the kind in which psychoanalysis is interested—just as antibodies constitute evidence for the presence and causal role of a virus. However, resistance is not a fight directed against the original noxious agent (the conditions of threat), but involves both anticipation of attempts to interfere—or anticipation of events or circumstances that might interfere—with a defensive strategy, and the prevention or circumvention of these.

Resistance and defense should be differentiated, as often they are not in research and clinical work. The terms "defense" and "resistance" are not synonyms.

Research purporting to test psychoanalytic theory usually fails to differentiate between "repression" and "countercathexis"—between ways of *expelling* something from awareness and ways of ensuring that what has been expelled is permanently prevented from returning to awareness. It usually fails also to differentiate between the *contents* of a mental state and the *significance* of the contents of a mental state as targets of repression and countercathexis.

What does "avoiding awareness of the *significance* of the contents of a mental state" mean? One may avoid awareness of (a) experience of the *phenomenology* of a mental state (what it is like to be in that state) rather than merely knowledge of the contents of that mental state; (b) what the *connection* is between the contents of different mental states (these mental states may be contemporaneous, or precede or follow one another); (c) what the *connection* is between some remembered mental state and what one is doing or saying now; (d) knowledge of what being in a mental state having certain contents *implies* about oneself or another, or about the likelihood that various other mental states will occur; or (e) knowledge of the *causal efficacy* of a mental state or set of mental states (that it is causally related, for example, to ego-alien symptoms and ego-syntonic character traits).

Defense is not, as a good deal of research purporting to test psychoanalytic theory suggests it is, something that happened once in the past. It is possible to observe the results of a currently active process of defense in vivo, in the present—as an analysand, for example, not only reports amnesia for whole periods of his life ("I can't remember anything before age twelve"), but amnesia for everything said in the session of the day before, or later in a session for what he said earlier in that session.

How does repression occur? Erdelyi (1985, pp. 149ff.) speaks of omissions, of ellipses, of attenuations, of displacements of accent. If an analysand does not remember or is not aware of a fantasy, and struggles against becoming aware of it, we speak confidently of the unconscious and repression. But the concept of repression is more complicated than the usual research of this type depicts it to be.

What about the analysand who mentions something casually without apparent awareness of its significance, in the manner of the miscreant who hides the stolen letter by crumpling it up and throwing it down in plain view, so that whoever sees it will say, "But that couldn't be what I'm looking for!"? Here the analysand enacts both detective and thief. How do we identify the rule-governed operation called *the defense of the purloined letter?*

What about the analysand who buries a significant utterance under an avalanche of worries, busy thoughts, preoccupations? What about the analysand who allocates his attention in such a way that he just doesn't notice certain things? What about the analysand who when something is called to his attention says, "oh yes, I was thinking about that when I came in, but I forgot about it!"? What about the analysand who can go for years beginning each day's session with, "I can't remember anything about yesterday's session," but who has no difficulty remembering it if the psychoanalyst tells him what happened during it? What about the analysand who blurts things out in an altered state of consciousness, in a trance-like state, but acts almost immediately as if he had never said them at all, although he has no trouble remembering saying them when reminded by the psychoanalyst of what he said? (Did Freud give up the notion of the etiological role of hypnoidal states in favor of the notion of defense too readily?)

We seem to have a vast number of ways to keep things "out of mind," and a vast number of different states called "out of mind." I am not satisfied that order can be brought to all this by introducing (as some ego psychologists are wont to do) abstract gatekeepers or filters—devoid of content—between unconscious and preconscious, between preconscious and consciousness, between what is conscious and the communication of what is conscious.

Research that would be relevant to the psychoanalytic theory of defense would see itself as studying a continuously active process when it studies repression or countercathexis—a process that inexorably extends its sphere of influence. Repression takes in or encompasses more and more. If something is a target of repression, then anything connected to it, and anything connected to anything that is connected to it, and so on and on, may become a target of defense.

Repression "swallows" more and more—and in fact repression, which is a fantasy rather than a contentless operation on mental contents, may be an imagined swallowing. The inability to remember, then, is the effect of a fantasy in which contents have been swallowed and disappeared. Here is a strategy that can be used to mitigate conditions of threat, at the same time that it is a wish-fulfilling fantasy of a pleasurable swallowing.

The linkages between the various contents swallowed are determined by primary, not secondary, process, a supposition often ignored by those who wish to assimilate psychoanalytic theory to learning theory or cognitive science. That is, the links are formed in fantasy, in the world of imagination.

Similarly, a countercathexis becomes ever more elaborate; it is put together from, and propped up by, whatever is at hand. But if defensive strategies are improvisatory and jerry-rigged from whatever is at hand, "defense" denotes a difficult-to-delimit set of mental activities. One cannot simply tell from "looking at it" that something observable is a manifestation of defense or of what defense it is a manifestation. Repression should be differentiated not only from the other traditional "mechanisms of defense" but from an indefinite number of defensive strategies more or less jerry-rigged from whatever is at hand (Bibring et al. 1961).

The notion of defense mechanisms does not seem to capture the quality of improvisational inventive problem solving often conveyed by an analysand. Erdelyi (1985) speaks of the ego as the seat of "*sophisticated strategies* for avoiding or distorting reality" (p. 130). (Are these strategies really concerned with reality, or with products of imagination, namely wish-fulfillments?) He also speaks of "makeshift stratagems" (p. 132). If we wish to consider fantasies attempts at problem solving, of dealing with unattainable wishes, conflicts, and dilemmas, then Schafer (1968b) has stated one problem that a fantasy is produced to solve: insuring "the maximum of instinctual gratification possible under conditions of danger" (p. 55).

The efforts of the analysand often seem to be a perfect jumble of ad hoc tactics and strategies, taking opportunistic advantage of whatever is at hand, whatever is found, whatever material is available (Freud 1900, p. 237). A bricolage is not constructed according to known rules, operations, or procedures.[3]

What strategies for achieving gratification under conditions of threat are possible? One may as well ask, "What are the limits of human imaginativeness and inventiveness?" These may turn almost any object, event, or state of affairs that exists or occurs in the ad hoc nonlawful circumstances in which such a problem is encountered into a resource that can be used in attempts to solve it.

Perhaps, it will turn out that different *defensive strategies,* broadly conceived, can be characterized by their position on five dimensions: repression, the tactics or measures that enter into a defensive strategy, countercathexis and resistance, gratification and threat, and duration.[4]

a) *Repression.* To what extent is repression a part of a defensive strategy? Just what is repressed? How extensive is the realm swallowed by repression?

b) *Tactics.* What measures other than repression are employed to mitigate conditions of threat?

c) *Countercathexis and resistance.* How is what is repressed kept repressed? How are attempts to interfere with a defensive strategy, or events or circumstances that might interfere with a defensive strat-

3. For a discussion of the term "bricolage," see Chapter 2.

4. For a similar formulation, see Waelder (1960), who mentions three coordinates: repression, countercathexis, and gratification.

egy, anticipated, and prevented or resisted, or forestalled or circumvented?

d) *Gratification and threat.* What kinds of threat are mitigated by a defensive strategy? To what extent does a defensive strategy make gratification possible? How does it make gratification possible? Just what kind of gratification does it make possible?

e) *Duration.* To what extent is a defensive strategy occasionally evoked, in response to specific kinds of threat, and to what extent maintained permanently in place?

One Possible Formulation of the Core Theory of Psychoanalysis

How would I—taking *The Interpretation of Dreams* and "Formulations on the Two Principles of Mental Functioning," for example, as a point of departure—formulate the core theory of psychoanalysis? It is this core theory that I am proposing empirical research should attempt to test. What is the role of imagination in it? How are imagination—and, in particular, fantasy that is constructed by means of primary process—and defense related?

Wishing and Imagining

The mental process of interest in psychoanalysis begins with a sexual wish. The sexual wish results in a fantasy, which fulfills the wish.

There is a special relation between wishing and imagination that does not hold for wishing and action.[5] Rational action is explained in an intentional psychology by citing a combination of wish and instrumental belief that are concurrently efficacious. For carrying out an action, a wish is necessary. But, also, given that wish, an instrumental belief is necessary. And, other things being equal, the two together will suffice to produce an intention to act.

For imagining an action, wishing is enough. This is true even for those wishes that, in conjunction with belief, might lead to intentions to act. It is also true for those wishes that are in conflict with, or that compete, with other wishes. But, most importantly to psychoanalysis, it is true for wishes that would not ordinarily or could not lead to action, wishes that some unalterable condition, for example, were changed–that you were different, or that I were differ-

5. For a beautiful, perspicacious statement about the relation between wish and imagination, see Wollheim (1979, pp. 58–60).

ent, or that something that has happened in my past had not happened. For, without any particular belief coming into the matter at all, I can still imagine you being different, or my being different, or that what has happened has not happened at all.

Wishes and beliefs produce intentions to act, and actions lead to change in the external world. But wishes and imagination may not be connected or connected immediately with any change in the external world at all. That, of course, is why psychoanalysis poses such conundrums for the philosophy of science and for most conceptions of scientific methodology.

Fantasy, Primary Process, and the Truth of Beliefs

That the fantasies of interest to psychoanalysis are produced by primary process distinguishes them from anticipatory reality-oriented imaginings.

For rational action, it is essential for a subject to have some conception of truth. That the subject considers, more or less justifiably, that his belief is true determines that together with his wish it produces an intention to act.

Imagination, however, may be quite comfortable with beliefs in whose truth the subject not only has no confidence but about the truth of which he has no concern. It is not simply that considerations of truth may not influence imagination, but that "imagination has *the power* to reject such considerations: the considerations are not influential, but this is because they are not permitted to be influential" (Wollheim 1979, p. 52).

Wishes and Beliefs

What are some relations that may hold between wishes and beliefs?[6]

a) In rational action, the content of an *instrumental* belief determines whether or not an action that will satisfy a wish is possible, and, if possible, what action will lead to its satisfaction. Psychoanalysis is not distinctively concerned with instrumental beliefs.

b) In attempting to understand action, we may say that a *causative* belief gives rise to a wish ("I believe he wishes to hurt me, and therefore I wish him dead").

c) A *presuppositional* belief stands as presupposition to a wish—"I wish to revenge myself on my father, but I can only do that if he is alive. Therefore, I believe he is alive." (When the Rat Man permitted

6. Here I make use of Wollheim's explication of the relation between wishes and beliefs as the internal constituents of neurotic symptoms (1971, pp. 144–147).

truth to matter, he believed his father dead, but, under the influence of primary process, he believed him—imagined him—alive, and therefore available for purposes of revenge.)

d) *Wish-fulfillments* support and sustain beliefs. "This must be so because I have this wish." (The Rat Man held on to wishes directed to his father in his imagination, so that he could maintain his belief that his father was still alive.) "This must have happened, because I imagine this wish fulfilled." (Freud thought Dora wished to be pregnant and that she imagined she was pregnant, so that she could maintain her belief that she really did have intercourse with K, the friend of her father.)

Of course, the beliefs that are maintained are "themselves the objects, and the products, of desire; of, that is, further and more fundamental desires" (Wollheim 1971, p. 147).

Can such relations among wishes, beliefs, and imaginings be explained by the kinds of theories that appeal to cognitive scientists, which give primary to rule-governed computations on mental states? The answer to this question is crucial in determining to what extent psychoanalysis is the same kind of representational theory of mind formulated by cognitive scientists such as Fodor (1968, 1975, 1981b, 1983) and Pylyshyn (1985). I shall consider the question further later in this chapter.

Wish-fulfillments and Effective Imagination: A Cinematic Model

Wish-fulfillments, of the kind that dreams are thought to be, are not hallucinations but the products of effective imagination. There are degrees of imagining. Wish-fulfillments are products of effective imagination, the extreme degree of imagination.[7]

The closest that one can come to this idea is in remembering what it was like to watch a movie as a child. It is a mistake to think the child does not know he watches a movie, or that he does not know who he is. But he inhabits the movie and usually some central character in it with such completeness that he feels what that character feels. Perhaps at times he inhabits some secondary character in the movie, a secondary character who judges the central character favorably or unfavorably; the child responds as that secondary character does to the hero or heroine.

It is like a long swoon into an inner world. Questions of the truth of this or that fact or circumstance or of the reality of the whole spectacle are not permitted to disturb the child's immersion in it or to

7. The concept of effective imagination is Wollheim's. Wollheim, in works I have previously cited, has also explicated effective imagination in terms of theatre. I have found his explication useful, and have borrowed from it, in developing the cinematic model described in the text.

interrupt his open-mouthed trance. The effects are enormous. He laughs hysterically, yearns, despairs, cries, trembles in terror, is unbearably excited, joyful, fulfilled, exultant, triumphant, stirred to new effort.

The next closest that one can come to the idea of effective imagination is to remember the vividness, peremptoriness, power, and effects of the sexual fantasies of adolescence. Thinking of the child in the movie and the adolescent in the grip of his own created excitement, one accepts the description of dreams as the products of effective imagination.

The cinematic model is particularly persuasive to me, because as a psychoanalyst I have often felt I was being assigned a part to play in some script I only gradually came to know. Both psychoanalyst and analysand are imagining, and each is in the imagination of the other. The power of the imagination comes home to me as I find myself with each analysand saying things, especially saying things in a certain way or with a certain intonation, and accompanied by shifts in position, hesitations, and even gestures, things different from analysand to analysand, which, although ambiguous, of course, are capable of being interpreted by each analysand in such a way that they fit right in with that analysand's script. I often have the experience that something the analysand is saying or doing is conveying a scene to me, or a series of scenes, as if I were indeed watching the screen on which his fantasy unfolds. Is the psychoanalyst's state of mind such that he imagines himself into the internal audience of the analysand's effective imaginings?

So-called countertransference also testifies to the power of the imagination. Countertransference may involve the psychoanalyst's assigning the analysand a part in some fantasy of the psychoanalyst's own—occurring somewhat independently of, and perhaps distracting the psychoanalyst from, the objectives of the psychoanalysis. Countertransference may also involve the psychoanalyst's awareness of the mental states the analysand's activity arouses in him, which for the psychoanalyst are clues to the content of the inner drama of the analysand in which the psychoanalyst, like it or not, plays a part, and to which, at the same time, he bears witness.

Effective Imagination and a Corporeal Conception of Mind

Psychoanalysis belongs among representational theories of mind.[8] Of great interest to psychoanalysis is that subjects are able to repre-

8. In referring to a representational theory of mind, I more or less follow Fodor (1981b). See p. 26, for example, for one formulation by him of the characteristics of a representational theory of mind. See also Dennett (1978); Fodor (1968, 1975, 1983); Pylyshyn (1985); and Searle (1983), as well as Chapter 7 of this book.

sent symbolically their own mental states, and in particular are able to represent symbolically their own psychological activities (thinking, feeling, perceiving, remembering, believing, wishing). In other words, a subject is able to form a higher-order mental representation of himself (the subject) carrying out a psychological activity that in turn is directed to a mental representation of an object or state of affairs. Of even greater interest perhaps to psychoanalysis is that a subject is able to represent symbolically not only his own body (and, as we know, that representation need not correspond to the facts of anatomy) but his own mind as well—what Wollheim refers to as "the mind's image of itself" (1969).

It is part of effective imagination that the fact that we are merely imagining is suppressed; the dream is an example (Wollheim 1979, p. 54). It is also a part of effective imagination that it is governed by a corporeal concept of the mind itself (Wollheim 1969, 1979, 1982).

The analysand, in fantasizing, may have a spatial or corporeal concept of his mind something like the following. "My mind is a place, in a geography." "My mind is a house, where things are put and kept, and where things happen in different rooms." "My mind is a body, with a gastrointestinal tract or a gastrointestinal-urinary tract, that takes things in from the world, digests them, keeps them inside, excretes and loses them. These things inhabit my mind, and do things in and to it, or are extruded and lost to it." "My mind is a penis, which throbs and pounds, can't wait to shoot things out, 'takes off,' penetrates incisively, wilts or can't get anywhere."

The corporeal concept of the mind is a concept of the mind as possessing powers; its states and contents have an exaggerated efficacy, adjusting "the world so that it conforms to its objects" (Wollheim 1979, p. 54). Given a corporeal concept of mind, we have, for example, "omnipotence of thoughts," belief in the magical powers of the mind, and what Freud calls "the overvaluation of psychic phenomena."

A New Look at Primary and Secondary Process

This entire description of effective imagination and its differences from rational action corresponds to descriptions of primary process and its differences from secondary process. In short, primary process might be defined as: (a) the construction of wish-fulfilling mental representations (wish-fulfillment has primacy compared to any other end for a primary process), by means of condensation, displacement, pictorialization, and iconic symbolization; and (b) effective imagining, in which the truth of beliefs is not permitted to matter, in which the fact that one is merely imagining is suppressed, and which is dominated by a corporeal concept of mind that results in an overvaluation of psychic phenomena (the belief that thoughts are omnipotent and that the mind has magical powers). If action is a manifestation of primary process, it is an enactment, an exteriorization of an inner drama, not a rational act, resulting from a conjunction of wish and reality-accommodated instrumental belief.

Affects and the Cinematic Model

In the cinematic model, an internal scriptwriter-director determines or initiates what it is that we imagine; we feel that we determine or initiate it. To say there is an internal audience is to say that what we imagine has an effect on us that is intrinsic to the imagining itself and not merely a consequence of it. The adolescent's sexual excitement is *in* the fantasy.

It is important to distinguish between affects that are *in* a fantasy and affects resulting from reflecting on a fantasy. Reflecting on the content of the fantasy, which is the fulfillment of a sexual wish, and the excitement the internal audience experienced in response to this content, leads to evaluative judgments. Insofar as the judgments themselves are made under the influence of primary process, they too are embodied in fantasies.

One may think of anxiety, guilt, or shame as responses to the fantasy, but outside of it. "How could I get excited by something as yukky as that!?" "What does that imply about me?"

Post-fantasy evaluative responses may themselves be expressed in fantasies, constructed by primary process. "What, in light of what I believe, would happen to me if I actually did something like that?" (So, the Wolf Man as a child, in the light of his sexual researches and his preoccupation with castration, interpreted a scene of his parents copulating as implying that having such passive pleasure with his father would necessarily entail his being castrated.) "What if my father had come in while I was doing that?" "How would my mother react if she knew what went on in my mind?"

But it is also possible that the internal audience identifies with the central character, the hero of the fantasy, as he reacts to another actor who responds unfavorably to him. So, in a scene in which the hero takes pleasure in exhibiting himself, one of the other actors turns away in disgust, belittles him, or mocks him. In this case, the wish-fulfillment and the conditions of threat are both part of the same fantasy, with the effects of the fantasy shifting as an internal audience identifies empathically first with the central character's pleasure and then with his response to another actor who regards him unfavorably.

The adolescent's guilt, anxiety, or shame, as he notices or reflects that he had this fantasy, or that he became sexually excited in it, or both, are a consequence of his imaginative activity, but also a part of it. The adolescent's guilt, anxiety, or shame may be *in* his sexual-wish-fulfilling fantasy, and not a simple reaction to that fantasy, separate from it. Guilt may be the feeling the central character, with whom the internal audience identifies, has as he is scolded or shunned by another actor in a fantasy that is otherwise a wish-fulfillment; anxiety the feeling he has as he is threatened with harm by another actor in a fantasy that is otherwise a wish-fulfillment; and shame the feeling he has as he is laughed at, belittled, or ridiculed by another actor in a fantasy that is otherwise a wish-fulfillment.

Emotions are mental states in which feeling is a relation to a mental representation, just as wishing and believing and perceiving are. Emotions are not simply contentless physiological reactions to stimuli. It is of course the case that, just as intentions to act may cause action, so emotional states may cause physiological changes (both are mental-representation-to-world causal relations). If an emotion appears without contents in a psychoanalysis, the contents are somewhere; it is a matter of discovering what the analysand has done with them.

Is there a tendency in psychoanalysis ("affects are discharge phenomena") and cognitive science to regard emotions as analogous to *action* rather than as mental states? "[E]motional states can be caused by cognitive states, just as the movements of my fingers as I type this sentence are caused by my thoughts and intentions" (Pylyshyn 1985, p. 269).

More typically, in cognitive science, emotions are bypassed. "The structural model [in psychoanalysis] is not easily accommodated to modern information-processing models of cognition, which typically (as one would expect of models based on the computer metaphor) avoid the problems of emotions and morality" (Erdelyi 1985, p. 131).

Fantasy and Defense

If the response to a wish-fulfillment in fantasy is anxiety, guilt, or shame, which are painful affects, then the subject, who seeks both to experience pleasure and to avoid the experience of pain, faces the problem of fulfilling wishes under conditions of threat. He may attempt to solve this problem, in fantasy, governed by primary process, so that some degree of wish-fulfillment is possible while the threat is to some extent mitigated.

Are defense mechanisms as a *kind* of mental process actually different from what is being defended against? I am not so confident that

I can distinguish them in the clinical situation. Defenses are primitive, magical, "inappropriate," and function according to the primary process (Gill 1963, p. 105), just as forbidden wish-fulfillments are and do. Could it be that defenses are fantasies, too? "Whistling in the dark" (Bibring et al. 1961), which is typical of the way analysand and psychoanalyst talk about "defense," is a fantasy, not a computation. An analysand does not simply set a "mechanism" into motion. He is imagining doing something. He is imagining doing something to mental states or contents. He is imagining what sorts of things he is doing and what sorts of things these are he is doing them to.

How can one separate the anxiety-reducing and wish-fulfilling aspects of a mental phenomenon? Any mental phenomenon seems to function both to gratify and to make secure, in both an expression of impulse and defense, provides discharge for what it also provides protection from. It rarely if ever makes sense to interpret a mental phenomenon as if here was a disembodied "mechanism of defense," pure and simple, no more than it makes sense to interpret a mental phenomenon as if here was a pure manifestation of impulse uncorrupted by defensive aims. In fact, there often appears to be a hierarchy or layering, such that what is defense at one level (or from one point of view) is what is defended against at another (Gill 1963, pp. 120–127). Homosexuality may defend against the dangers of heterosexuality, and then heterosexuality in turn may defend against the dangers of homosexuality. And where is the wish-fulfillment and where is the defense? In the same fantasy.

Another way, perhaps, of expressing this problem is to note that a mental phenomenon such as a neurotic symptom may seem to be the product of a wish, a fear, and what the analysand does to gratify the wish and at the same time to reduce the fear. But at the same time, a complex or network of wishes and beliefs, of different kinds, and in different kinds of relations to one another, seems to be implicated—seems to constitute the structure of the phenomenon (Wollheim 1971, pp. 144–147).

Alas for the theorist and therapist, there is no easy way to classify or separate wishes, fears, and what the analysand does about his fears. If an analysand says, "I am a guinea pig or dog, stretched out on this couch, and you are probing my mind, and it hurts, and I hate you for doing this to me," the analysand's fear is his wish, his wish is his fear, and what he does about his fear, which is also his wish, is to produce a fantasy so that he gratifies his wish and avoids what he fears simultaneously, without letting himself quite know what he accomplishes in the world of his imagination.

The view of the defenses as fantasies has been anticipated by others; I mention Schafer (1968b) as an example. He conceptualizes the mechanisms of defense as wishes—that is, as "dynamic tendencies having mental content" (p. 51). *Isolation* is an effective imagining of keeping certain things from touching—"the intrapsychic enactment of not touching, hence not engaging in the anal-sadistic acts of masturbating, soiling, killing, playing, observing the primal scene, making babies, and so forth" (p. 54). *Undoing* is an enactment of "blowing away." "'[T]o blow away' I take to include a reference to fantasies about the creative, curative and annihilating powers of flatus, breath and wind in general" (p. 54). *Projection* (imagining oneself as attacked by an outside force) gratifies the wish to be penetrated. Fantasies of spitting out, defecation, and being devoured may also be implicated in projection. Schafer adds to this account of projection: "the patient's search for his emotional life in the therapist's; the passive experience of his impulses; the relation of anxiety and guilt to both his active and passive experiences; the oral, anal and homosexual significance of the shift of emphasis [from one's own impulse, thought, or feeling to the other's]; and the pleasure and dread of merging with the therapist" (p. 56). *Denial* is a psychic act involving an assault on oneself, on another, and on the relation between the two. *Introjection* "realizes the wishful fantasy of incorporating the object orally for purposes of regaining omnipotence and effecting destruction or preservation of the object" (p. 57). *Repression* may be a fantasy about swallowed ideas. Gaps in memory may be a fantasy about female genitalia. A sense of defect, of incompleteness, of a need to have the psychoanalyst fill something in, of voids in the mind, all may be fantasies of being a woman or of castration. To repress a name or memory of a person may be a killing of the person in fantasy.[9]

When the venue changes from rational action to effective imagination as a way of gratifying wishes, including wishes to avoid painful states, the objective is to gratify wishes through effectively imagining them as gratified. In order to gratify as many wishes as possible, such devices as condensation are employed by the scriptwriter, where one element in the script can simultaneously allude to many events or objects because, for example, it has played a part in each of them, or resembles each of them in some way.

As we have noted, an evaluative judgment as a response to a wish-fulfillment may be expressed in a fantasy outside the fantasy that is a

9. Holland (1973) is aware of this way of looking at the defenses; it certainly keeps, as Suppes and Warren (1975) noted, his "ego's algebra" from being as neat and mathematical as he might have wished it to be.

wish-fulfillment, but also may occur *in* the fantasy that is a wish-fulfillment. In the latter case, an example was that, in a complication of a pure wish-fulfillment, the internal audience in the fantasy that is a wish-fulfillment identifies with the central character's reaction to being regarded unfavorably by another actor *who has a necessary part in the scenario of wish-fulfillment.* Just as the internal audience's identification with a sexually excited central character has the power to create pleasurable sexual excitement in the subject who has the fantasy, so its empathic identification with the central character when he is made to feel anxious, guilty, or ashamed by another actor has the power to create anxiety, guilt, or shame in the subject who has the fantasy.

Such painful affects will motivate the subject to carry out defensive strategies—for example, to revise the fantasy in various ways to mitigate the painful affects and at the same time safeguard a measure of pleasurable wish-fulfillment.

These revisions may have external manifestations, since fantasies, as dispositional mental states, have the power to cause conscious mental states of various kinds, including intentions-to-act. An imagined undoing in a fantasy, for example, may cause an actual enactment (such as the Rat Man's clearing the path so that his lover will not be hurt by what he imagines his anger at her has placed in the path to hurt her).

But the symbolic operations on such a mental representation as a fantasy will not necessarily be observable by the research investigator from a mere scrutiny of the subject's actions. Both fantasy and symbolic operations on it may have to be inferred. That psychoanalytic theory refers to hypothetical mental states and processes, which cannot be observed, does not make it unusual, as scientific theories go. It does pose some difficulties for experimental/quantitative research purporting to test psychoanalytic theory, which on the whole has not been noticeably successful in dealing with these difficulties to date.

I shall not attempt to enumerate the inexhaustible supply of strategies and devices available to the scriptwriter in his rewriting. He may keep the story, but attempt to tone down (defense: make it less exciting!) or otherwise change the response of the internal audience. He may draw the audience away from an empathic feeling with the central character, or induce it to sympathize instead with the feelings of an actor responding favorably to the central character. He may do something about an actor who regards the central character unfavorably. He may also abandon the script, locking it up in an attic or safe (repression as an act of effective imagining!). He may instead close the movie house, or allow the film to unfold before an empty house, or refuse admittance to critics of any kind (repression as an act of effective imagining!).

Might defenses be regarded as contentless mechanisms—operations on mental representations? I once was entranced by the possibility of formulating psychoanalytic theory in a canonical formal language for the sake of rigor and explicitness (Edelson 1977). In particular, I thought for some time it might be possible to apply the notion of rule-governed transformations (like those used in Chomsky's work), acting upon latent or unconscious propositions (or, in the present context, unconscious fantasies), however these are mentally represented, to produce manifest or preconscious or conscious propositions, however these are mentally represented (see Chapter 2 of this book).

My *analogy* ran something like this:

a) the operations of the dream work are like linguistic operations insofar as they are rule-governed transformational operations;

b) unconscious ideas or impulses (I now would say unconscious fantasies) are like deep structures in linguistics insofar as they are the objects of such transformations;

c) the manifest content of the dream—images under a linguistic description—are like surface structures in linguistics insofar as they are the results of such transformations;

d) the subject's capacity to dream is like language capacity insofar as the difference between knowledge of the dream work and the performance of making a particular dream is like the difference between knowledge of transformational-generative grammar and making a particular sentence; and

e) the creativity of dreaming is like the creativity of language, in that with finite means an infinite set of novel forms can be generated.

As part of a justification for such an approach, I referred to some comments by Freud in the Schreber case (Freud 1911a, pp. 63–64), in which he represented the principal forms of paranoia as resulting from transformations of, or operations on, the linguistic representation of the proposition "I (a man) *love him* (a man)."[10]

I continue to believe that the transformational-generative model is important for psychoanalysis, despite the particular reservation I now have and am about to express.

Just consider the complicated transformations Freud uses in his explanations in "A Child Is Being Beaten" (1919). Freud traces the

10. As did Suppes and Warren somewhat later, in a somewhat different context, but for similar reasons (1975, p. 406).

vicissitudes of the historical development of the fantasy "A child is being beaten."[11] The first phase of the beating fantasy, in a case of a female child, is represented by the phrase: "My father is beating the child whom I hate." The content and meaning of the fantasy is: "My father does not love this other child, he loves only me." The next phase of the beating fantasy (as constructed in the psychoanalysis) is represented by the phrase: "I am being beaten by my father." The content and meaning of this fantasy, which is an expression of the child's guilt, is: "No, he does not love you, for he is beating you." This fantasy is not only a punishment for forbidden genital wishes, but a regressive substitute for a genital relation with father. In the third phase, accompanied by sexual excitement, the content becomes: "Strange and unknown boys are being beaten by someone, perhaps a teacher, and if I am there at all, I am probably looking on." The child is now identified with the boys who are being beaten, and the gratification is, therefore, masochistic rather than—as in the first phase—sadistic.

What I want to emphasize here is that it is clear Freud is *not* thinking in terms of rule-governed operations that are contentless mechanisms (not themselves mental states, representations, or contents), which simply operate on mental states, representations, or contents. He is thinking rather of the kind of changes in a fantasy that are like those an author might make in a script. What we have here is like a theme and variations, with each variation motivated by different considerations, or, in other words, different wishes and beliefs, and often providing in the imagination both protection and gratification simultaneously.

The important point is, *defenses are effective in reducing anxiety (or guilt or shame), as well as gratifying wishes, because they are themselves, just as wish-fulfillments are, effectively imagined doings.*

The Unanticipated Consequences of Defensive Strategies

A frequent mistake in both research and clinical work is to regard all the results of a defensive strategy as motivated, that is, as something sought in using the defensive strategy.[12] Sometimes the consequence of a defensive strategy is an unanticipated consequence of it, a "side effect," so to speak. Effective imagination results in changes in

11. See, especially, pp. 185–191.

12. A similar mistake is made by those who, because some anatomical structure has a present function or a use, regard it as necessarily having been selected in the process of evolution because of the adaptive value of that function or use. On this point, see Gould (1977, 1980, 1983).

the subject, in the phenomenology of his mental states, in his dispositions, and therefore in his actions.

Michelangelo, according to Freud, defending himself against the loss of his mother, incorporates her. She is now inside him. As is the way with effective imagination, she is capable there of doing things to him, pleasant and unpleasant, and he is also capable of becoming her. Becoming her, and it is part of his imagining that he now possesses her beliefs and wishes, he loves boys as she loved him. It was no necessary part of what he intended that he now becomes homosexual in his life—an unanticipated and perhaps not-altogether-welcome-to-him effect of the efficacy of his imagined gratification of his wish to hold on to her (Wollheim 1974c).

A similar point can be made about the student who imagines that when he writes he excretes. When, of course, he imagines closing the sphincter and becoming constipated, he is simply managing internal contents, which for one reason or another he enjoys holding on to or controlling. He may indeed want to avoid killing his teacher with noxious matter. He does not necessarily self-punitively intend that he should flunk the course because he could not get a paper in on time.

The Pathological Effects of Defensive Strategies

I have emphasized here a conception of psychoanalysis as a science of imagination.

If psychoanalysis is a science of imagination, pathology becomes the effect of the contents of unconscious fantasies. These fantasies influence other mental states, dominating and infiltrating and shaping them. Pathology may manifest itself in disturbances in the realm of imagination itself, in the analysand's reaction to and rejection of the products of imagination—specific wishes and wish-fulfillments — but also in the analysand's reaction to and rejection of imagination itself.

This conception of psychoanalysis rejects the currently popular idea that the primary disturbance and focus of therapeutic effort are to be found in the realm of object-relations or interpersonal relations. The primary disturbance is in, and the therapeutic action of psychoanalysis depends on influencing, the analysand's relation to his own inner life.

What analysand does not fear his own imagination, does not treat it like a beast to be tamed, caged, or destroyed? (Indeed, sad to say, there are treatises on psychoanalytic theory and technique and the objectives of psychoanalytic therapy that appear to take the same attitude toward the "id.") How many analysands report inner emptiness—the screen of imagination is what stands blank, without script, scene, or character? How many analysands report being trapped before the screen of imagination, enthralled, unable to leave the movie theater, going through the same scenes over and over, the world outside the movie theater pallid and unreal, the world inside the movie theater intense, colorful, the only reality? How many analysands report that they are no longer in control of the actors or the play: scenes flash into their minds they have not determined or initiated; the action takes a surprising turn; an actor appears in search of a part; an actor refuses to perform his role as directed, forgets his lines, does not appear on cue; the projector breaks down; the screen is rent by an ugly distracting tear?

An analysand may, of course, under some frightful circumstances, want to live in fantasy alone. Then, he imagines, the world cannot reach or hurt him. Then, he can have the world he wants. He can express feelings without danger of rebuff or ridicule, gratify wishes without obstacle or attack, even communicate what he wants to communicate to anyone without that one knowing what he has communicated and hurting him in response. Here, it seems he might live without risk.

But he cannot rid the imagination of the demons he introduces into it. His fantasies, which have causal power, have unexpected and unwelcome consequences. He discovers that he has sacrificed what gratification he might sometimes otherwise have achieved by acting in and changing the actual world.

If it is effective imagination that solves the problem of achieving wish-fulfillment under conditions of threat, we should not regard defense as some sort of well-defined contentless computation, from a finite list of such computations, operating on a mental state. Effective imagination does its work by means of primary process. So it makes our dreams—and it would seem our defenses as well. It is primary process, and the ubiquity and power of primary process, rather than unconscious conflict or infantile sexuality, which are its progeny, that may turn out after all to be Freud's single greatest discovery.

What Does Current Experimental/Quantitative Research Have to Do with the Psychoanalytic Theory of Defense?

Experimental/quantitative research purporting to be relevant to psychodynamic/psychoanalytic theory is frequently preoccupied with measures of what is easiest to measure—vicissitudes of interpersonal relations, or cognitive processes (memory, perception, learning). In my opinion, it has yet to investigate the causal efficacy (even with respect to symptoms, dreams, and parapraxes) of wish-fulfillments (effective imagining), in particular, those wish-fulfillments resulting from sexual wishes. Such research also does not, although it should, investigate the causal efficacy of both pleasurable and unpleasurable affects, *as these are both evoked in and manifestations of sexual wish-fulfillments.* Nor does it, as it should, investigate, above all, the primary processes used in constructing such fantasies, and also in revising them according to strategies of problem solving that are designed to address the problem of fulfilling wishes under conditions of threat.

10 Psychoanalytic Contributions to a Theory of Sexuality[1]

The Domain of Psychoanalysis

What Is Distinctive about Psychoanalysis?

Re-reading Freud's remarkable work, *Three Essays on the Theory of Sexuality,* I came across this passage in the 1914 preface to the third edition.

> It is impossible that these *Three Essays on the Theory of Sexuality* should contain anything but what psycho-analysis makes it necessary to assume or possible to establish. It is, therefore, out of the question that they could ever be extended into a complete 'theory of sexuality', and it is natural that there should be a number of important problems of sexual life with which they do not deal at all. But the reader should not conclude from this that the branches of this large subject which have been thus passed over are unknown to the author or have been neglected by him as of small importance. (Freud 1905, p. 130)

This is just one of many passages by Freud consistent with my contention that psychoanalysis is not a general psychology. It is not even a complete psychology of mind. It is a branch of a psychology of mind. It accepts that its primary and distinctive task is to investigate the relation between desire and imagination. More specifi-

1. In preparation for Symposium "A Dialogue Between Psychoanalysts and Philosophers about the Psychoanalytic Theory of Sexuality and the Psychoanalytic Theory of Dreams," midwinter meetings of the American Psychoanalytic Association, December 1987. I have repeated in this chapter enough of the discussion about causality, which was formulated first in writing Chapter 15, to maintain continuity.

cally, psychoanalysis investigates the way in which sexual desire produces not actions but imaginings and, in particular, sexual wish-fulfillments and sexual fantasies; the way in which such imaginings become enduring mental dispositions; and the protean ways in which these dispositions are manifested.[2]

It does not make sense to ask grandiosely, "What is the psychoanalytic theory of mind, the psychoanalytic theory of sexuality?" It makes sense to ask modestly instead, "What can psychoanalysis contribute as a body of knowledge to our understanding of the mind's workings, and specifically to our understanding of sexuality and its effects, that is not likely to be contributed by, and perhaps cannot be contributed by, other schools of thought, given their preferred methods of inquiry and the conceptual frameworks within which they work, which determine what their interests, questions, and programs of research are?"

Similarly, it makes sense to ask, "What causal factors should be especially associated with psychoanalysis as a body of knowledge—in contrast to other psychologies—not because psychoanalysis denies the role of other kinds of causes, but because it has something distinctive to say about just these factors and because it is, given its particular mode of inquiry, in the best position to investigate and elucidate their causal roles?" Certainly, sexuality is one such causal factor.

In the 1945 British movie *I Know Where I'm Going,* Joan Webster (played by Wendy Hiller) is a young woman who knows where she's going, but her plans to marry a wealthy man are disrupted when she falls in love with Torquil MacNeil (played by Roger Livesey), the impoverished Laird of Kiloran. (Kiloran is an isle in the Hebrides.)

In one scene, Joan is talking to Catriona (played by Pamela Brown), who lives in an old, run-down castle and has to shoot rabbits to eat. Joan is lying in bed after a terrible and chastening adventure; her ambition has nearly sent her and others to the bottom of the sea. Catriona is stoking the fire.

Joan: You must think I'm awful.

Catriona: I don't think anyone is awful.

Joan: Not even when I'm breaking my neck to marry a rich man?

Catriona: Well, what's wrong with that?

Joan: But I thought you didn't care about money.

Catriona: Who says so! I'd swim to Oban for ten pounds, Glasgow for twenty.

2. For one explication of this view of the task of psychoanalysis, see Wollheim (1969, 1971, 1974a, 1974b, 1974c, 1979, 1982, 1984).

Joan: What about Torquil?

Catriona: He'd do it for fifteen.

Joan: But I thought that you and Rebecca Crozier and Torquil were perfectly happy without money.

Catriona: What else can we do?

Joan: Well, you could sell [this place], and Rebecca could sell Acroneish and Torquil could sell Kiloran.

Catriona: (thoughtfully) Yes. (pause) But money isn't everything. Now go to sleep.

So it is with sexuality as a causal factor in psychoanalysis. It isn't everything—but it certainly is important.

The Characteristics of Sexuality

What characteristics make sexuality such an attractive plausible candidate as a putative cause of phenomena in the domain of psychoanalysis?

1. Its physicality. The origins of sexuality lie in bodily processes, and its expression is intertwined with and dependent on bodily states. There is no way to talk about sexuality without mentioning anatomy and physiology: bodily postures, bodily changes, and bodily parts, regions, or zones. To accept sexuality as a major cause of mental phenomena is to insure, at the least, that even as a psychology of mind psychoanalysis cannot ignore or forget the role of the body in mental life.

2. Its sociality. Society takes an enormous interest in sexuality. Every society strives, through processes of socialization and sanctions, to monitor, regulate, and determine the form of expression of sexuality. What can be explained about sexual phenomena as the result of such social goals and processes is not in any central way a proper focus for psychoanalytic inquiry. But to accept sexuality as a major cause of mental phenomena is to insure, at the least, that even as a psychology of mind psychoanalysis cannot ignore or forget the role of the social in mental life.

3. Its plasticity. The quest for sexual, sensuous, or voluptuous pleasure takes many forms. It finds its fulfillment in many different kinds of acts and is directed to, or makes use of, many different kinds of objects. What was once perhaps merely exciting becomes pleasurable. Fore-pleasure becomes end-pleasure. Preliminary acts, acts that in their origin were primarily instrumental, become pleasurable in themselves.

4. Its antiquity.[3] That sexuality is already present from infancy on

3. Wollheim's term (1971, p. 133), used in a similar discussion of those characteristics of sexuality that make it credible as a cause.

gives it plenty of time to accomplish its work and to achieve far-reaching and complicated effects.

5. Its vulnerability to fixations and to disorders of development in general. Just because of the plasticity and antiquity of sexuality, deviations from outcomes that processes of socialization expect, hope for, and strive to bring about are all too possible and frequent.

6. Its imperiousness. The power of sexuality to grip, obsess, and drive a person, leading the person in pursuit of sexual pleasure to override all obstacles and other considerations, is well known, and makes it plausible to suppose that, similarly, sexuality may have multiple strong effects outside the sexual realm.

What Does Psychoanalysis Want to Explain?

What is it that psychoanalysis wants to explain as having its origins in, or as determined by, sexuality? Mental representations—which symbolize objects, states of affairs, or events—comprise the *contents* of mental states. Psychoanalysis raises questions about the inexplicability of the contents of particular mental states.

These contents may be inexplicable in a number of ways.

1. They do not seem to have any causal links to the contents of other mental states to which a subject has access. For example, they do not seem to be linked to the contents of any wishes and beliefs the subject is aware of.

2. The objects, states of affairs, or events in external reality, which are their putative cause or which would be the right kind of cause of them, do not seem to exist or ever to have existed (for example, dreams, hallucinations, nonveridical memories).

3. The actions one might expect, given the nature of the contents and the particular mental state to which they belong, do not occur. For example, a person, although he has the relevant wishes and beliefs, may be unable to form the right kind of intention to act, or the intention, once formed, does not result in action.

4. In some cases, it is the absence of certain contents—when, given a particular causal constellation, these contents might be expected to be present—that calls for explanation: for example, a person whose memory, as a psychological capacity, is unimpaired wants to remember, tries to remember, but is unable to remember certain contents.

Neurotic symptoms, dreams, and transient and circumscribed psychological dysfunctions such as the parapraxes, as well as more extensive and enduring psychological dysfunctions, are paradigmatic of such *causal gaps*. It is a distinctive contribution of psychoanalysis to explicate the role of the sexual, especially as it is represented in unconscious mental contents, in bringing about these causal gaps.

Conceptual Contributions of Psychoanalysis

Extensions of the Conventional Definition of Sexuality

The first contribution of psychoanalysis to a theory of sexuality is to extend the conventional view of what is sexual in two ways. Most important, psychoanalysis distinctively supposes that just these two extensions, and not other facts about sexuality, are of critical importance in arriving at causal explanations of neurotic symptoms, character, dreams, and parapraxes. This programmatic supposition will, of course, determine the content of specific causal conjectures.

Psychoanalysis reacts to the conventional view of what is sexual—that "sexual" implies genital union, resulting in orgasm, where the genitals are those of two persons of opposite sex, and the two persons are both adult. It goes in two directions to extend our understanding of what is sexual.

The first direction has to do with object and aim. The quest for sexual pleasure may find its fulfillment in something other than the union of the genitals of two adult persons of opposite sex. Here psychoanalysis makes use of the facts of sexual inversion and perversion to overthrow what is conventionally comprehended by "sexual." The object of a quest for sexual pleasure may not be another person or may not be another person of the opposite sex. The aim of a quest for sexual pleasure, the act that brings about sexual pleasure, may not be genital union.

The second direction leads to infantile sexuality. The quest for sexual pleasure is found not only among adolescents and adults but among infants and children as well. Sexuality in adolescents and adults has a personal history. This history determines the nature, structure, or character of adult sexuality, and therefore the nature of its effects.

The theoretical role assigned to these two extensions is as important as the extensions themselves in identifying the distinctive contributions of psychoanalysis to a theory of sexuality. A non-psychoanalytic psychology might accept one or both extensions of what is comprehended by "sexual"—the existence of perverse sexual tendencies and even the existence of a sexual life in infants and children—without any inclination to suppose that these characteristics of what is sexual have the causal efficacy psychoanalysis supposes they have.

What is *sexual?* Perhaps sexual is one of those concepts better left somewhat indistinct and imprecise. Psychoanalysis appears to answer the question in part by making sexual, without regard to the age of the subject, any kind of activity with any kind of object that satisfies the following criteria. It produces a distinctive kind of excitement and pleasure. Furthermore, if this activity (and the object it involves) is not conventionally associated with sexuality, it is causally related to the combination of activity, object, and outcome conventionally accepted as sexual. This causal relation depends on its playing a role in the origin of, acting as a precursor of, or eventually becoming an aspect of what is conventionally accepted as sexual. This causal relation also depends on its being capable under certain cricumstances of replacing or serving as a functionally equivalent substitute for what is conventionally accepted as sexual.

Excitement and Pleasure: Implications for Theory

Psychoanalysis makes a second contribution to a theory of sexuality in blurring the distinction between the excitatory state and the consummatory act, and it makes this contribution against its will, so to speak—in contradiction of its own metapsychological models and principles. Freud's entire *Three Essays on the Theory of Sexuality* is a refutation of the reflex arc model in his overly attended to Chapter 7 of *The Interpretation of Dreams.* (Chapter 6, which is an account of the mechanisms comprised by the dream work, is the true masterpiece of that masterpiece.) The reflex arc model, with its associated constancy principle, conceives pleasure as a matter of getting rid of, or discharging, excitation, and assumes the mental apparatus operates to reduce to a minimum the level of excitation brought about by internal and external stimuli.

The reflex arc model requires a sharp differentiation between pleasure and excitement. Stimuli cause excitation at one end; the excitement is discharged at the other end. But it is clear from many passages in the *Three Essays* that one and the same act can be a source of pleasure and at the same time create an influx of new excitement. This may be another way of saying that pleasure leads to a wish for more pleasure. The memory of a particular kind of pleasure itself can reinforce the disposition to seek more of that kind of pleasure. In any event, the excitatory state and consummatory pleasure are not sharply differentiatable beginning and end points in a sequence of states or events.

> On the one hand these activities [touching and looking] are themselves accompanied by pleasure, and on the other hand they intensify the excitation. (Freud 1905, pp. 149–150)

> And everyone knows what a source of pleasure on the one hand and what an influx of fresh excitation on the other is afforded by tactile sensations of the skin of the sexual object. (Freud 1905, p. 156)

All the facts—the inexorable persistence of the quest for sexual pleasure, and the overriding of disgust, guilt, shame, and fear that is frequently associated with it—suggest that pleasure is a positive state, compatible with the presence and even increase of excitation, rather than a state defined by the reduction or absence of excitation. Man in his sexual preoccupations, obsessions, and frenzies resembles the animals who, with electrodes implanted in the septal region of the limbic system (where, apparently, there exists a pleasure-producing center) delivered as many as five thousand self-stimulatory shocks in an hour, or the patient with electrodes placed in the septal region who stimulated himself fifteen hundred times in three hours, and protested vigorously each time the unit was removed, begging to self-stimulate a few more times; in one session this self-stimulation resulted in orgasm (A. Freedman et al. 1975, pp. 154–157).

All this, of course, is not to say that we cannot distinguish between desire (and excitement seems related to desire) and the satisfaction of desire (and pleasure seems related to the satisfaction of desire). What is characteristic of a desire is that it can be satisfied or frustrated. What is characteristic of a belief is that it can be true or false (correspond or not to some state of affairs in the world external to or existing independently of the believer).[4] To be unable to distinguish between desire and its satisfaction—between excitement and pleasure—would make it impossible to conceive of psychoanalysis as a representational theory of mind, with desires and beliefs as typical of entities in its domain. However, the distinction between desire and satisfaction does not call for a reflex arc model in which excitement precedes pleasure and ceases with the occurrence of pleasure, any more than it would make sense to say that there can be no desire that arises from the satisfaction of desire or that desire cannot have as its object a state of desire.

Psychoanalysis does raise questions such as the following. What determines what in particular excites a person? What determines the particular weight given to different sources of sexual excitement for that person? Why is one person excited by a stimulus, mental activity, or mental state that is not at all exciting to another? What are the possibilities open to a person in responding to sexual excitement? What determines what a person does about sexual excitement?

4. See, for example, Fodor (1987).

Source, Object, and Aim

Psychoanalysis makes a third contribution to a theory of sexuality by drawing a sharp conceptual distinction between the *source* in bodily processes of a quest for sexual pleasure; the sexual *object* (i.e., to what kind of being or thing does the subject relate in order to obtain sexual pleasure); and the sexual *aim* (i.e., what kind of things are done to, by, or with an object in order to obtain sexual pleasure). The end of the quest is, ultimately, and in the last analysis, to obtain some particular kind of sexual pleasure. Objects and acts are means to that end, and what objects and acts are involved determine what kind of sexual pleasure it is.

Bisexuality is a universal disposition, in the sense that there exists in everyone propensities, of varying strength, which may change from one occasion to another or from one time of life to another, to choose an object of the opposite sex and to choose an object of the same sex. Even in homosexual object-choice, we have a variety of possible combinations. A man, for example, may seek a male who is young, boyish, or feminine, or a male who is mature, muscular, or masculine. He may seek in a male the ideal boy, girl, man, woman, or the idealized brother, sister, father, mother. He may imagine himself a man seeking a man who reminds him in important ways of a woman, or he may imagine himself a woman seeking a man who conforms to his conception of what is quintessentially masculine.

Object and aim vary independently of each other. The same quest for sexual pleasure—by *different* persons, or by the *same* person on different occasions or at different times of life—can be directed (in *action* or *fantasy*) to different objects (oneself or another, a person of the same or opposite sex, a person of the same or a different generation, a member of a different species, an inanimate object, an entire body or a single body part). So, also, the same quest for sexual pleasure—by *different* persons, or by the *same* person on different occasions or at different times of life—can involve (in *action* or *fantasy*) different kinds of acts (involving different parts of the subject's body and different objects).

There is no fixed relation, no necessary connection, between object and aim. Another way of saying this is that preference for a certain kind of object does not determine (inexorably and without exception cause) the choice of a particular kind of activity, and that preference for a certain kind of activity does not determine the choice of a particular kind of object. This variability is in marked contrast to animal instinct.

The conceptual distinction between source, object, and aim, the related postulation of a universal bisexual disposition, and the observation that object and aim vary independently of one another, are connected to a fundamental psychoanalytic conjecture. The conjecture is that what source is dominant, what object is chosen, and what aim is preferred, in the same person at different times of life or on different occasions, or from one person to another, will be found to be causally relevant in one case after another, although just what particular causal relevance they have (just what particular causal role they play) may differ from case to case.

The concept of source remains incompletely explicated. "Source" sometimes is used to refer to physicochemical processes in the body, to which persons respond by seeking a particular kind of sexual pleasure. Differences in source then imply different physicochemical processes, or differences in the anatomical region of the body in which a physicochemical process occurs. Of course, to the extent that "source" refers to physicochemical processes, acquiring knowledge about the source of the quest for sexual pleasure is not a task of psychoanalysis.

The concept of source, which has to do with the bodily origin of excitement or desire, seems to merge with the concept of erogenous zone, an anatomical region of the body that is the site of activity leading to the satisfaction of desire or sexual pleasure. But the concept of erogenous zone itself remains incompletely explicated.

Sometimes what is implied by "erogenous zone" is a bodily surface—mucous membranes or skin—or a sense organ (the general idea seems to be "external stimulus receptor") the direct stimulation of which leads to sexual pleasure. However, it is also said that almost any organ of the body can become an erogenous zone. I suppose one can unite these ideas by noting that almost any organ of the body is a receptor for internal stimuli, especially stimuli emanating from the activity of the brain.

Another complication arises around the concept of erogenous zone. On the one hand, sites of the body, with richly innervated mucous membranes and, therefore, like genitals, capable of being directly stimulated and so perhaps capable of acting as a source of end-pleasure, are erogenous zones. Here, I refer especially to orifices, such as mouth or anus, which can come to substitute in the subject's mind, and therefore in the subject's acts, for genitals. On the other hand, anatomical regions (eyes, skin, muscles), which are instead *instrumentally* important as ways to the object—ways of finding, coming into contact with, and controlling or possessing the object, so that it can be used in the quest for sexual pleasure—are also regarded as erogenous zones. Here again we touch on the discovery that what is exciting and what is ultimately pleasurable are not sharply differentiatable, and that what may have begun as merely exciting, as forepleasure, can become end-pleasure, sought for its own sake and not as a means to any other end.

In any event, the central idea in psychoanalysis is that in a quest for a particular kind of pleasure, a subject's act that produces that kind of pleasure crucially involves some body site (some region or part of the subject's own body) and makes use of some object in the subject's situation. Again a problem arises, now in explicating the distinction between body site and situational object.

I take it that mind is the system with which psychoanalysis is concerned and that, therefore, from that point of view, the subject's own body as a physical object and its various regions or parts are in the subject's situation. The subject brings some region or part of his or her body into relation or contact with, or causes such a region or part to interact with, some object in his or her situation. (It is also possible that some such region or part is acted upon by the situational object without regard to any intentions of the subject.) The object in the subject's situation may be, for example, the body or a body part of another person of the same or opposite sex or of the same or another generation, a member of another species of the same or opposite sex, an inanimate object—or some part of the subject's own body. The object's being a part of the subject's own body is a special case perhaps, since in this case it is possible for pleasure to arise simultaneously in more than one region or part of his or her body, which may help to account for the prominent causal role played by vicissitudes of autoerotism in the domain of psychoanalysis.

We have, then, a site, a region or part of the subject's body; an external object; and the relation between them. The relation is the activity that brings about sexual pleasure—for example, penetrating/being penetrated; touching/being touched; looking/being looked at; controlling, mastering, possessing, dominating/being controlled, mastered, possessed, dominated; or inflicting pain/submitting to pain. Employing a cinematic model of mind, we may ask about any quest for sexual pleasure: What site, part, anatomical region, or erogenous zone of the subject's body is dominant, is the scene of the action, the stage upon which the drama unfolds? What object—alter not ego, other not I—is brought into the scene? What part does that object play? What screenplay does the subject write? What action does he or she direct? What does the *I* do to or with the *other* in this scene?[5]

5. Review Chapter 9 for a discussion of a cinematic theory of mind.

Activity and Passivity

Psychoanalysis makes a fourth contribution to a theory of sexuality by choosing the distinction between activity and passivity with respect to sexual aim as the distinction that is theoretically relevant for psychoanalysis. It is in this sense only that psychoanalysis regards the polarity masculine-feminine as explanatory. However, the distinction between activity and passivity with respect to sexual aim is in some ways problematic.

Anatomy is destiny. The distinction between masculinity and femininity, when it is a matter of anatomy and physiology, or a matter of values and social role prescriptions, is on the whole not a matter of central importance with respect to distinctively psychoanalytic contributions to a theory of sexuality. The notion that anatomy is destiny does not make anatomical or physiological markers of gender theoretically important in psychoanalysis. This notion is part of the effort in psychoanalysis to trace backwards from unconscious sexual fantasies to their causes, rather than to investigate the effects of unconscious sexual fantasies, when they themselves are regarded as causes. I have some doubt that the attempt to find ultimate causes here can be successful and still remain within a distinctively psychoanalytic conceptual framework. Anatomy presumably is just one of many factors (culture and processes of socialization being other factors) that cause a propensity to have one kind of unconscious sexual fantasy rather than another.

The notion that anatomy is destiny should not distract us from the distinctive emphasis that psychoanalysis places upon the observation that gender (as defined by anatomical and physiological markers), social role, and preference for activity or passivity in sexual aim vary independently. Different persons—or the same person on different occasions or at different times of life—though in gender masculine or feminine may be either masculine or feminine in their social role behavior, and either masculine or feminine in their preference for activity or passivity with respect to a sexual aim. Gender does not determine (inexorably and without exception cause) the nature or acceptance of a sexual role, or a preference for activity or passivity with respect to a sexual aim. Adoption of a socially defined sexual role does not necessarily determine preference for activity or passivity with respect to a sexual aim (though it is possible that acceptance of a particular sexual role may imply acceptance of a prescription governing that preference). Preference for activity or passivity with respect to a sexual aim does not determine which sexual role is adopted.

Differences between men and women. It is not clear that such distinctive concepts of psychoanalysis as source, object, aim, or activity and passivity provide an adequate apparatus for investigating differences in development, mental functioning, or character between men and women as groups. Given the obvious necessity for the language of biology and sociology, and the need for statistical comparisons to test conjectures related to this problem—probably not.

Unproblematic uses of the distinction between activity and passivity. In three instances, the distinction between activity and passivity is easy to draw. All three have to do with sexual aims that are in origin instrumental and connected with finding, coming into contact with, and controlling or possessing the object. These aims are looking, touching, and mastering the object. The important anatomical regions, which ultimately becomes zones of excitement and pleasure, are the eyes, the skin, and the muscles.

In each case, there is an active and passive form of the sexual aim: looking/being looked at; touching/being touched; and controlling, mastering, possessing/being controlled, mastered, possessed. Sadism/masochism (inflicting pain/suffering pain), as the active and passive forms of a sexual aim, are thought to derive from, and to remain associated with, the active and passive forms *controlling, mastering, or possessing/being controlled, mastered, possessed.* Psychoanalysis supposes that the history of these sexual aims, the vicissitudes of both the active and passive forms of sexual aims, in either a male or female, and the way in which a person ultimately deals with and expresses both the active and passive form will have causal significance.

Problematic uses of the distinction between activity and passivity. When it comes to the mouth, anus, and genitals as erogenous zones, the duality active-passive becomes more problematic. From the point of view of the subject (or the central person in a fantasy) to be active is to be the agent in a particular activity (that is, with respect to a particular sexual aim)—the one who penetrates; the one who sucks; the one who makes, holds on to, or deliberately expels solid or liquid contents. To be passive, from the point of view of the subject (or the central person in a fantasy), is to be the object in a particular activity (that is, with respect to a particular sexual aim)—the one who is penetrated; the one who is sucked; the one who opens up, just lets go; or who is spattered or sprayed with expelled contents.

The complications here are formidable. But one fact is certain. The behavior does not carry its own theoretical classification marker. Observing the behavior alone does not automatically determine what kind of behavior it is; this is true of most phenomena in the domain of psychoanalysis. One needs to know, to classify behavior in theoretically relevant ways, what is going on in the subject's mind. How does he or she classify the behavior? What does he or she imagine it is like? What does he or she imagine his or her orifice or the orifice of the other, whatever that orifice might be, is like—vagina, anus, or mouth? What are the role and feelings of the central person in the fantasy that is causally connected to the behavior?

In our understandable reluctance to offend, in our wish to avoid the odium of appearing to be pornographic or unprofessional, we have become accustomed in our theoretical discourse to a formalized and latinate language. But this very language makes it difficult for us to see certain difficulties the distinction active-passive runs into. By referring in two instances, for example, to very different acts as "mouth-genital contact" or "fellatio," we obscure the difference fantasy makes. In the first instance, one participant imagines eagerly sucking the other's penis, as if sucking the nipple of a sumptuous breast, while the other participant imagines lying back, relaxed and effort-free, and letting the semen like milk flow out gently and easily. In the second instance, what both participants imagine is a steel poker rammed forcibly and deeply into a vagina-like throat. Does it make theoretical sense to classify in both instances the one whose orifice receives the other's penis as "passive," and the one whose penis enters the other's orifice as "active?" Similarly, shall we call "passive" both the one who, from above, grasps a penis in an orifice and moves energetically up and down upon it, riding imagery in mind, and the one who, from below, relaxes an orifice to receive a penis, being-filled-up imagery in mind? Is the penetrated one who tightens vaginal muscles or the anal sphincter around the penis and squeezes it, as the subject imagines biting it off, to be regarded as "passive" merely because penetrated?

The language of theoretical discourse may have accustomed psychoanalysts, and others, including nonclinicians, who are interested in psychoanalysis, not to think of these acts in the vernacular, or "under a description" which makes use of the everyday and perhaps "obscene" language of the clinical situation. This state of affairs goes to the heart of how psychoanalysis classifies the phenomena in its domain, the verbal reports of the analysand. Its consequences are detrimental therefore not simply for the clinical enterprise but for the theoretical enterprise as well.

> An obscene word has a peculiar power of compelling the hearer to imagine the object it denotes, the sexual organ or function, *in substantial actuality*. . . . I would [call] special attention to the fact that delicate allusions to sexual processes, and scientific or foreign

> designations for them, do not have this effect, or at least not to the same extent as the words taken from the original, popular, erotic vocabulary of one's mother-tongue. (Ferenczi 1911, p. 137)

The theoretical language to which we have become accustomed may prevent us from hearing, through the imagery and metaphors an analysand uses in describing nonsexual experiences, allusions to sexual fantasies. It may make increasingly unlikely that a clinician will help an analysand express such fantasies more directly and obtain thereby evidence of their causal efficacy. It may have the consequence, ultimately, that psychoanalysts increasingly will take less seriously conjectures that such fantasies are causally relevant in understanding apparently nonsexual phenomena, and unconvinced themselves increasingly will be unable to convince others of the scientific credibility of core psychoanalytic conjectures.

A theoretically irrelevant sense of "active." Being active in the sense of trying to bring about, and planning and acting to bring about, sexual pleasure, as distinct from having it just happen to one, "accidentally," so to speak, is not the causally relevant distinction. Every quest for pleasure is, of course, in this sense, "active"; as it is sometimes said, "libido is always active." The theoretically relevant distinction active-passive refers specifically to the sexual aim, whatever it is.

On the Limits of Psychoanalysis and Difficulties in Accepting Them

Variation

Is there, besides the explanation of inexplicable mental states (causal gaps in mental life), another explanatory task that is both central to and distinctive of psychoanalysis? Just as evolutionary theory attempts to explain the variety of species, so psychoanalysis seems to take on the explanation of the enormous variety of human sexual experience—among cultures, persons, or on different occasions or at different periods in the life of one person.

In addition to the variation associated with choice of object or aim, or active or passive forms of a sexual aim, there is variation in the degree to which a choice of sexual object or sexual aim is exclusive (replaces all other objects or aims); the degree to which a person's preference for a certain kind of object or sexual aim is acceptable to that person; the degree to which a preference for a sexual object or sexual aim persists, the frequency with which it appears and disappears, the weakness or strength of a person's investment in such a preference, the ease or difficulty with which a particular person abandons a preference in the face of obstacles to gratification, or the ease or difficulty with which a particular person returns to a long since abandoned or rejected preference in the face of obstacles to gratification involving a later preference; the degree to which a particular kind of object-choice is linked to a particular kind of sexual aim; the sources of sexual excitement, and especially what evokes exciting unconscious sexual fantasies (external stimuli, mechanical stimulation of body parts, strong emotions, intense intellectual concentration, mental activity); what in particular is sexually exciting to a person (what kinds of fantasies); the degree to which affection or love and lust converge on the same object.

That there is such variation, that human beings appear to be omnipotential in their sexuality, and yet that so many persons by adulthood seem to have eliminated all but one or at least a very few possibilities in their sexual lives is also a fact that must be explained by a theory of sexuality. One does not have to account only for the occurrence of variation, but for the regularity with which it decreases in the course of life. Freud (1905) attributed this decrease in part to the action of disgust, pain, shame, and moral prohibitions. These are examples of the kind of negative causes in psychoanalytic theory that prevent other causes (for example, sexual desire or sexual fantasies) from generating the effects they have the power under certain circumstances to generate. Psychoanalysis sets itself to explain how these negative causal factors come into being, how they achieve their effects, and how one is to account for the variability of their occurrence, and the variability in what it is upon which they act, in the lives of different people and in the life of any one person.

Certainly, psychoanalysis uses these many kinds of variation in its explanations of causal gaps in mental life. But that all these kinds of variation can themselves be explained using the distinctive conceptual apparatus of psychoanalysis alone is, I think, unlikely.

History and Continuity: Infantile Sexuality

Psychoanalysis extends what it takes to be the conventional view of what is sexual—that it occurs in adolescence and adulthood. Psychoanalysis observes the sexual life of infancy and childhood as well. It shows that the same object-choices and activities that occur in the undeniably sexual perversions of adulthood have their counterparts in infancy and childhood, and are associated there too with the quest for sexual pleasure.

Psychoanalysis also emphasizes the continuity between the kinds of perverse object-choices and activities of infancy and childhood, on the one hand, and conventional sexuality in adolescence and adulthood, on the other, by demonstrating in clinical studies how one kind of aim or object-choice covaries with another. Frustration in one direction (e.g., of genital union) may be regularly followed by a quest characteristic of an earlier period of life (e.g., an upsurge of anal-sadism); and excitement of one kind (e.g., arousal by a heterosexual object) may regularly be followed by gratification of another kind (e.g., with a homosexual object).

The psychoanalytic theory of psychosexual stages is, I believe, only marginally a theory of development. Its major importance is in providing a classification of organizations of preferences for object and aim, which appear as themes in unconscious sexual fantasies, and in emphasizing that unconscious sexual fantasies as mental dispositions have a transformational (which may coincide with a temporal) history.

Unconscious sexual fantasies as mental dispositions are the proximal causes in current mental life with which psychoanalysis is distinctively concerned. The emphasis on infantile sexuality in psychoanalysis may be in part an expression of the desire to trace the causal history, to discover the origins, of these mental dispositions, which, although they have their primary role in psychoanalytic theory as causes, are in this context regarded as effects of other causes. What it is about a particular mental disposition that is responsible for its having the causal power it has, and for its bringing about the specific effects it brings about, surely comes into being through its history.

Psychoanalysis, in order to carry out its particular explanatory task, must discover just what it is that makes a fantasy qua fantasy causally efficacious, just what its causally relevant characteristics are. A fantasy's contents are among its causally relevant characteristics. Psychoanalysis, in order to carry out its explanatory task, must also discover what transformations the contents of a fantasy undergo in the various editions or versions of the fantasy. But psychoanalysis may not always be able to reconstruct just what causal influences over time have given rise to particular contents.

What psychoanalysis reconstructs from evidence obtained for the most part but not exclusively in the psychoanalytic situation are the sexual fantasies of infancy and childhood, which are thought to be constructed from experiential and constitutional influences (about the latter, psychoanalysis does not have much to offer to a theory of sexuality), or to be later versions of universal primal fantasies (a phylogenetic heritage). Recourse to universal primal fantasies is motivated by the wish to eliminate the need for the languages of other sciences to explicate the nature of ultimate causes in psychoanalysis.[6] My own belief is that psychoanalysis should restrain its passion for speculating about origins, lest this passion lead it into an infinite regress, to a mythology or, at the least, to the postulation of causal factors about which, given its theoretical apparatus, it can have little to say.

Psychoanalysis may have something to say about the sexual fantasies of infancy and childhood. Perhaps transformed, in revised versions or editions, these become the unconscious sexual fantasies, the mental dispositions, that have the status of causes in psychoanalysis. The sexual fantasies of infancy and childhood characteristically center on parts of the body that at different times of life are major sites of experience and so dominate the mind of the child. The form and content of these fantasies are influenced in various ways by experiences of too much or too little excitement; experiences of too much or too little of a particular kind of sexual pleasure; or experiences of sexual pleasure that are too early or too late. The form and content of these fantasies are influenced in various ways by the sexual researches of childhood, the attempts to solve puzzles and mysteries—notably, where babies come from, and the differences between the sexes—by constructing "Just-So" stories in imagination. The form and content of these fantasies are influenced in various ways by attempts to respond to incest taboos, to variations on oedipal themes, as well as relations with siblings, afforded by experiences in a particular family. The form and content of these fantasies are influenced in various ways by retrospective reflection on them and revisions of them in the light of subsequent experiences and imaginings.

Fantasy, of course, is not a mere imitation of experience. In every case, what the imagination makes of experience will depend on the way in which suspension of considerations of ordinary belief, the concept of mind as spatial or corporeal, condensation, displacement, iconic symbolization, and the requirements of visualization enter into imagination. In every case, what the imagination makes of experience will depend on the way in which the child's sexual desires determine what he or she perceives and remembers, therefore what he or she believes, and what construction the child places on what is perceived and remembered.

Sexual fantasies may be placed on a continuum from relatively enduring mental dispositions, relatively inaccessible to conscious-

6. See, for example, Laplanche and Pontalis (1968).

ness, to occurrent fully conscious daydreams. The latter are taken, along with dreams, symptoms, enactments, parapraxes, inexplicable moods and mental states, to be effects of the former. Unconscious sexual fantasies themselves may be regarded—rather than as single fantasies—as groups of fantasies related to one another by transformations. These transformations may be carried out through time as the fantasizer has new experiences and develops new capacities, desires, beliefs, guilts, shames, and anxieties—or may be carried out currently in response to the exigencies of defense. The result is the existence of different revisions, versions, or editions of an unconscious sexual fantasy.

These versions are regarded as "layered," in the sense that one is more superficial or deep, that is, more or less accessible to consciousness, than another, or in the sense that one is a defensive response to another. The conception of psychosexual stages is in part at least an attempt to account for this layering, construing, for example, some version of a fantasy as a temporally earlier version, and relative inaccessibility to consciousness as an expression of relative temporal remoteness in origin.

Pathology

Is there any reason for psychoanalysis to be concerned with the question of the extent to which a sexual deviation or aberration (aberrant, given some conventional view of sexuality) must also be considered to be psychopathological—other than pragmatic interest in whether suggesting that treatment is required for it is justified? Doesn't the effort to distinguish between what is normal and what is not immediately plunge us into conjectures that cannot be supported by evidence obtained by psychoanalytic (as distinct from statistical) methods of inquiry?

Certainly, we want to exclude from consideration here, as outside the domain of psychoanalysis, those cases in which sexual aberration is caused primarily by the existence of serious incapacities in other psychological functions, arising from genetic, anatomical, or physiological defects, or profound environmental deficit. Certainly, not all cases of any kind of sexual aberration can be explained in this way.

On the other hand, we have no difficulty with cases that are assimilable to the class of inexplicable causal gaps psychoanalysis takes as what it intends to explain. Here, being sexually aberrant makes no sense to a person in the light of that person's other wishes, beliefs, values, standards (i.e., a choice of object or aim is felt to be alien—"I can't figure out why, given what else I know about myself, I should want to do this kind of thing or prefer this kind of object; it seems inconsistent with what I know about myself.") But, of course, not all cases of any kind of sexual aberration can be so described. (And what if a person rationalizes his or her aberrant sexual preferences so that they do make sense to him or her?)

Freud is clearly of the opinion that not every deviation or aberration, not every perverse tendency, is in itself to be regarded as psychopathological. He is reluctant to regard every deviation from the normal as pathological, since such deviations may occur as part of the sexual life of adults who are normal in their primary sexual preferences. Furthermore, he observes, even those for whom deviations substitute for what is considered normal in sexual life, or serve as fixed conditions for carrying out the normal sexual act, may in every other way be functioning satisfactorily and nondeviantly. He repeats frequently that perversion is not necessarily pathological. He goes so far as to distinguish from perversion (deviation with respect to aim) a particular deviation with respect to object, homosexuality, which he labels "inversion" or contrary sexuality, because he regards homosexuality as a universal disposition in sexual life, and not incompatible with health (but then adopts inconsistently another attitude when he treats such deviations with respect to object as sex with animals or children).

Freud does attempt to define markers for what might be regarded as psychopathological. He chooses two criteria. One criterion has to do with the determination that a type of aberration is pathological, given its contents. Some aberrations—for example, having intercourse with a dead body—are too horrible or repugnant to be anything but psychopathological. The other criterion has to do with the determination that a particular instance of a type of aberration is pathological, given that it is characterized—as not all instances of this type of aberration are—by fixation and exclusivity.

The first criterion is unsatisfactory because it relativizes the definition of pathology to the values, attitudes, norms, and practices characteristic of a particular society. The second criterion, exclusivity, is unsatisfactory because it has the unwelcome result, for example, of making bisexuality healthier than exclusive homosexuality. Fixation in the developmental sense substitutes an evaluative standard apparently developmentally defined ("mature genitality") for one socially defined ("straight," "normal"), and it is just such a normative attitude toward sexuality that Freud is subverting.

Here I note a tendency of psychoanalysis to make its conception of psychosexual development the basis of a new evaluative standard of what is normal. This seems to be a tendency characteristic of every psychological theory of development. Piaget's theory of cognitive development, for example, is used in America to decide what kind of thinking—among different kinds of mental operations—is most mature, and therefore superior or better (not just different, or better for some particular purpose). This tendency to regard the end of a line of development as an ideal may have its origin in a utopian evolutionism—that is, in the idea that any result is evidence of the result's adaptive superiority. Sports and jerry-rigged solutions in nature teach us rather that there is no Engineer Who has contrived ideal outcomes, and the evidence from evolutionary theory is that many results are "side-effects" or correlates of adaptive solutions that are just dragged along and may have in themselves no adaptive significance.

For some psychoanalysts, the existence of sexual aberration is prima facie evidence of a failure to achieve ideal (adaptively optimal?) outcomes, a failure, for example, to resolve the oedipal complex, to accept the sexual role (including the valuation of parenting) of the parent of the same sex, to become free of castration anxiety or penis envy, or to achieve mature genitality. Since such failure is held to be pathogenic, a (sufficient?) cause of neurosis or character disorder, sexual aberration must be therefore inevitably associated with neurotic or character disorder.[7] (Freud, of course, appears on the contrary to have believed that such failure is a necessary but not sufficient cause of psychopathology.)

Similarly, sexual aberration may be regarded as prima facie evidence for an impairment of cognition. Its roots are held to lie in denial of the differences between the genitals of the sexes and of the distinction between different generations.[8] The existence of the genitals of the opposite sex is denied. The undervalued or overvalued genitals of one or the other sex are regarded with horror, disgust, shame, envy, or anxiety. Limitations with respect to choice of sexual objects dictated by differences between generations or by blood-ties go unrecognized—and the necessity of renouncing incestuous objects is rejected. These pathogenic factors are, of course, those postulated for any psychopathology; they cannot therefore be held to determine sexual aberration in particular.

Within an evolutionary conceptual framework, psychoanalysis is inclined then to consider pathology in sexual life as the result of fixation and regression, a regression that goes unopposed, or an inhibition or failure of development. Inevitably it seems, such a framework carries with it the suggestion that some forms of gratification are higher and more mature and others lower and more immature.

7. See, for example, Kernberg (1986).
8. See, for example, Chasseguet-Smirgel (1984).

Here again perhaps psychoanalysis shows a tendency to go beyond its own domain and to exceed the grasp of its conceptual apparatus. Sexual aberration or deviation insofar as it is a departure from what is conventionally regarded as normal or desirable is neither necessarily sick or evil. It is first of all social deviance. There is no society that does not monitor or regulate attempts to satisfy desire through action. The role-prescriptions of every society include prescriptions for sexual behavior consistent with the value system and the norms derived from these values that, shared by its members, constitute that society. Even the expression of aberrant sexuality (when, by whom, with whom, what specific acts) is prescribed.[9] Social deviance is at the least a miscarriage of processes of socialization. Its explanation requires first of all the conceptual apparatus of sociology (although psychoanalysis may make a contribution to the understanding of why a particular person remains unsocialized and perhaps in this one area of life only).

A diagnosis of psychopathology within the boundaries of the domain of psychoanalysis cannot be based, as it is in psychiatry, on a type of manifest behavior, or a cluster of manifest traits. Nor can such a diagnosis be made prior to an investigation of the mental life of a person, by relatively casual inspection, so to speak. (How much knowledge of a person's unconscious mental life would one have to have to diagnose sexual aberration as pathological?)

What do I conclude, then, about sexual perversion and psychopathology? The following set of circumstances might justify a diagnosis that a sexual aberration is psychopathological. If a sexual aberration, a choice of aim or object, is primarily, rather than a quest for pleasure, a technique for mitigating or avoiding any one of a number of kinds of anxieties, and as such is merely a substitution for another kind of pleasure, which otherwise—if not for anxiety—would have been preferred or sought, one might consider it, because it involves incapacity, psychopathological. A particular instance of sexual aberration may be then primarily a matter of security rather than pleasure, a technique of anxiety avoidance or mitigation, or more precisely a technique for achieving gratification under conditions of (internal) threat. The threat must be internal and by "internal" I do not mean simply a realistic anticipatory assessment of the likelihood of social disapproval and sanctions. (In addition, the pleasure that is associated with the gratification of perverse tendencies, like any pleasure, may be used to distract a subject from, to muffle, or to soothe over, dysphoric affects.)

9. See, for example, Liebert (1986).

Inability to carry out the aberrant act or to carry out an act with the aberrant object, no matter what the nature of the obstacle, produces anxiety rather than the frustration one might expect from a failure to satisfy desire. To carry out such an act or to carry out an act with such an object will therefore be experienced as obligatory, driven, or compelled—rather than a matter of freely chosen recreation or a form of play. "This does not seem to have to do with what I want, and no reflection on undesirable consequences affects its power over me." Strenuous efforts to override, mitigate, get rid of, or inhibit the development of, intense feelings of horror, disgust, shame, or guilt about what one finds exciting or pleasurable—and the costs of such efforts—are likely to be evident in such a case.

The person suffering from this kind of sexual aberration would prefer and would choose another kind of sexual pleasure if the way to that kind of pleasure were not blocked by anxiety. The preference for the sexual aberration is at the cost of something else of more value to the person, whether he or she is aware of this or not. What may be of more value is another kind, a nondeviant kind, of pleasure. Observations in a particular case that nondeviant desire—excitement aroused by a nondeviant object or an exciting fantasy involving a nondeviant sexual aim—regularly precedes attempts to gain pleasure deviantly provide evidence that such is the state of affairs. What may be of value are highly valued goals, with which a sexual aberration is not consonant; it interferes seriously with their achievement. (What if a person is unaware of any connection between his sexual proclivities and his failure to achieve important goals?)

This way of dealing with the question "When is sexual aberration psychopathological?" has the following satisfactory consequences.

a) Heterosexuality as well as sexual aberration may be considered psychopathological if in an individual case it is primarily a technique of anxiety avoidance or mitigation. In other words, the classification of what is psychopathological will cut across rather than duplicate the classification of what is socially deviant.

b) This way of dealing with the question "When is sexual aberration psychopathological?" leaves open whether in a particular case the sexual aberration is merely social deviance or is rather to be regarded as psychopathological. That is, not all sexual aberration is automatically to be regarded as psychopathological, which is in accord with the observation that sexual aberration may exist in persons who are not only like nondeviant members of their society with respect to the adequacy of their psychological functioning and the value of their achievements but who may not show any other manifestation of social deviance.

c) This way of dealing with the question "When is sexual aberration psychopathological?" is consistent with psychoanalytic views about psychopathology in general—that is, about forms of psychopathology that do not manifest themselves as symptomatic sexual aberration.

d) This way of dealing with the question "When is sexual aberration psychopathological?" is consistent with the view that any instance of seuxal deviation may arise as the result of any number of causal factors, or as the result of different combinations of causal factors. These causal factors are not necessarily social or psychological. Even when the factors are psychological, the scenario or causal story or explanation of an instance of a particular kind of sexual aberration may differ from case to case, although to be sure nomothetic themes involving castration anxiety, penis envy, and various kinds of failure to resolve the oedipus complex may predominate.

Freud comments in this connection that sexuality is the weak point with respect to socialization and psychological functioning. An individual may be able to meet all other demands placed upon him and still be unable to regulate his sexual desires and their expression in accordance with his own values and standards. This gives another meaning, besides that the neurotic symptom is an expression of the quest for sexual pleasure, to Freud's conjecture that while there can be disorder in the sexual life without neurosis, there can be no neurosis without disorder in the sexual life. If a person is having difficulties managing other aspects of life, he is certainly going to be having troubles with sexuality as well. "Disorder in the sexual life" may connote the existence of a perverse tendency, which blocks the way to another kind of pleasure, because the imagining of this other kind of pleasure arouses anxiety. Imagining the perverse kind of pleasure mitigates or eliminates that anxiety. The perverse tendency therefore persists despite any shame, guilt, or anxiety that it itself arouses.

The diagnosis of sexual aberration as itself psychopathological must be sharply distinguished from the attempt to explain psychological dysfunction or incapacity as the result of the vicissitudes of sexuality, which will of course include the fate of perverse tendencies. Here, psychoanalysis calls attention to the way in which vicissitudes of sexual desire may render a person unable to work or play; to enter into mutually satisfactory personal relations; or to carry out psychological functions in nonsexual realms. Such vicissitudes may impair a psychological capacity such as perception, memory, cognition, language, emotion, or volition, or interfere with bodily functions. Here, psychoanalysis is distinctively concerned to discover the mechanisms by which the sexual life of a person (e.g., the sexual preferences of, or what sexually excites, a person) can come to influence mental functioning or mental states that the person does not consider have anything to do with sex. In the realm of nonsexual mental functioning or mental states, psychoanalysis includes psychoneurotic phenomena, and also nonpsychoneurotic phenomena such as choice of occupation, vicissitudes of friendship, preferences for recreational activities, choice of a place to live, failures and achievements. Justifying what is included in this realm and determining more rigorously what are the boundaries of this realm are significant parts of the research program psychoanalytic theory sets for psychoanalysis.

Unconscious Sexual Fantasy as a Causal Mental Disposition

Dispositions as Causes

All the facts about sexuality on which psychoanalysis focuses, especially the varieties of sexual experience, the efflorescence of combinations of object and aim, activity and passivity, as well as the multiplicity of causal scenarios that can lead to any one kind of outcome, suggest what kind of causal mental factor must be operating. It is imagination that can account for such variety. Sexual wishes produce their effects through the mediation of unconscious sexual fantasies, which are complex narratives, mixtures of wish-fulfillments and responses to wish-fulfillments, often wish-fulfillments achieved under conditions of (imagined) threat.

These fantasies are also mental dispositions, which really exist "in the mind" or, if you prefer, as properties of the mind. I am not talking about a figure of speech, a convenient conceptual fiction, or a placeholder for some future real explanatory entity. As dispositions, unconscious sexual fantasies may exist at any point in time without observable manifestations, as a virus may coexist with its host for long periods without manifesting its existence in the form of symptomatic illness, or as a stick of dynamite may exist for long periods of time without exploding. Like the virus and the stick of dynamite, an unconscious sexual fantasy by virtue of its causally relevant properties has causal powers. Under certain conditions these powers will be evoked, resulting in manifestations that are effects of the fantasy and are linked to it by specifiable steps, operations, or mechanisms.

The causal role of sexual desire is mediated by the sexual wish-fulfillments and fantasies sexual desires or wishes produce through imagination. That is, such sexual wish-fulfillments and fantasies issue forth from the relation between desire and imagination, and come themselves to constitute mental dispositions, having protean manifestations. These manifestations are evoked: periodically and spontaneously (the sexual fantasy has a characteristic intrinsic propensity to manifest itself, which is independent of situational vicissitudes); by situational objects or events, which might be said to "remind" the person of the fantasy and therefore excite him or her; or by the willed mental activity of the subject (the person wants and intends to activate the fantasy, in order to experience excitement or desire and its satisfaction in a wish-fulfillment). If sexuality achieves its effects through the mental dispositions—notably, the unconscious sexual fantasies—it creates, then its nature or structure, what determines how its causal powers will be expressed in mental life, is given by what it is like to have such fantasies and by the contents of these fantasies.

We are guided here by a generative, powers, or propensity account of causality (Harre 1970, 1972), as some of the work and comments of Robert Shope (1970, and personal communication), for example, also suggest.[10] In this conception of causality, things have causal powers, by virtue of their nature and structure, that can be evoked in suitable circumstances. Things are active, not passive. They have the power to bring about effects, and respond to influences and stimuli in the way they do because they are constituted the way they are. They possess causal powers by virtue of their internal constituents and the relations among these. When some change of condition occurs, changes in internal constituents or in the relations among them set into motion causal mechanisms or processes that produce the observable effects and indeed the actual chain of links between cause and effect. A structure is implicated in a sequence of events, traverses a number of steps, undergoes a series of transformations. In other words, there is a real connection—a causal process or mechanism—existing between, and linking, cause and effect. A cause is not a thing independent of its effects. It cannot be described without referring to the effects it is capable of generating or producing. That capability is part of its nature.

Investigating the nature of sexuality as a putative cause is similar not to the formulation of the laws of mechanics but to the investigation of a virus as a putative cause. One wishes to demonstrate that sexual fantasies exist; what their constituents (contents) are; how these are structured, organized, or related to one another; and how the nature of a sexual fantasy so defined evokes causal mechanisms (primary pro-

10. See, especially, Chapter 15 of this book.

cess, defense mechanisms, dream work)—and, in any particular instance, which causal mechanisms—to produce its effects.

The Central Theoretical Problem for Psychoanalysis

This is *the* central question psychoanalytic theory must answer for its causal claims to be taken seriously. How is an unconscious sexual fantasy causally efficacious? How is it causally linked to its effects? By what mechanisms do we start with an unconscious sexual fantasy and reach a putative effect?

Two major stories are offered as solutions or partial solutions to this problem. The first story has tended to depict fantasies as appearing rather "late" on the scene, expressing ego-id-superego conflicts. Various versions or editions of these fantasies, which vary in the degree to which they are inaccessible to consciousness, are constructed by both primary and secondary processes. The second story about the causal powers of fantasies has tended to depict fantasy as appearing rather "early" on the scene, perhaps before the differentiation of ego, id, and superego, and having its origin in wish-fulfillments. I cannot tell whether or not the way in which fantasy is depicted is necessarily tied to one of the stories told about the causal powers of unconscious fantasies. The two stories about causal powers do not seem to me to contradict each other; and it seems to me that one could believe both stories without inconsistency.

1. *The epistemological story.*[11] An unconscious sexual fantasy may produce its effects by virtue of its power to influence what a person perceives, remembers, notices, or pays attention to. Perceptions and memories are picked out by it, in part at least because of similarity relations (here is where metaphor comes in), and assimilated to it. An unconscious sexual fantasy has an epistemological effect; it serves as a model by virtue of which reality is interpreted or known. "What I am experiencing or perceiving is just like something in my fantasy. I suppose something of the same sort is happening here."

Unconscious sexual fantasies influence what is perceived, how it is interpreted, what is remembered. Perceptions and memories have the power to fix belief. Beliefs have the power to influence action and arouse feeling.

If a person is unconscious of a fantasy evoked on an occastion, he may experience on that occasion an inexplicable mood, emotion, or mental state that can only make sense in the context of the fantasy. Such moods, emotions, or mental states have the power to color and

11. See, for example, Arlow (1969a, 1969b).

determine the interpretation of current experience. Similarly, relationships between persons may be governed by the way in which their interpretations of each other's words and actions are influenced by shared or reciprocal fantasies.

2. *The relations-to-internal-objects story.*[12] Here, I paraphrase Wollheim's account.

Sexual desires, like other desires, most notably desires that cannot by their very nature be gratified in reality, may be represented as fulfilled by effectively imagining them as fulfilled. Effectively imagining them as fulfilled means having the belief that the desire is being satisfied and having at least some of the pleasure that would result from gratifying the desire in reality. Imagination is able to bring about these two conditions under certain favorable circumstances.

First, truth-directed considerations of belief are suspended. They are not allowed to exert influence.

Second, imagination under these circumstances operates under or is governed by a spatial or corporeal concept of mind, so that mental operations and the contents of mental states are believed to have the same efficacy physical processes and objects in a real space are believed to have.

Third, the person having the fantasy is engrossed in the imaginative activity. The fact that he or she is merely imagining is suppressed. This suppression is assisted when imaginative activity occurs under a spatial or corporeal concept of the mind. Then, the mind is believed to have exaggerated powers. Imagining swallowing an object results in the belief that the object, with all the properties (beliefs, attitudes, feelings, motives) attributed to it, is now somewhere inside the mind, where it can manifest its causal powers. Here is the source of "overvaluation of psychic phenomena," which derives from "magical belief" in the "omnipotence of thoughts." It is important to distinguish between reality-accommodated fantasies about, for example, possible future courses of action and their consequences; fantasies about external objects (all reality-accommodated fantasies will be about external objects); and fantasies about internalized objects.

Fourth, if the unconscious fantasy is like a drama on a stage (or, as I would say, on a movie screen), then an internal audience is a constituent of the imagining. If the internal audience is empathically identified with the central person in the drama, and that person is excited and then experiences pleasure, the internal audience will be excited and then experience pleasure. If the internal audience is favorably disposed, and therefore sympathetic, to the central person,

12. See, for example, Wollheim (1969, 1971, 1974a, 1974b, 1974c, 1979, 1982, 1984).

and the central person experiences pleasure, then so will the internal audience. In either case, the result of the imaginative activity is pleasure.

Fifth, as the result of identification with an internalized object (that is, an object imagined to have been orally incorporated), the person who fantasies about the doings of an internalized object comes to imagine that he or she possesses the beliefs and desires he or she supposes the internal object to have.

Sixth, one may reflect upon and react to one's own fantasies with shame, guilt, disgust, or anxiety. Ordinarily, such affects will be isolated or displaced from the fantasy context and attached to conscious experience; so it will determine interpretation of that experience. (Here is where story one joins story two.)

I would add that shame, guilt, disgust, or anxiety may also be what the central person is feeling because other persons in the drama, who perhaps are required to be in it if the wish is to be gratified, are responding, so it is imagined, unfavorably to the central person's doings. The internal audience, empathically identified with the central person, will also feel shame, guilt, disgust, or anxiety.

I would add further that anxiety, guilt, shame, disgust motivate revisions of the fantasy, saving whatever of wish-fulfillment can be saved and still permitting mitigation or avoidance of anxiety, guilt, shame, disgust. This clearly involves rewriting the screenplay or changing the way the actors are directed to play their parts. Transformations leading to new editions of the fantasy include displacement, condensation, the use of iconic symbolization, translation of concepts into nonverbal and therefore inevitably ambiguous imagery.

A group of fantasies is constituted by transforms of one "deepest" central fantasy, which is either most inaccessible to consciousness or earliest in development, or both. The task for psychoanalysis is to explicate with increasing clarity and precision the way in which such groups of fantasies influence psychological functions such as perception and memory and through them cognition, affect, and action; the way in which empathic and other kinds of identification, and especially identification with internal objects, come about; and the way in which changes in various circumstances (the degree of engrossment, the extent to which the concept of mind governing imagining is spatial and corporeal) influence what imagination produces and therefore the causal powers of its products.

Important questions remain. These questions suggest with what problems a distinctively psychoanalytic research program might concern itself, and they may also require perhaps the assistance of philosophers of mind to answer. What endows imagination itself, as a psychological function, with causal powers? What is its causal role vis-à-vis other psychological functions? How is it affected by various kinds of mental states? How does it affect various kinds of mental states?

The phenomenology of a mental state—what kind of mental state it is, what it is like to be in that kind of mental state, and the contents of that mental state—produce its effects. Unconscious fantasies are especially suitable to be fundamental causes in psychoanalysis, for then, like all mental states in our commonsense intentional psychology, such as beliefs and desires, they possess their causal powers by virtue of their phenomenology (Wollheim, previous citations). Which effects of imagination depend upon the fact that it is imagining that is going on? Which effects of imagination depend upon the fact that a mental representation (or sequence of mental representations), the object to which imagining is directed, has just the contents it has, symbolizes just the state of affairs or object(s) it represents? To what extent do contents themselves, surviving through transformations, supply the causal links between cause and effect?

What causal powers of a wish or desire lead to attempts to find fulfillment in effective imagining (attempts at wish-fulfillment) rather than reality-accommodated action? What features of the final observable outcome—for example, a symptom, a dream, in general, an inexplicable mental state, better described as an *attempt* at wish-fulfillment than as a wish-fulfillment—are the manifestations of such attempts? What features are "side-effects"—not wish-fulfillments—of successfully carrying through such attempts? What features—not wish-fulfillments—are the effects of negative causes, which act to prevent these attempts from succeeding?

Finally, what is the difference in causal roles between mental dispositions and occurrent mental states?

However one assesses Wollheim's account, this last question is an important one for psychoanalytic theory. Wollheim suggests, as I understand him, that mental dispositions can only play a causal role by virtue of the occurrent mental states that are their manifestations. These occurrent mental states are what have a phenomenology and it is this phenomenology—what it is like to be in such a mental state and what the contents of the mental state are—that determines the causal powers of occurrent mental states. Occurrent mental states, which are caused by mental dispositions, mental activity (attention, e.g.), or other mental states, in turn cause mental states, cause new mental dispositions, and strengthen existing mental dispositions.

Assessment of Psychoanalytic Contributions to a Theory of Sexuality

The Kind of Theory It Is

Psychoanalysis is concerned with explanation rather than prediction. It is concerned with showing how certain nomothetic themes are worked out in individual cases. In doing so, it is guided by the assumption that different causal scenarios may be responsible for the same kind of observable outcome in different cases.

It is not concerned with the formulation of general laws, which are without exception true. The "other things being equal" clause cannot be made to work by coming closer and closer to realizing it in ideal experimental situations. "Other things being equal" will in general not be applicable. In any particular case, other things are not equal, because a number of ad hoc, nonlawful (unpredictable) events will intervene, affecting the outcome of any particular nomothetic theme or combination of such themes. Some of these events will be the expressions of causal factors for which psychoanalysis has no distinctive theoretical language; other sciences must explain the effects of these factors. Some of these events will be the expressions of causal factors that psychoanalysis can demonstrate may interfere with the manifestation of the causal powers possessed by the particular theoretical entities it postulates.

The psychoanalytic theory of sexuality includes not only substantive causal conjectures, inferences, or explanations, the scientific credibility of which must be assessed, but includes as well a set of proposals about the ways to proceed if one wishes to find out something about sexuality. The theory offers something perhaps more important than its substantive causal claims. It offers conceptual tools for conducting an inquiry into the nature of sexuality and, in particular, some of the causes and effects of what we call "sexual." The distinctions between source, object, and aim, and between activity and passivity are examples.

In assessing the psychoanalytic theory of sexuality, it is important to distinguish among: (1) description and theoretical concepts; (2) the questions that are raised, and answered, using these concepts; and (3) the answers to these questions, answers in the form of causal conjectures, inferences, or explanations.

The concepts are freely chosen. They represent hunches about ways of cutting nature at its joints. They represent decisions about what things or events shall be grouped together and called by the same name.

These concepts thereby determine what shall be considered a fact, how a fact shall be described, and what shall count as a critical or central and what as a peripheral fact. They draw attention to the possible explanatory status of certain groups of things or events; they suggest what factors should be given causal priority in attempts to account for or explain the properties of such groups of things or events.

The standard by which concepts are assessed is usefulness, not truth. Do they lead to a program of inquiry? Does the program have momentum, a progressive forward propulsion? In other words, do the concepts inspire and enable the investigator to ask interesting, novel questions, the answers to which in turn lead to formulating still other novel and interesting questions?

Truth enters in when, as part of our assessment of the psychoanalytic theory of sexuality, we ask about the answers to the questions, "Do these causal conjectures, inferences, and explanations stand up to empirical tests of their credibility?" The status of such propositions with regard to their scientific credibility is inevitably provisional. This is so because that status may change—as some new way of explaining the facts appears, and some new verdict is rendered about which explanation (among rival explanations) is now favored by an always growing body of evidence.

Two Problems

Psychoanalysis seems to face two different problems.

The first has to do with documenting facts about sexuality, and thereby demonstrating the existence and nature of sexuality, including the mechanisms it evokes that are capable of linking it causally to its putative apparently nonsexual effects. The second has to do with providing evidence to support the claim that, in fact, sexuality is causally connected to these apparently nonsexual effects.

Here we distinguish between, on the one hand, demonstrating the existence and nature of infantile sexuality or of unconscious fantasies, for example, and, on the other hand, demonstrating that infantile sexuality or unconscious fantasies actually play the causal role claimed for them.

It is possible to accept that infantile sexuality or unconscious fantasies exist and still to question that they cause the particular effects psychoanalysis claims they do. Adolf Grunbaum (1984, 1986) provides a case in point. If I understand him correctly, he is willing to concede the empirical adequacy of the demonstration that unconscious conflicts exist but not the empirical adequacy of the demonstration that they have the causal status psychoanalysis assigns them. However, as I will go to some pains to argue in Chapter

15, demonstrating empirically that a cause exists, that it has a particular structure to which it owes its power to produce certain effects, and that the mechanisms exist that are capable of linking this cause to its putative effects may in itself be part of making an argument for the claim that the cause in fact does produce its putative effects.

Controversy: The Quest for Pleasure vs. the Quest for the Object

Freud thought that whether the instinct (the quest for sexual pleasure itself) was idealized (the object not much mattering), or the object was idealized (the instinct here tending to be regarded as somewhat disreputable), depended on the culture of and its values. Currently, it is object-relations theories that are popular in psychoanalysis.[13] They tend to be opposed to emphasizing the vicissitudes of perverse sexual tendencies, the existence and causal role of childhood sexuality, and the distinction between source, object, and aim, and between activity and passivity, as especially causally relevant in the domain of psychoanalysis.

Does this state of affairs signal the opposition of rival paradigms, which will ultimately be assessed as having different degrees of success in explaining the phenomena of interest to psychoanalysis? Do object-relations theories involve rather a redefinition of just what phenomena are of interest to psychoanalysis? Do we have here simply two notational variants, which, although they use different words, will end up having the same empirical consequences and the same range of explanatory achievements? I don't know.

But I have the following thoughts, which are related to these questions.

1. The preoccupation of object-relations theories with pre-oedipal and presumably largely nonverbal experiences—and especially with the *actual* wordless relation between mother and infant—as especially causally relevant seems to me a radical departure from an emphasis on psychic reality, the importance of verbal symbolizations in constructing psychic reality, and the primacy of fantasy and its causal powers in psychoanalytic explanations. Even when fantasy is important in an object-relations theory, there is an inevitable shift away from fantasy in favor of the actual vicissitudes of interpersonal relations, which results from assigning causal priority to the actual experiences with the mother. Even when fantasy is said to be important, it turns out that the qualities of the actual relationship with the mother determine finally what the effects of fantasy will be and in particular whether their potential pathogenicity will be mitigated or

13. See, for a review of such theories, Greenberg and Mitchell (1983).

not. The inevitable slide away from the mind's workings to interpersonal interactions directly contradicts, as far as I am concerned, what is most distinctive about psychoanalysis.

Given the particular optimism, Social Darwinism, and therapeutic zeal that were added to psychoanalysis when it crossed the sea to America, it is not difficult to understand at least some part of the enthusiasm with which psychoanalysts in their role as psychoanalytic therapists feel for object-relations theory. For if the quality of the actual mothering received, rather than unruly sexual desires (the last stubborn preserve of the uncivilized in us), is given causal priority, then bringing about change through a corrective emotional experience with the psychoanalytic therapist, or more widely by changing the way that mothers mother, loom as exciting possibilities.

2. I note the absence of any serious reference to anality and its vicissitudes in a writer like Fairbairn. If it is typical, it represents a serious lacuna in such theories as his. In attempting to explain what I try as a psychoanalyst to explain, I would not like to have to do without anality, anal pleasures, anal themes, anal-sadism, and anal theories, preoccupations, and fantasies. Perhaps the problem for such theorists is that there is no person as object in the anal scenario (for them, interest in a nonhuman object for pleasure is simply a degradation product of defects in real relationships), although of course the mother is around when bowel training is at issue. There are similar lacunae when it comes to phallic exhibitionism, as expressed also in urination, and its relation to shame. And so on. The oral themes are repeated with unvarying relish and often utilizing quite stereotyped formulas. It is difficult to introduce much causal variety or individual nuance in an account of the way nomothetic themes have worked out in a particular person's life when one is limited to commenting on supposed events, inner or outer, in preverbal mental life; elaborating on oral themes; or empathizing with an analysand's feelings about current events.

3. I am uneasy that the analysand's theory about his difficulties—my parents did this to me, it would all have been different if I had had ideal mothering or fathering—sounds so much like that of the object-relations theorist (or vice versa). I suspect that the neglect of sexuality is associated with taking what the analysand says at face value, an attitude that is at odds with psychoanalysis as a depth psychology.

So also if the analysand talks about problems with intimacy, lack of a sense of identity, or dissatisfactions with self, the phenomena referred to are taken to be explanatory (transparent presentations of causes) rather than what is to be explained. In my experience it is difficult indeed for an analysand to take seriously, attend to, and report even relatively conscious derivatives of unconscious fantasies (daydreams, masturbation fantasies). Taking these references to intimacy, identity, and self at face value only supports this particular reluctance on the part of the analysand.

Of course, I do not comment here on what changes in technique and in the nature of the dialogue might be called for, when environmental deficiencies and trauma are essential causes of a patient's difficulties, and impairments of capacity rather than causal gaps in mental states are characteristic of the pathology presented. I certainly do not mean to imply that preoedipal experience is not clinically or theoretically important in development. I do mean to question whether it is part of the core of psychoanalysis. Are psychoanalytic method and process capable of investigating or shedding light upon it?

4. There are, however, some interesting (technical? theoretical?) problems lying around here. Ferenczi (1909) comments that it is not the idea introduced by the hypnotist that is directly causal. The idea does not go into the subject's mind where it does the causal work. Rather, the role of the idea is in evoking a mother-complex or a father-complex (what I have been calling a mental disposition); the disposition does the causal work. It is the complex that is evoked that determines whether or not the voice of the hypnotist is heard as the soothing crooning voice of a mother or the intimidating commanding voice of a father and should be responded to accordingly.

Yet do the actual qualities of the hypnotist's voice not have something to do with which complex is evoked by it? It is a neat trick sometimes to decide whether the actual qualities of the psychoanalyst's voice evoke some particular disposition in the analysand, which then manifests itself; or the analysand's disposition determines which qualities of the psychoanalyst's voice will register; or the analysand's disposition manifests itself in ways that evoke a complementary or reciprocal unconscious fantasy in the psychoanalyst, which is manifested in part by an alteration in the qualities of the psychoanalyst's voice.

Five Nomothetic Themes

The following five nomothetic themes are distinctively psychoanalytic. The fate of psychoanalysis as a body of knowledge will depend, I believe, to a large extent on how successful it is in providing empirical evidence for those of its causal explanations that make use of these themes.

1. Causal gaps, inexplicable mental states, in both sexual and non-

sexual areas of life, are manifestations of the causal powers of unconscious sexual fantasies.

I believe that a major theoretical conjecture, distinctively psychoanalytic, is that unconscious sexual fantasies are manifestations of the causal powers of the imagination. Under favorable conditions, the imagination will respond to desires or wishes by attempts at effective imagining in its most extreme form—attempts, in other words, at wish-fulfillment. Other causal powers oppose these attempts, more or less successfully. The unconscious sexual fantasy and its transforms are the effects of an attempt at wish-fulfillment, which negative causes have more or less interfered with. A causal gap, or inexplicable mental state, is a manifestation of an unconscious sexual fantasy.

2. Neurosis is one outcome of attempts to gratify perverse sexual wishes in the imagination (through acts of effective imagining or, in other words, wish-fulfillments). These attempts at wish-fulfillment are opposed. Features of a particular outcome of such an attempt at wish-fulfillment reflect the fact that the attempt has occurred under conditions of imagined threat.

3. Sexual deviation in some cases may prove to be an outcome of the use of sexuality in a quest for security—anxiety mitigation or avoidance—rather than primarily in a quest for pleasure.

4. In some cases, the cooptation of a zone, region, or part of the body by a quest for perverse sexual pleasure may cause a relative impairment of some particular psychological function, and sometimes simply a less than optimal use of that function (a failure of achievement requiring sublimation), when the function is subserved by that zone, region, or part of the body.

5. Unconscious sexual fantasies may provide the model for—cause some of the features characteristic of—responses to and interpretations of external reality, in nonsexual and sexual areas of life, and whether or not dysfunction or achievement is the outcome.

The scientific credibility of explanations making use of these nomothetic themes is at issue in any assessment of psychoanalytic theory. If some of these themes turn out to yield scientifically credible explanations, even in some persons, all well and good. If none of these themes turn out to yield scientifically credible explanations, in any person at all, then another paradigm will replace psychoanalysis. Considering the explanatory power such themes have conferred on psychoanalysis to date, I doubt that the latter outcome is in the offing.

PART 2
The Case Study Method in Psychoanalysis

The Hermeneutic Turn and the Single Case Study in Psychoanalysis[1]

Science, above all, teaches us to doubt—to question our conceptions of the world. It is an institutional embodiment of Freud's reality principle. It requires us to justify our beliefs about the world on empirical grounds. To provide such grounds, we pit our own hypothesis against a rival hypothesis and, when the data we obtain favor our own hypothesis over its rival, we consider what alternative explanations would account for our having obtained these data even if our own hypothesis were false.

Two kinds of deviation from the values and norms of science must especially concern social scientists. The scientistic deviation refuses to accept the demands of a thoroughgoing skepticism. Although this rebellion is disguised by an apparently zealous conformity to "science," it is revealed when an investigator neglects the "clinical" or human demands of social science research. He ignores, in other words, personal and social forces operating when such research is done which influence what data are and can be obtained.

The hermeneutic deviation also refuses to accept the demands of a thoroughgoing skepticism, but rebellion here is expressed by "dropping out"—abandoning the quest for scientific knowledge altogether. Under some circumstances, perhaps, this deviation has its beginnings when concern with the human aspects of social science research becomes an end in itself.

The scientistic deviation expresses itself in the overinfatuation of some social scientists with the form of conventional research designs. A research design, however, is merely a means to enable an investigator to

1. This chapter was originally published as Edelson (1985b).

argue that provisional acceptance of a test hypothesis over rival and alternative hypotheses is justified; the argument is the substance. A particular research design is rationally valued only if it does its job in the study of persons or social systems, no matter how successful it has been in other kinds of studies. In the social sciences, the intrapersonal processes of the investigator, his values and biases, his conscious and unconscious wishes, fears, anxieties, and defenses against impulses and dysphoric affects, influence what information he is able to obtain from the human subjects with whom he interacts and how he interprets the information he does obtain. Unavoidably, he affects his subjects and they affect him. The intrusion of the personal and the social in social science research is a major source of alternative hypotheses, which would account for his having obtained the data favoring his own hypothesis, even if this hypothesis of his were false. The possible influence of such factors casts doubt on the conclusion that his data necessarily support his hypothesis. Ignorance or willful neglect of the clinical aspects of social science research will lead the investigator to overestimate his own hypotheses and to interpret the data he obtains uncritically.

In this chapter, I shall have very little more to say about the scientistic deviation in social science research. My concern here is rather with the second deviation, which strongly asserts itself currently in my own—the psychoanalytic—community. Some psychoanalysts at least, and their allies in the humanities and social sciences, regard the emphasis in science on testing hypotheses not as an expression of the reality principle but rather as some eccentric offshoot of an aberrant logico-philosophical movement ("logical positivism" is the usual epithet) that is out of touch with what most scientists do. This attitude leads quickly to the conclusion that testing hypotheses is not what "we" are up to—because psychoanalysis is a special science, with its own rules of inference and evidence (whatever these might be) and its own mode of understanding ("Verstehen"), or because it is not science at all but a hermeneutic endeavor, which in its objectives resembles more "reading" (i.e., interpreting) a text than explaining a state of affairs. In the context of a discussion of responses, evasive or otherwise, to the clinical demands of social science research, it should be noted that likening the psychoanalyst's activity in making clinical interpretations to the humanist's activity in interpreting a text quite plays down the reciprocal interactions of psychoanalyst and analysand as well as the crucial contribution to clinical interpretation made by the analysand's own activity.

The recoil from the discipline of hypothesis-testing may take a less extreme form, apparent, for example, in a tendency to deprecate work in the context of justification and to emphasize disproportionately discovery and the invention of explanations as quintessential scientific activities. Those who are enthusiastic about the psychoanalytic case study join its detractors in relegating such case studies to the context of discovery, since both groups believe that, however richly suggestive the ideas generated and however satisfyingly complex the explanations proposed, case studies lack design-controls and cannot, therefore, provide scientific support for the credibility of psychoanalytic hypotheses.

In this chapter, questions are raised about all these versions of the hermeneutic turn in psychoanalysis—and by extension, in any of the social sciences.

Some Pernicious Epistemological Justifications of the Case Study

Recently in *Hypothesis and Evidence in Psychoanalysis* (1984), I took the position that psychoanalysts can, and should, conform to the canons of procedure and reasoning that characterize science and distinguish it from other human endeavors, especially when they use clinical data from the psychoanalytic situation to provide support for psychoanalytic hypotheses. It will certainly be held by some colleagues that, as a psychoanalyst, I have given away too much when I accept the view of science held by those who question the scientific status of psychoanalysis, and especially those (e.g., Grünbaum 1984) who believe that data obtained in the psychoanalytic situation cannot be used to *test* psychoanalytic propositions.

In my experience, this kind of response from colleagues comes eventually to rest on an appeal to epistemological justifications for the proposal that psychoanalysis is and should be exempt from scientific canons. The methods of study and manner of reasoning in psychoanalysis are and must be different, so the argument goes, because the way in which psychoanalysts acquire knowledge, as well as the kind of knowledge they acquire, differ qualitatively from the way in which natural scientists acquire knowledge and the kind of knowledge they acquire. First, the phenomena and our ways of knowing them are subjective rather than objective. Second, our interest is in knowing or understanding the meaning(s) of phenomena and not in determining their status as cause or effect. Third, we seek to understand (rather than explain or predict) phenomena, and in all their complexity, wholeness, and uniqueness, not merely as exemplars or instances of generalizations about relations between or among a set of relatively few variables that are themselves the colorless result of a process of conceptual analysis or abstraction from multifaceted phenomena.

These differences, the argument concludes, justify a preference for the case study and all its relatives over the experiment and all its relatives. Just contrast the former's richness, narrative complexity, presentation of a unique individual who is to be understood as a whole, and emphasis on the subjective and the qualitative, with the latter's preoccupation with deductive and inductive schemata, reliance on quantification and prediction in testing hypotheses, concern with empirical generalizations involving simple relations among a few variables, and efforts to control (i.e., hold constant or eliminate) the influence of many factors in order to demonstrate the influence of an isolated few.

In this chapter, I shall examine what is meant by references to subjectivity, meaning, complexity, and uniqueness (which are contrasted with objectivity, causality, abstraction, and generalization), and I shall reject these epistemological justifications of the case study, and the view based on them of psychoanalysis as a hermeneutic enterprise, a special "science," exempt from ordinary canons of scientific method and reasoning. I shall conclude, perhaps unexpectedly in light of what has gone before, that the case study is indeed a valuable method in psychoanalysis and, specifically, is capable of providing data that will test psychoanalytic hypotheses—just as clinical methods, in general, are valuable in the same way in social science—but that the use of such a method not only does not imply rejection of, but should and indeed can conform to, the canons of method and reasoning that are characteristic of science no matter what its subject matter. The psychoanalytic case study is particular, and case studies in general, should not only *not* be relegated to the context of discovery but can and indeed should be used to make various kinds of arguments that justify on empirical grounds provisional acceptance of some hypotheses as more credible than others.

Science[2]

Certain presuppositions, stated below, underlie ordinary canons of scientific method and reasoning. It is difficult to conceive that those who want exemption from the canons are indeed prepared to discard the presuppositions.

2. The view of science presented here should be compared and contrasted with the view of science presented in Chapters 13 and 15.

1. *The actual world is independent of our descriptions or knowledge of it, our values, preferences, and emotional responses to it, and our attempts to understand or explain it.* To deny that there are brute facts indifferent to our wishes and feelings, to talk as if there is no actual world, only symbolic forms constructed by symbol-making organisms, is to show oneself somewhat deficient in a sense of reality. Of course, our values, wishes, and emotions influence what questions we ask, what problems we address, and how we interpret states of affairs, specifically, how we classify them and what significance we attribute to them, but that this is so does not constitute an insurmountable obstacle to testing and revising our beliefs about the actual world. Here, for one, we are able to depend on our senses, our instruments for obtaining information, which accommodate to brute reality and do not merely assimilate what is observed to already existing schemata, as well as on canons of scientific method and reasoning.

2. *The scientist, therefore, adheres to a correspondence theory of truth.* A scientist asserts that a certain state of affairs obtains in the actual world. If the state of affairs does obtain, the assertion corresponds to reality; it is true. If it is not true, then it is false. The truth-status of a sentence expressing such an assertion is independent of our state of knowledge or our capacities for acquiring knowledge. If a sentence is true, it is true whether or not we know it is true, or are able or not to construct an argument concluding that, given the empirical data and the way these data were obtained, this sentence is more credible than its rival(s) and, therefore, should be provisionally accepted as true.

Canons of scientific method and reasoning are necessarily, then, directed to two problems: given presuppositions 1 and 2, how to obtain evidence relevant to evaluating the truth of a sentence asserting that such-and-such a state of affairs does obtain in the actual world; and, given presupposition 3, which follows, how to argue that such evidence, so obtained, provides more support for that assertion than for its rival(s).

3. *All hypotheses are underdetermined by data, facts, or evidence. There is no criterion for deciding that some hypothesis is the best (the one, the certainly true) hypothesis. There are only criteria for deciding that some hypothesis, which is a member of a set of hypotheses, is a better (more credible) hypothesis, given the available evidence, than the other members of that set of hypotheses. In general, a hypothesis will not be accepted provisionally as credible just because positive instances of it are observed; it will be so accepted only if evidence favors it over rival hypotheses.*

Any data obtained are entailed by, and therefore in that sense explained by, any one of an indefinite number of rival hypotheses about the actual world. It is always the case that any one or any combination of a number of causally relevant factors may have been responsible in the research situation for generating particular data. A scientist is always faced with the problem, then, of obtaining data that favor his hypothesis over another, and of eliminating from consideration possible plausible alternative explanations, other than the truth of his hypothesis, of how he came to obtain these data.

It follows that scientific knowledge is never certain and that scientific belief in the truth of a hypothesis is always provisional—relative to the rival hypotheses that have been challenged by the hypothesis; the available data that have been brought to bear to support the hypothesis over its rivals, and the alternative hypotheses, also capable of accounting for these data, that have been eliminated—and relative, inevitably, therefore, to the conceptual tools and methods for acquiring knowledge we have, and the background knowledge we presume to be true, when a hypothesis is tested.

Data do not support a test hypothesis unless those data can be argued to favor a test hypothesis over some rival hypothesis about a domain and have been obtained in a way and under circumstances justifying the conclusion that the truth of the test hypothesis is a more credible explanation for the fact that these data have been obtained than plausible alternative hypotheses. Contrary to what is assumed by many clinicians, observations, no matter how great their number, cannot be said to support a hypothesis just because they are consistent with or entailed by it.

All of these statements about what is involved in testing a hypothesis, and what kind of data may count as supporting a hypothesis, follow from the fact that hypotheses are underdetermined by data. Lack of interest in testing a hypothesis betrays excessive credulity about, or overinvestment in, one's own conjectures about the world and an inability to imagine what rivals they might have or what plausible alternatives might account for the data obtained in attempts to "confirm" them. Such lack of interest in a psychoanalyst (or social scientist) could well be a manifestation of the primacy of wish-fulfillment over considerations of reality.

4. *The canons of scientific method and reasoning are independent of subject matter. These canons do not depend in any way upon, and are not altered by, what domain an investigator chooses for study; they are not affected, in other words, by what kinds of observable or*

hypothetical entities, and what kinds of properties attributed to these entities, constitute the subject matter of any particular science.

Scientific activity may be said to occur in either the context of discovery or the context of justification, depending on what objectives, and therefore what skills and attitudes, are given priority.

Scientific activity occurs in two analytically distinguishable phases, no matter how great the preference of a particular scientist for one kind of activity over another is, or how rapidly the oscillation from one kind of phase to the other occurs—as when a scientist devises a model, evaluates it, revises it, evaluates it again, revises it again, always with the objective of coming closer to what is true of the actual world. Science cannot do without either discovery or justification.

In the context of discovery, establishing a frame of reference for investigation (selecting a domain, identifying the kind of entities constituting the domain, deciding what kind of properties are of interest that may be attributed to such entities) and inventing powerful iconic or discursive causal explanations of facts of interest (or both) have priority. Decisions in the context of discovery are not constrained in any simple way by what is true of the world. These decisions are matters of conceptual invention—the generation of conceptual schemes which may or may not fit the world as it is. While there are many circumstances and heuristics which appear to facilitate acquisition of insights leading to the formulation of new hypotheses, it is quite difficult to state precisely what these are. There are certainly no algorithms for "discovering" or generating *true* hypotheses. Nor is there any way to evaluate the scientific credibility of a hypothesis on the basis of its origins. A hypothesis may come to a scientist as the result of a dream, or following immersion in—or painstaking examination of—a body of data. Neither route automatically decides that the hypothesis will survive, or fail to survive, tests of its scientific credibility.

In the context of justification, taking a skeptical or critical attitude toward causal explanations and subjecting them to rigorous empirical tests have priority. Here, the question is not how powerful or otherwise satisfying an explanation is, but how accurate it is. Accuracy is evaluated by addressing two questions: How warranted is the conclusion that a causal explanation is better (i.e., more credible) than some rival? How warranted is the conclusion that the truth of this causal explanation accounts better for the fact that certain favorable data have been obtained by an investigator than alternative explanations account for it?

Subjectivity

In criticizing subjectivity, meaning, complexity, and uniqueness as epistemological justifications for exempting the case study from canons of scientific reasoning and method, I do not, of course, intend to deny that, for example, insight into one's own motives as well as special interpersonal skills may be required to obtain data in case studies (whether of persons, groups, or organizations). Nor do I intend to devalue the use of case studies to achieve objectives other than acquiring and evaluating scientific knowledge, for example, illustrating a technique, a concept, or a clinical phenomenon, or sharing clinical work with colleagues. (See Edelson 1984, pp. 70–73 for a discussion of various objectives a case study may have.)

What I am arguing is rather that such epistemological justifications as these do not work to exempt the case study that purports to make a contribution to scientific knowledge from canons of scientific reasoning and method. These canons apply in acquiring and evaluating scientific knowledge of any kind, no matter what its subject matter, and therefore apply also in case studies in psychoanalysis in particular, and clinical studies in social science in general, if these kinds of studies are to be used, not only in the context of discovery, but also, as I think they can and should be, in the context of justification to assess the credibility of claims about the actual world.

What is meant by references to subjectivity when warrant for exemption from scientific canons is sought? Here are some possibilities.

1. *There is no actual world that is independent of any particular observer's descriptions or experience of it. What we have are sets of unique symbolic representations of "reality," and these constitute the only reality we shall ever know.*

Currently, in psychoanalysis, this assertion takes the form of emphasizing an analysand's language, and his rules for using it in different ways in thought and communication (a hermeneutics), rather than those elements of the actual world to which the analysand refers, or what in the actual world is responsible for his capacity to use language at all or his various uses of language on particular occasions. If any "explanation" occurs here, in answering, for example, such questions as "Why does the patient say that?," it consists of references to syntactic (form-determining and form-transforming) and semantic (content-determining and content-transforming) rules, which are then applied to decode or translate one verbal production (what is to be explained) into another verbal production (the explanation). When there is no concern, as here there apparently is not, with what constraints or conditions actually determine what syntactic and semantic rules have been in fact used on a particular occasion by an analysand, then an infinite number of translations are possible, with no way to assess the truth of a claim that such-and-such *is* (not just *may be*) in this instance the translation of such-and-such. If this description captures at all what is meant in some references to subjectivity, then out of considerations of consistency it makes little sense to talk of science or scientific knowledge at all. This is not to say that the attitudes, beliefs, and activities just described might not have considerable value of another kind.

It is possible that what is at issue behind this kind of reference to subjectivity are simply an investigator's preferences:

a) for a particular kind of data—the verbal reports of a subject over classifications or ratings of the subject's behavior by others;

b) for assessing and attributing causal status to how objects or states of affairs in reality are represented by, look to, or are interpreted by, different subjects (or the same subject at different times), over assessing and attributing causal status primarily or solely to intrinsic features of a subject's situation.

These preferences do not constitute grounds for abandoning canons of scientific method and reasoning, or make it impossible on logical grounds ("in principle") to conform to such canons.

2. *In psychoanalysis and the social sciences, the investigator's stance must be subjective rather than objective if he is to acquire knowledge* of *his subject rather than* about *his subject.*

a) *The investigator must be involved with, rather than a detached observer of, the subject. He acquires knowledge of, rather than about, the subject through empathic identifications with him, or by noticing the various internal states stirred in himself, as a participant who interacts with, is influenced by, and reciprocally responds to the subject he studies.*

b) *The investigator inevitably brings to bear all his own biases, preferences, and values, and he responds with his own largely unconscious anxieties and his largely unconscious defenses against the impulses and anxieties aroused in him, as he interacts with the subject. These aspects of his own mental life determine how he influences the subject and, therefore, what the subject says to him and shows him, what he selects or is able to receive of what the subject says to him and shows him, and how he interprets what is said or shown, when it is ambiguous (as it usually is).*

My response, when the above statements capture what is meant by subjectivity, is as follows.

a) We may suppose that the investigator who uses empathic identifications or reciprocal responses as a means to knowing is in the context of discovery. What he concludes from observation of his own internal states, what hypothesis about the subject he forms as he makes such observations, remains to be tested against plausible rival and alternative hypotheses.

However, we may also regard the investigator's reports of his internal states as the same kind of data as the subject's reports of internal states. For example, if a kind of investigator-state, or report of such a state, is associated or correlated with a kind of subject-state, or report of such a state, that itself becomes a fact to be explained, or a fact that might provide support for one hypothesis over another.

b) The way in which aspects of the mental life and propensities of the investigator determine what data are accessible to him, what data he selects or notices, and how he interprets such data is a strong argument for the implementation of the canons of scientific method and reasoning. For here we describe, in fact, a plausible alternative account of how the investigator might come to obtain observations that seem to support his conjectures about the subject, even if these conjectures are false.

To be a skilled observer requires the ability to acquire the kind of insight into one's own mental life and propensities that will expand one's observing capacities. The canons encourage the investigator to become aware of his characteristics as an instrument for making observations, and to discover what systematic and random influences upon his observations arise from his own nature.

In addition, these canons check the investigator's overinvestment in his own conjectures, counter his credulity where these are concerned, and provide him with the means to give primacy to the reality principle rather than to the pleasure principle. For such canons, above all, impose a conceptual discipline upon the investigator to seek out, to imagine, and explicitly to formulate rivals to the belief he seeks data to support, and alternatives to the explanation he favors in accounting for the data he obtains; he must wrestle with these. (This chapter is not the place to show in what way such conceptual discipline is part of clinical skill as well.) The value of a case study, with respect to its contribution to scientific knowledge, is diminished when it makes no mention of rival or alternative hypotheses and how these are assessed in relation to the hypothesis the data of the case study supposedly support. Freud's extended consideration of rival and alternative hypotheses in *The Interpretation of Dreams*, on the other hand, clearly enhances the value of that work.

Is it really so that, because of biases, values, and unconscious meanings, there can be no agreement about what the facts are? Judges can be used to classify or rank order the contents of protocols of psychoanalytic sessions (Luborsky 1967, 1973; Luborsky and Mintz 1974), judges who are blind to the purposes of the investigator and who do not necessarily share his biases. Such judges, no matter the differences in their intrapsychic processes, can be shown to be capable of agreement among themselves in carrying out such tasks.

Finally, if an investigator is smuggling in evaluative assessments, preferences, or proposals as matters of fact, if he makes value-statements about the states of affairs he observes, or comes to conclusions from his investigations that are essentially value-statements, even though he disguises such statements as truth-claims, then it is possible to make use of conceptual analysis to expose this deception, to show that his observations and conclusions can be expressed by the kind of sentences to which it is not appropriate to apply the predicates "true" or "false." To this extent, his work is not directed to acquiring and assessing scientific knowledge. (I do not comment here on the problem of basing policy proposals, action, or technology on scientific knowledge.)[3]

3. *We work within a subjective frame of reference, not in the objective frame of reference of the natural scientist. What we study is "inside" the subjective life of the subject. There is nothing objective "out there" to measure; and objective methods, in general, are unavailing applied to subjective phenomena.*

a) *We do not study the subject as a physical object, as a natural scientist would. We study wishes and beliefs, the whole range of kinds of psychological entities or intentional states. These are individuated by subject, by content (object, state of affairs), and by psychological mode (a subject's psychological state, or attitude or relation to the content)—and not, for example, by spatiotemporal location. These psychological entites or intentional states are not physical objects.*

b) *We do not study the intrinsic properties of states of affairs, events, or objects, as the natural scientist does, but rather how they seem to, how they are interpreted by, what they signify or mean to the subject. Inevitably, therefore, we study irrational as well as rational phenomena, and the rationality and logical arguments of the natural scientist have no relevance in any effort to understand such subject matter.*

If this is what is meant by subjectivity, my response is as follows.

3. Chapter 8 has a discussion of this problem.

a) Neither the canons of scientific method and reasoning nor the character of scientific explanation is dependent on the subject matter of science, the choice of a domain for study, what kind of entities are selected for study, or what kinds of properties are attributed to such entities. Edelson (1984, especially pp. 86–101) and Searle (1983), for example, have shown, in somewhat different formulations, how psychological entities or intentional states can be individuated, and that the choice of a domain of such entities for study, or of the kinds of properties to be attributed to them, does not need to affect in and of itself the scientific status of that study. There is no reason to suppose a priori that there are no regular relations between and among kinds of wishes and beliefs (or other such entities), or that such entities cannot be caused or causative. There is no a priori logical basis for concluding that the causes of the properties of, or the causal relations between and among, wishes and beliefs (or other such entities), or the causal efficacy of such properties or relations with respect to other phenomena, cannot be studied according to the canons of scientific method and reasoning, however difficult such study might prove to be.

Measurement in this subjective realm is possible. If an investigator is able to classify kinds of psychological entities or intentional states, for example, according to the kinds of contents characteristic of each, as in everyday life we certainly in a rough and ready way already do, and as every psychoanalyst certainly does, he is making use of a kind of measurement. If an investigator is able to order members of a set of psychological entities or intentional states, for example, according to their relative strength or intensity, as in everyday life we certainly in a rough and ready way already do, and as every psychoanalyst certainly does, he is making use of a kind of measurement. Such nominal and ordinal measurements are not to be despised. Even if in this realm we never achieve the level of equal-interval measurement, where, for example, each entity can be said to have so-and-so many equal units of some property, nominal and ordinal measurements are sufficient to carry out much scientific work.

b) Suppose that the relations of interest, those which lead to fruitful explanation, are just those which depend on the way in which the subject himself classifies the states of affairs, events, or objects that are the contents of his psychological or intentional states, or that assignment of a particular psychological entity or intentional state of a subject to a class depends critically on whether or not the investigator believes that that entity or state is for that subject a member of that class. This sort of subjectivity does not constitute an intrinsic obstacle to work according to scientific canons. What is important here is that the investigator be able to specify in what way his hypothesis that the subject classifies a particular entity or state in a particular way is supported by evidence.

To do this, the investigator must keep very clear the difference between his data and his inferences or hypotheses about the subject. That such-and-such is a content of a psychological or intentional state of the subject, according to the subject's own statement, may be accepted as data. If an investigator wishes to be very cautious, he may accept as data only that the subject reports that such-and-such is a content of a psychological or intentional state of his.

The psychoanalyst's assignment of the content of an analysand's verbalizations to a class of contents characterized by a state of affairs in which one object *takes in or merges* with another object, or a state of affairs in which one object *hides away or keeps in* another object, or a state of affairs in which one object *penetrates* another object with a third object—or the psychoanalyst's inference that for the analysand the content is so classified—may be based upon such evidence as the particular choice of words used by the analysand in describing his psychological or intentional states. The evidential status of the analysand's choice of words constitutes one reason for the psychoanalyst's careful attention to nuances associated with the exact words used by the analysand; he would not and should not be satisfied by some rough approximation to, or translation of, these words by someone other than the analysand.

These considerations about classification apply also to assigning relative strengths or intensities to psychological or intentional states.

What also seems to be at issue behind both these kinds of references to subjectivity (3a and 3b above), besides the two preferences previously mentioned, is an investigator's unproblematic (with respect to scientific canons) preference for regarding performances or behaviors as effects of internal states, or as manifestations or realizations of dispositions or propensities, over regarding performances or behaviors in and of themselves as both the causes and effects of interest.

4. *Objectivity as a characteristic of scientific method depends on the possibility of intersubjective agreement, at least in principle, among observers. But such intersubjective agreement is impossible in psychoanalysis. The psychoanalyst can read anything he wants into what the analysand says. One psychoanalyst will hear as "anal efforts to control" what another psychoanalyst hears as "compensatory strivings to overcome a sense of inferiority." So, Freudian, Adlerian, and Jungian psychoanalysts all claim that the data sup-*

port their hypotheses. Surely, what we have here are different stories, each with its own value, and no way to choose between them on grounds of veridicality.

Meehl (1983) gives this particular problem of subjectivity extended consideration, in one of his characteristically brilliant papers. Toward the conclusion of the paper, which has much else of value in it, he does argue convincingly that there are ways to achieve intersubjective agreement about the inferences made from the analysand's reports.

The general problem is just what degree of intersubjective agreement can be achieved about what the data are. When Meehl refers to the subjectivity of inference in psychoanalysis, however, he means in part by "inference" how data are classified; the problem becomes the possibility of intersubjective agreement about such classifications. In his example, he argues there is a way of achieving intersubjective agreement about what kind of theme is expressed by the patient's utterances in a particular psychoanalytic session—that is, intersubjective agreement about how such a set of utterances should be classified according to type of theme. However, the classification of themes Meehl presents (i.e., the kind of facts he picks out to be explained) seems to me to be dependent on psychoanalytic theory in just the way I argue in what follows it must not be.

My own response to the problem of achieving intersubjective agreement is as follows.

a) The difficulty in achieving intersubjective agreement in psychoanalysis follows from the preoccupation with seeking confirming instances, without regard to the problem of excluding plausible alternative explanations of the occurrence of such instances.

b) More attention must be paid to keeping clear the difference between data and inferences based on data. Data include, for example, that such-and-such is the content of an analysand's psychological or intentional state, according to his own statement; or that such-and-such a content is assigned to one or another class or rank—where the classifications or bases for ranking involve reference to nontheoretical properties of such contents; or even that one kind of content, so classified, co-occurs regularly with another kind of content, or that the strength of one kind of psychological or intentional state co-varies with the strength of another kind. Inferences from data include interpretations or hypotheses about the analysand entailed by a conjunction of particular psychoanalytic hypotheses (making use of theoretical concepts) and particular data. The interpretations of psychoanalysts clearly are not first-order facts and should not be, as they often are, presented as such.

If incompatible inferences about an analysand, based on different theories, are made by different psychoanalysts, then the same kinds of scientific arguments about the relation between rival or alternative hypotheses and evidence I shall describe in the penultimate section of this chapter must be used to decide which inference or hypothesis is to be provisionally accepted over another.

In connection with the need to keep clear the difference between data and inferences or interpretations, note that the classification "taking in or merging," "hiding away or keeping in," and "penetrating" is not the same as the classification "oral," "anal," and "phallic." The former classification is based on rather commonsense notions, and while one must have some kind of theoretical knowledge to use it, that knowledge is, for example, knowledge of metaphors and figures of speech; it is not knowledge of the psychoanalytic theory of psychosexual development. Therefore, this classification may be considered nontheoretical, not because no theory is required to use it, but because carrying out a classification of psychological or intentional states according to these notions does not require knowledge of psychoanalytic theory, which the investigator wants to use these data to test. (Here, we have some way of understanding the emphasis on metaphors and figures of speech in Freud's writing and in psychoanalytic practice.)

That this classification is nontheoretical, in the sense described, is so, even if considerations based on psychoanalytic theory have led to a decision to use such a classification. In the same way, the usefulness of characterizing physical objects in terms of location in space and time is certainly suggested by Newtonian theory, but knowledge of that theory is not required, although knowledge of some theory is required, to assign a physical object to a particular spatiotemporal location.

It is quite otherwise with a classification such as "oral," "anal," and "phallic." It is a mistake in my opinion to consider a statement such as "the analysand's utterance or state is anal" as a typical data-statement in psychoanalysis. For, if the distinction nontheoretical-theoretical is not kept in mind, an investigator is in peril of guaranteeing acceptance of a theory by assuming it to be true prior to any empirical justification for accepting it as true, and smuggling it into the very constitution of the facts that are held to test it. Including as a premise in an argument the very conclusion that is supposed to follow from a set of premises results in tautology.

Clearly, none of this denies that all data are both theory-laden and corrigible. What is asserted here is that data must not be laden with the very theory one is using them to support, and that what are regarded as facts at any time are those nontheoretical assertions about the actual world it is easiest, at that time, to achieve intersubjective agreement about.

Meaning

Psychoanalysis is not natural science. It is a hermeneutics, an explication of the meaning of what the analysand reports, which is regarded as a text to be interpreted (in the sense of "decoded" or "translated"). The relations to be explored are between signifiers and what they signify, symbols and what they represent, signs and what they refer to, allude to, or stand for—not between causes and effects.

Psychoanalysis is the study of a signifying or symbolizing function. If that study involves explanation at all, it is closer to the kind of "explanation" exemplified by answers to questions about the meaning of words, phrases, concepts, or proverbs than it is to causal or functional explanations exemplified by answers to questions about the origin or source of some state or condition, how some change or difference in state or condition has come about, or how something works. What is at issue are logical connections (in a broad sense), denoted by such terms as "means," "symbolizes," "is a paraphrase of," "implies," or "is by analogy like"—primarily conceptual connections between linguistic entities (including descriptions of nonlinguistic entities), or their counterparts, rather than intrinsic connections obtaining in the actual world between or among nonlinguistic entities themselves. (See Sherwood 1969 for a discussion of explanation in terms of significance, as distinct from explanation in terms of origin, genesis, or function.)

The case study is not a scientific argument. It is not an account of causal connections. It is a story, which reveals the complex coherent network of signifiers and signifieds that makes sense of the analysand's symptoms, dreams, and parapraxes. It translates and decodes, rather than causally explains, these.

My response to this characterization of psychoanalysis and the case study follows.

A careful reading of Freud's work reveals that he sought persistently and throughout causal and functional explanations of mental phenomena. To save space, I shall not cite in this chapter all the relevant passages in that work; in any event, Shope (1973), in his excellent paper on Freud's uses of the term "meaning," has already cited most of them; additional citations can also be found in my comparison of Freud's theory of dreams and Chomsky's theory of language (Edelson 1975, and Chapter 2 of this book). Certain passages (occasional rather than preponderant) allude, often metaphorically, to symbolizing activities in human life. I think it could be argued that these indicate an effort on Freud's part to clarify by analogy aspects of the subject matter he is studying, including in some instances aspects of the clinical activity of the psychoanalyst—while at the same time perhaps he paid too little attention to disanalogies—rather than indicate any abandonment on his part of the explanatory objectives he so clearly pursues. There is no more reason to suppose that just because Freud refers to language, symbols, representations, and symbolic activity (part of his subject matter), he has rejected, or should have rejected, canons of scientific method and reasoning, than to suppose that just because Chomsky studies language (his subject matter), his theory of linguistics cannot be a theory belonging to natural science and that he cannot be seeking causal explanations in formulating it.

Of course, that Freud sought causal and functional explanations is no reason to suppose psychoanalysis continues to do so, but I would argue that clinical psychoanalytic theory as we know it, which is what is applied as far as I know in most current instances of psychoanalytic treatment, exemplifies causal and functional explanation, and that any other characterization of the clinical theory of psychoanalysis involves a rather bizarre misreading of it or is in the service of a radical revision of it. In the latter case, there is no particularly good reason to call the revision a version of psychoanalytic theory. We cannot be sure that two such "versions" of psychoanalysis are in fact comparable, in the sense that they stand as rivals such that in principle at least investigators should be able to find evidence warranting acceptance of one over the other.

For psychoanalysis, the *meaning* of a mental phenomenon is a set of unconscious psychological or intentional states (specific wishes or impulses, specific fears aroused by these wishes, and thoughts or images which might remind the subject of these wishes and fears). The mental phenomenon substitutes for this set of states. That is, these states would have been present in consciousness, instead of the mental phenomenon requiring interpretation, had they not encountered, at the time of origin of the mental phenomenon or repeatedly since then, obstacles to their access to consciousness. If the mental phenomenon has been a relatively enduring structure, and these obstacles to consciousness are removed, the mental phenomenon disappears as these previously unconscious states achieve access to consciousness.

That the mental phenomenon substitutes for these states is a manifestation of a causal sequence. An example of one such causal sequence: a wish leads to an image of a dreaded state of affairs, which the subject believes will follow upon the gratification of the wish. Therefore, imagining the gratification of the wish and believing that gratification to be possible and imminent produce anxiety. Anxiety may lead to changes in states of consciousness and to various operations (e.g., the dream work, primary process, defense mechanisms) upon these psychological or intentional states, which determine the form and content of the mental phenomenon which comes to substitute for them. The psychological or intentional states, in conjunction with, or by means of, the various operations upon them, cause the mental phenomenon to come into being (i.e., make it happen).

For psychoanalysis, the *meaning* of a mental phenomenon is also the purpose (e.g., the wish-fulfillment) the mental phenomenon serves, and here we may have reference to the functions it serves both in keeping rejected contents out of consciousness in order to avoid dysphoric affects such as anxiety and in providing unconscious motives, drives, or impulses a means of expression (i.e., a path to satisfaction).

Meaning, in the first sense, requires causal explanation of a mental phenomenon (i.e., an answer to the question "What makes it happen?"). Meaning, in the second sense, requires functional explanation of a mental phenomenon (i.e., an answer to the question "What end does it serve as a means?"). With reference to meaning in the first sense, especially, Freud writes of that which fills a gap, repairs a discontinuity, or finds its place in psychic continuity (i.e., fits in with other psychological or intentional states in conscious mental life, or belongs to a sequence of such states that has been interrupted). With reference to meaning in either sense, Freud writes of that which, by virtue of psychoanalytic explanation, gives intelligibility or significance to a mental phenomenon that does not make sense. (For a similar account of meaning in psychoanalysis, see Shope 1973.)

The following are especially important.

a) Semantic explanation, which refers to meaning or significance (see, e.g., Sherwood 1969), is not, in psychoanalysis at least, a noncausal kind of explanation. Suppose the question is: "Why does the analysand fear the snake so?" Suppose the answer to that question is: "A snake stands for, or symbolizes, a penis." It is easy to see that by itself this is no answer at all; for one thing, it leads immediately to the question: "Why does the analysand fear a penis so?" The question is about an inexplicable mental phenomenon (i.e., "fearing the snake so") and its answer depends on an entire causal explanation of the kind described above. "A snake stands for, or symbolizes, a penis" makes sense as an answer only if it is understood as shorthand for a causal explanation of that kind. Correspondingly, "the child stands for, or symbolizes, the boss" is not a satisfactory answer (it does not even sound right) to the question, "Why does this father beat his child?"

b) Causal and functional explanation are, given the above account, two different perspectives on, or two different ways of talking about, the same relations between or among events or states of affairs. Explanation in terms of purpose or function (reference to a means-end-schema) should not be contrasted as teleological (i.e., determination by ends) to causal explanation (i.e., determination by antecedents). No *future* state of affairs is required by functional explanation to explain what obtains or occurs in the present. Causal efficacy resides in the representation, in the *present,* of possible future states of affairs; in a *present* image of, orientation to, or anticipation of a possible future state of affairs; in a *present* desire or wish that an imagined state of affairs obtain in the actual world; in a *present* intention to make that imagined state of affairs happen; in a *present* plan for bringing that imagined state of affairs into being, a plan which is based on beliefs about means-end relations.

An unconscious image of an end-state-of-affairs modeled on past memories of the gratification associated with that kind of state of affairs can dominate psychological life, even in the absence of conscious intentions to make such a state of affairs happen. The subject repeatedly imagines that kind of state of affairs, acts to bring it about, interprets whatever he perceives as that kind of state of affairs, and then acts and responds in terms of this interpretation. The image is endowed with the power to compel the subject to repeat over and over again his attempt to resurrect that kind of state of affairs.

c) Empirical generalizations describing statistical associations or correlations entail, and therefore may be used to "explain" or predict, singular facts. They are also themselves facts the scientist wishes to explain. To do so, he has recourse to causal mechanisms. In the absence of an explanation that spells out a causal mechanism, an empirical generalization remains a description of an accidental relation, rather than, e.g., a scientific law.

Causal explanation (e.g., as explicated by Harre 1970) traces how one event or state of affairs leads to, produces, or generates another.[4] To elaborate such explanations, a scientist may call on a regress of

4. For an expanded discussion of the account of causality and causal explanation given here and in the next section of this chapter, see Chapters 13 and 15.

causes and effects, e.g., may identify an antecedent event or state of affairs as an even more remote cause, or interpolate between any cause and effect of interest other cause-and-effect-links which constitute steps from the one to the other.

Causal explanation may also make use of a move from whole to part or from part to whole. How the parts or constituents of a kind of enduring structure are organized or relate to each other, and what properties such constituents have (usually different from properties the structure itself has), may be shown to produce or generate facts about such structures. Conversely, facts about a structure may be shown to result from, or involve, responses by the structure to causally relevant events or states of affairs in its situation; situation and structure then, form the larger whole of which the structure itself is a part.

So, physicists explain macro-characteristics of physical objects (e.g., temperature, the change from water to ice) in terms of the properties of, and relations among, micro-constituents of such objects. Freud explains psychological phenomena as the result of both the properties of, and relations among, intrapsychic constituents (e.g., wishes and beliefs) of a psychological system and the impact of that system's relations with situational objects (including its own body). Talcott Parsons explains sociological phenomena as the result of the interaction between internal relations (inputs and outputs) among the subsystems of a social system and that system's external relations to conditions, resources, and obstacles in its situation.

Freud wishes to account for the properties of mental phenomena, including why they should have such-and-such contents, by causes which are also mental. The mental phenomenon that does not make sense is both caused by, and is the realization of, a set of relations among hypothetical (i.e., unconscious) psychological entities or intentional states, which do make sense, somewhat as the properties of water or ice both are caused by, and are the realization of, the movement of molecules under different conditions of volume and pressure, molecules which have properties unlike those of water or ice.

A major contribution of psychoanalytic theory is not to rest content with empirically observable properties or regularities (co-occurrence or co-variation), but to spell out the mechanisms by which the properties of and relations among a set of psychological entities or intentional states cause (make happen, produce, generate) and are realized by a phenomenon with somewhat different properties than they have. A single case study is often used to spell out causal mechanisms; this is a particularly important feature of Freud's case studies. What is called a psychoanalytic narrative by some is in part at least an effort to explicate causal mechanisms, to show how one kind of state of affairs or event can lead to, produce, or generate another, or how relations among constituent entities can cause or be realized by the properties of a certain kind of structure.

Complexity and Uniqueness

The case study captures the subject completely, presents the subject vividly, matches with fidelity the complexity of the subject, and preserves what is unique about the subject—in a way that no objective-quantitative study can.

But scientists do not presume to capture "all the phenomena." They seek to study a specified domain or domains, constituted by certain kinds of elements, which are characterized in certain ways. An interest in vivid presentations more often than not disguises a reliance on rhetorical persuasion as a strategy for compelling belief, in place of logical argument about the relation of hypothesis and evidence. An emphasis on complexity more often than not disguises a shrinking from the necessary work of abstracting out aspects of reality—properties of interest (ultimately variables)—in order to investigate just the relations between or among these. Discouraging talk of countless variables (and the different kinds of relations, and the possibly very complicated network of relations, between or among them) is no substitute for beginning to specify what these variables are and to map their interrelations.

Those who emphasize the *complexity of psychological phenomena* perhaps appreciate too little the complexities of physical phenomena. We may say, for example, that such an event as striking a match causes such an event as fire. A moment's reflection reveals the complexities here. Such necessary conditions as the presence of oxygen, and the dryness of the match and of the surface against which it is struck, must be distinguished from causes. A cause makes the difference; it makes something happen. The necessary conditions contribute to the production or generation of the effect, but they cannot by themselves cause it.

The concept *cause* implies a capacity or power to produce, bring about, make happen, or generate, and also implies a causal mechanism, which explicates the nature of this capacity or power. But a great many things are, if not causes, *causally relevant*. Anything is causally relevant if it makes a difference. A kind of event or state of affairs is causally relevant—even if it cannot be shown to, or does not, determine an effect (i.e., bring it about without exception)—if it increases the probability of that effect from what the probability of it would be in the absence of that kind of event or state of affairs. (A kind of event or state of affairs can also be causally relevant, but in a negative sense, if it decreases its probability.) Causally relevant events or states of affairs may not have the power or capacity to produce or generate an effect, even if they are necessary conditions without which the effect will not occur. These necessary conditions are often part of, or used by, a postulated causal mechanism; without them, therefore, a cause cannot produce its otherwise expectable effect.

In psychoanalysis, Freud distinguishes not only between dispositions or propensities (e.g., fixations), favoring circumstances (e.g., fatigue or debility), instigating causes (frustration and regression), and essential causes (unconscious conflict), but also between the mental cause of a mental phenomenon such as a dream (e.g., its motive force) and the sources of the manifest contents of that dream (recent indifferent impressions, somatic stimuli, symbols), those materials which are made use of in generating, but which themselves do not have the power to cause, the dream.

Concentration and abstraction, and in this sense a simplification of the phenomenal world, are essential to science. No scientific theory ever seeks to explain everything there is to be explained. That would be impossible. No set of hypotheses can encompass all properties, for some of these belong to different kinds of elements and are at different levels of analysis. Unfocused indiscriminate impressionability, passive promiscuous immersion in phenomena, in and by themselves cannot bring forth scientific thought. No one can advance scientific knowledge who bemoans that dissecting out certain features of a person (or a cell, for that matter) is destructive to the person's complexity, or that no person (or cell), event, state of affairs, or structure is exactly like any other (which, of course, is always true). Striving to get it all, to get more and more of it, not to miss anything, as if the more phenomena, the more truth, to capture the look as well as the physics of the rainbow, are not so much anti-scientific as in many ways for scientific work just beside the point.

Discussions of the *uniqueness of the individual* more often than not miss the point that scientists are interested in relations—ultimately, relations between or among variables. Every statement about relations between or among kinds of states of affairs, properties, or amounts of properties is a generalization over a set of instances in each one of which values of different variables coincide. The set of instances may be generated by a group of persons, each instance by a different person. In the case study of a person, the set of instances are generated by that one person; each instance, for example, is a different exemplar of a kind of mental phenomenon or performance produced by that person, or is a different time-slice of that person. The phrase "individualizing or idiographic versus generalizing or nomothetic science" marks a pseudo-issue, as Holt (1962) has decisively and devastatingly argued. I add only the following to his discussion.

The same kind of reasoning about hypothesis and evidence may be used in studying a domain whose elements are collectives (societies, organizations, groups, or families), a domain whose elements are persons, a domain whose elements are cultural objects (books, movies, games, or athletic contests), or a domain whose elements are physical objects. A single person, a single cultural object, or a single society, organization, group, or family, each with its distinctive elements and their properties, may each constitute a domain that is the subject of a single case study. We may study the interpersonal transactions or interactions bounded by a particular single collective; or the intrapersonal events or entities bounded by—the perceptions, beliefs, memories, feelings, wishes, dreams, symptoms, dysfunctional performances, or achievements of, or objects produced or occasions occupied by—a particular single person.

In each kind of domain, a limited set of properties of the elements of the domain, and the relations between or among these properties, are what is of interest. Each element, generated by the single case regarded as a domain of such elements, together with the property or properties the element exemplifies, is the focus of a potential act of observation. The methods of making and recording observations, and the mode of reasoning about the relation between hypothesis and the evidence provided by such observations, are not fundamentally different in the study of an intrapersonal domain from those used in the study of a population of persons. In both, for example, groups of elements allocated to different conditions, or samples of different kinds of elements, may be compared. In both, each element observed instantiates a particular value of every variable of interest.

A hypothesis is always a hypothesis about a particular domain. The question raised when a hypothesis is tested is whether it can be provisionally accepted as true about that domain, rather than some rival hypothesis. One may also raise a question about the scope of a hypothesis. Different arguments about the relation of hypothesis and evidence are required, depending on whether the question is about the truth of a hypothesis about a particular case (e.g., "Does a postulated relation hold for a particular subject?") or the question is about generalizability (e.g., "Under how many conditions to which the subject is exposed or for how many such subjects does the postulated relation hold?").

Statements asserting that certain relations between particular concepts or variables hold for a single case, that in a single case one kind of state of affairs is causally relevant in bringing about another kind of state of affairs, or merely that a single case has a certain kind of property, are just as much hypotheses as statements making similar claims about populations of such cases. Indeed, contrary to an attitude sometimes apparent in critics of case studies, it is not necessarily a trivial achievement to provide evidence making a causal hypothesis about a single case scientifically credible, even if it is not also possible to provide evidence from a study of that case which would justify generalizing the same hypothesis to other cases. (A series of cases, however, might provide evidence justifying such generalization.)

The Case Study as a Scientific Argument

It is pernicious to relegate case studies to the context of discovery, for it discourages rigorous argument about the relation of hypothesis and evidence in such studies. My impression is that, in fact, in most case studies that have had an impact on the scientific community, there is a challenge, even if it is latent, to at least one rival set of concepts and hypotheses, a challenge which indeed motivates the case study. There is also an argument, however informal, that the data of the case study give greater support to the challenging than to the challenged set of hypotheses. The scientific value of such a case study ultimately depends upon the adequacy of that argument.

Such an argument usually presupposes that the author of a case study is able convincingly to provide warrant for the following assertions.

a) He can reliably identify instances of nontheoretical states of affairs. Consider Luborsky's painstaking explication of criteria (1967) for deciding whether what is reported is to count as an instance of momentary forgetting, or Freud's discussion (1900) of what shall count as a dream.

b) He can make such an identification without using—or depending upon procedures, instruments, or tests which presuppose the truth of—the test hypothesis he is proposing, or any other hypothesis which entails or follows from that test hypothesis. Consider the background theoretical knowledge, which is not being tested, that is presupposed by the use of a thermometer, a microscope, or free association in obtaining the facts which are to serve as evidence for hypotheses which themselves do not depend in any way on this background theoretical knowledge. (For a discussion of what is presupposed by the method of free association, see Edelson [1984] and Chapter 12 of this book.)

The contribution of the case study to scientific work is increasingly held to be both limited and minimal (e.g., Kazdin 1980, especially pp. 9–32). The prevalence of two arguments especially about the relation of hypothesis and evidence in such studies has done much to discredit them. These two arguments should not be used. One argument is that an empirical instance of a hypothesis, which follows logically from it, confirms the hypothesis, and that is sufficient to enhance its scientific credibility. Indeed, the argument, implicit if not explicit, is that the greater the number of positive instances of a hypothesis obtained, the more credible the hypothesis. Another equally dubitable argument concludes from the mere concomitancy of two states of affairs, events, or processes reported in a case study that one is the cause of, or causally relevant in the production of, the other.

What arguments, then, can and should an author of a case study use? In general, a case study shall be considered to provide some support for a test hypothesis—to make it to some extent scientifically credible—only if the author of the case study argues, and is justified in arguing, that all three of these assertions about it are true.

1. The test hypothesis has the power to explain the kind of observations made and reported in this case study, and some rival hypothesis does not. The observations are predictable or expectable if the test hypothesis is true, but are not predictable or expectable if the rival hypothesis be true instead.
2. In this case study, the test hypothesis ran a real risk of being rejected in favor of that rival hypothesis.
3. The test hypothesis accounts better than some alternative hypotheses for the fact that the observations apparently favoring it were obtained by the author of the case study.

It is my thesis that arguments such as the following about the relation between hypothesis and evidence can and should be used when the data are obtained by case study (or, generally, by clinical methods in social science) just as much as when they are obtained by experimental or objective-quantitative methods. (For a discussion of the contributions of Platt, Hacking, Popper, Campbell and Stanley, and Glymour to the formulation of one or another of these arguments, see Edelson 1984.) Indeed, if it were not possible to make these arguments when the data are obtained by case study, much of my objection to what I have called pernicious epistemological justifications of the case study would lose its cogency.

Explanatory Power and the Vanquished-Rival Argument

Authors of case studies should not, as they often do, confuse explanatory power and scientific credibility. Explanatory power is necessary, but it is not sufficient, to establish credibility. Von Eckhardt (1982) makes the distinction clear and claims (incorrectly, I believe) that Freud in his case studies ignored it to the detriment of the scientific value of these studies. The argument "because data-statements confirm a hypothesis (i.e., follow logically from it), the hypothesis is scientifically credible" is, as has been previously pointed out, an unacceptable argument.

Even if an author of a case study should succeed in arguing conclusively that his hypothesis "explains" (i.e., entails) his data, that does not, if he wishes also to argue for its scientific credibility, absolve him from the responsibility to test it in his case study. He tests it, both by exposing it to at least some risk of rejection, and by pitting it against rival and alternative hypotheses. Any attempt of this kind on his part, limited though it may be, will do much more than mere submission of a record of confirming instances of the hypothesis to convince a reader that the hypothesis is to some degree worthy of provisional belief, and to inspire a reader to want, and perhaps even to carry out, additional tests.

What is especially important is presentation of some version of the *vanquished-rival* argument. The author of the case study shows that, given the evidence, his hypothesis is more credible than one or more of its rivals. The rival hypothesis, which is held or entertained prior to the attempt to obtain evidence to support the test hypothesis, expresses a belief about the domain of interest different from that expressed by the test hypothesis. The author of a case study may argue that:

a) the rival hypothesis entails that a particular event or outcome, the occurrence of which is consistent with or deducible from the test hypothesis, will not occur, and so that it did occur favors the test hypothesis over the rival hypothesis;

b) the rival hypothesis entails that a particular event or outcome, the occurrence of which is likely, probable, or expectable given the truth of the test hypothesis, is rare, improbable, or unexpected given the truth of the rival hypothesis, and so that it did occur favors the test hypothesis over the rival hypothesis; or

c) the rival hypothesis does not entail that a particular event or outcome will occur or will not occur (it has nothing to say about such an event or outcome), while the test hypothesis entails that the event or outcome will occur, and so that it did occur favors the test hypothesis over the rival hypothesis.

An important subspecies of the vanquished-rival argument is the Bayesian probabilistic *impact-of-new-information argument.* The author of a case study may argue that the way in which new information, obtained in a study, changes both the prior probability attached to the proposition that a test hypothesis is true, and the prior probability attached to the proposition that a rival hypothesis is true, favors the test over the rival hypothesis. (Prior probabilities are based on information available before the study.) In other words, given new information, posterior probabilities are now attached to the two propositions, and these posterior probabilities favor the test over the rival hypothesis. This argument is implicit in much clinical reasoning, including that which goes into working out a differential diagnosis, or deciding upon, or changing, a patient's diagnosis, given new information about the patient. (Note that this argument does not involve simply the different probabilities some event or outcome has under different hypotheses, as in the vanquished-rival argument (b) above, but rather the probabilities attached to hypotheses themselves.)

The Risky-Prediction Argument

Suppose that the failure of a particular event or outcome to occur under conditions of investigation would be held to favor a rival hypothesis over a test hypothesis. Suppose that, given prior background knowledge, the probability that the particular event or outcome under conditions of investigation will indeed fail to occur is very high. To predict, then, nevertheless, that under conditions of investigation, because the test hypothesis is true, the particular event or outcome will in fact occur is to subject the test hypothesis to a correspondingly very great risk of rejection. If, therefore, the particular event or outcome subsequently under conditions of investigation does indeed occur, its occurrence provides correspondingly very strong support for the test hypothesis.

It is important, if an author of a case study is to be able to make this argument, for him to anticipate and identify with the critical skepticism his hypothesis will inevitably encounter by subjecting it to a real risk of rejection. If the cards are not to be stacked in favor of his hypothesis, he must choose a particular case, or obtain observations in studying that case, in a way that makes rejection of the hypothesis not only possible, but, indeed, more probable than acceptance of it.

Any wish to accept a hypothesis must be checked by making it difficult to accept it, by setting it up to be rejected. The author of a case study will, therefore, specify which observations, if he had obtained them, would have led him to reject his hypothesis—and how he proceeded in such a way that just these observations not only might have occurred but (though they did not occur) were very likely to have occurred. He will show that he avoided particular steps which might have prevented their occurrence, and indeed, took particular steps to increase the likelihood that, if they could be obtained, they would be obtained.

Any one of the following arguments may be used to justify a claim that the hypothesis in a particular case study survived a real risk of rejection: (1) the least likely case argument; (2) the rare cause argument; (3) the improbable outcome argument; (4) the convergence argument; and (5) the bootstrap argument.

The Least Likely Case Argument

If a hypothesis is a generalization of the kind "all dreams are wish-fulfillments," then a counterexample falsifying the hypothesis is "this dream is not a wish-fulfillment." A case that might be chosen for study, then, is one involving a dream which appears not to be a wish-fulfillment, given existing knowledge (which does not include the test hypothesis).

For example, the dream in the case study is made up of images of misery and pain. On the commonsense background assumption, which at the time of testing the hypothesis is widely accepted, that human beings do not wish for misery and pain, the expectation is that evidence obtained in a study of such a dream will show that the dream is not a wish-fulfillment. The choice of this kind of dream for study increases the probability that the test hypothesis will be rejected. If, even so, the evidence obtained does show that the dream is a wish-fulfillment, that improbable outcome counts more toward establishing the credibility of the test hypothesis than such an outcome would have counted if the dream chosen for study had been made up of images of joy and pleasure.

Similarly, in testing a hypothesis that all societies are rule-governed, a society whose members are apparently all wild and lawless or a primitive society whose members as children of nature are thought to behave spontaneously might be chosen for study. Or, in testing a hypothesis that an unconscious conflict involving homosexual impulses is causally necessary to produce a certain kind of psychopathology (e.g., paranoia), a patient who is apparently unconflicted about his homosexual behavior, but who is also paranoid, might be chosen for study.

The Rare Cause Argument

In testing a causal hypothesis, if the supposed cause of an effect is known to occur relatively rarely in a population, a case study in which these two events co-occur is a more convincing instance of the hypothesis than a case studv in which the supposed cause is known to occur relatively frequently in the population.

A hypothesis describes a relationship, for example, between a childhood event, such as the one postulated by Freud in the Rat Man case (a child is punished by his father for masturbating) and a mental phenomenon in adulthood (e.g., a neurotic symptom). In what follows, assume the frequency of the mental phenomenon in the population remains constant. If the childhood event occurs relatively frequently in the population, then a case which manifests the mental phenomenon is, quite probably, a case in which the childhood event also occurred, even if there is no causal relation between the two. The probability of obtaining a confirming instance of the hypothesis, in other words, is relatively high, even if no causal relation between the childhood event and mental phenomenon actually exists.

However, if the childhood event occurs relatively rarely in the population, then a case which manifests the mental phenomenon is, quite probably, a case in which the supposed childhood event did not occur, unless the two are causally related. Here the probability of obtaining a negative instance of the hypothesis is relatively high, if the two are not related. When an author of a case study postulates a cause which he knows occurs relatively rarely in a population, he can argue with some justification that he has thereby increased the probability of rejecting his hypothesis, and that, therefore, obtaining a positive instance of his hypothesis in his case study counts more toward establishing the credibility of that hypothesis than if he had postulated a relatively frequently occurring cause.

In passing, note here that the report of a rare case of something is to be distinguished from a report in which a rare cause is postulated to explain something. A rare case is one that is known to occur—or that a review of the literature shows to have occurred—very infrequently. Since there is so little opportunity to observe such a case, it may be thought worthwhile just to describe "the facts" in detail, even if the state of knowledge is such that no hypothesis presents itself to account for these facts and no hypothesis's credibility seems undercut by them. Facts, however, are always relative to a choice of elements and ways of characterizing elements and, therefore, presuppose some sort of assumptions about what in the phenomena is worth noting. The scientific value of such a report tends to be overrated, therefore, since it is unlikely that the author will include just those facts that will turn out to be relevant to a consideration of hypotheses which are at some later time of interest.

The Improbable Outcome Argument

The improbable outcome argument may take the form of (a) the unexpected prediction argument; or (b) the precise deduction argument.

The unexpected prediction argument. When an author of a case study makes a prediction, he asserts that a state of affairs has obtained, does obtain, or will obtain, or that an event has occurred, is occurring, or will occur, when in fact he cannot know that the assertion is true. It is not based on direct experience. It does not follow logically from any background knowledge available to him. It does follow logically from the hypothesis he tests by making the prediction. If this hypothesis is true, then his prediction must be or is likely to be successful; if the prediction should turn out to be unsuccessful, then the hypothesis must be or is likely to be false.

The probability that his prediction will turn out to be successful is, given his background knowledge, low indeed. Essentially, then, the author of the case study tests his hypothesis by deducing from it what is, on the basis of established knowledge, and excluding the hypothesis in question, an improbable state of affairs or event.

The predicted state of affairs or event will probably not occur. Its failure to occur, which is expected, would suggest rejection of the test hypothesis. Therefore, if this improbable state of affairs or event does occur, the hypothesis has survived a real risk of rejection. This unexpected occurrence, then, counts more toward establishing its credibility than the occurrence of a state of affairs or event observed in the course of studying a case, which also follows from and is therefore explained by the hypothesis, but which was not predicted.

That a state of affairs obtains or an event occurs, when the author of a case study knows it obtains or occurs, is completely probable. He cannot claim that the hypothesis has run any risk of rejection, unless he has predicted the state of affairs or event before knowledge of it. What has already been observed in a case study may be regarded, if no existing knowledge accounts for it, as in that sense "unexpected," but it is certainly not improbable in the sense of unlikely to occur. (It has in fact already occurred.) The author of the case study, in referring to such an "unexpected" observation and its relation to his hypothesis, has only demonstrated that this hypothesis has more explanatory power than other hypotheses in the body of knowledge previously available to him. He has not, however, exposed that hypothesis to any risk of rejection. To do that, he, not knowing an event has occurred or will occur, must predict that given his hypothesis it will occur or has occurred, when existing knowledge leads to the prediction that it is impossible for it to occur (or that it should occur only very, very infrequently).

An "unexpected" observation may simply be one that a case study documents has not been reported before. A case study may be written to record such an unusual observation and to announce what hypothesis the observation has instigated the author to formulate in order to explain it, or what hypothesis is called into question by it. The unexpectedness (in the sense just described) of this kind of observation contributes nothing at all to making the intended-to-be-explanatory hypothesis credible, though it may cast real doubt on the credibility of a hypothesis with which such an observation is incompatible.

There is no way to guarantee that an author of a case study will take a critical attitude toward his own hypothesis and will do everything he can to subject it to some risk of rejection. That is one reason that both the possibility of replication and the fact of replication are so important in scientific work. Even when an observation in a case study is described as unexpected because its occurrence when predicted was improbable, a reader cannot be certain that it was actually predicted. An author of such a study may write his story as if he predicted first and observed later, although in fact he observed first and "predicted" later.

Similarly—when multiple independent predictions are made from a hypothesis or a set of hypothesis, and no existing knowledge entails that they should be jointly successful, or a predicted outcome is measured in a variety of ways—an author of a case study may not reveal, or may report but ignore, how many of his predictions in fact did fail; he may report only positive findings. Indeed, any researcher may not calculate, or may calculate but ignore, what proportion of his positive findings would in fact have been expected to have occurred by chance even if his hypothesis were false. Achieving some successful predictions among many unsuccessful ones does not justify rejecting a plausible alternative hypothesis—that the proportion of successful predictions was itself a matter of chance.

Taking a pessimistic view of the capacity of people to be skeptical of their own ideas, one may say that how determined an author of a case study or any research is to be critical of his own hypothesis can be decided ultimately only from the fate of tests of his hypothesis carried out by others. However, since whether an assertion is actually a prediction, in the sense given to that term here, determines in part at least whether a hypothesis has run a risk of rejection, it is important for an author of a case study at a minimum to document that he has in fact made a prediction, when that is how he justifies his claim that his hypothesis has run such a risk. He should specify, for example, what he knew and did not know when he made the prediction, what procedures he followed in making the prediction, and just exactly what prediction he made. Did he predict from one *part* of a session, interview, or psychological test to another; from one *entire* session, interview, or psychological test to another; or from one kind of information to another kind of information? How precise was his prediction? His documentation should show, if true, that he took particular steps to elicit, discover, or search out observations damaging to his hopes, and that the observation he did make was not one that just anyone, without the aid of his hypothesis, could easily have stumbled over.

The precise deduction argument. A hypothesis does not, of course, run much risk of rejection if a prediction from a hypothesis, or an observation deducible from and therefore explained by a hypothesis, is one of a set of a few grossly discriminated states of affairs or events. Not much of what is possible can be excluded by a prediction or deduction when the possible outcomes an author of a case study specifies are few in number or vaguely discriminated or both, or if—as frequently happens—he does not even specify the states of affairs or events excluded by his hypothesis, while emphasizing what state of affairs or event follows from it. The freedom from risk of rejection is even more enhanced when the occurrence of any one of a relatively large subset of possible states of affairs or events is considered to be compatible with the truth of the hypothesis.

The degree of risk of rejection a hypothesis runs increases as the degree of precision of a prediction or deduction from that hypothesis increases. The degree of precision of a prediction or deduction increases as the degree of fineness of discrimination of possible outcomes increases. The degree of fineness of discrimination is measured by how many possibilities are discriminated, and how free from vagueness is the discrimination. Precision in a case study may depend, for example, upon how detailed is the description of conditions that must be satisfied for a state of affairs or event to be counted—or to fail to be counted—as a positive instance of a hypothesis. Degree of precision also increases with every increase in the size of the subset of possible outcomes that are negative instances of a hypothesis (those inconsistent with it) relative to the size of the subset of possible outcomes that are positive instances of the hypothesis.

The degree of precision of a prediction or deduction from a hypothesis may be used to argue that the hypothesis has indeed run a risk of rejection. The more precision, the greater the risk. The greater the risk, the more severe the test. A hypothesis surviving a more severe test than a comparable rival hypothesis has, other things being equal, more credibility than its rival.

Mathematicization and quantification acquire their value in scientific work both from the lack of vagueness or ambiguity in what is deduced or predicted from a hypothesis and from the enormous range of possible outcomes excluded by a prediction or deduction, which ensure that the tested hypothesis runs a high risk of rejection. There is no reason not to try to approximate in a case study achievement of such desiderata: (a) specifying an exhaustive set of mutually exclusive possible outcomes (states of affairs or events), which are discriminated from each other as sharply, with as little vagueness and ambiguity, as possible; (b) making this set of well-discriminated possible outcomes as large as possible; (c) specifying which of these outcomes follow from a hypothesis and which at the least cast doubt upon it; and (d) attempting always as much as possible to make the size of the subset of negative possible outcomes much greater than the size of the subset of positive possible outcomes.

The Convergence Argument

An author of a case study shows that he can make multiple independent deductions from the same hypothesis. The deductions are independent if and only if (a) the different observations, which are used to arrive at the same conclusion, are obtained from different sources, by different methods, or in different settings or life epochs, or involve different kinds of subject-performances, and when possible are obtained by different unrelated observers; and (b) no available knowledge entails that these different kinds of observations are actually of the same sort (have the same cause) and should coexist. The author may then argue that, since each deduction independently has exposed his hypothesis to the possibility of being rejected, he has, by making multiple independent deductions, increased the opportunities for such rejection, and therefore the risk the hypothesis has run in the case study.

The improbability of the assumption that the hypothesis is false, given a set of such observations, makes the argument more than merely an argument for the explanatory power of the hypothesis. In arguing about explanatory power, that a vast number of observations follow deductively from a hypothesis is what is emphasized. Here, the emphasis is on different kinds of observations obtained by different and independent routes. In each study or with each different method, different alternative hypotheses, which might have accounted for the data, will be regarded as plausible and may be eliminated. If so, it is more probable that the test hypothesis is true than that (1) the test hypothesis is false and (2) in each study some different alternative hypothesis, which was not eliminated from consideration, accounts for the occurrence of the event or outcome.

The Bootstrap Argument

A subset of independent but interrelated hypotheses is jointly supported if they converge upon the same outcome. For example, one hypothesis in conjunction with some data entails the occurrence of an event or outcome (that a variable has a particular value) and another hypothesis in conjunction with some other data entails the occurrence of the same event or outcome (that the variable has that value), and the event or outcome occurs (the variable is observed to have that value). Or, if a variable is theoretical, one hypothesis in conjunction with some data entails that it has a particular value, and another hypothesis in conjunction with this computed value entails the occurrence of an event or outcome (that another variable has a particular value), and the event or outcome occurs (this other variable is observed to have that value). In each instance, it is possible that the expected event or outcome may not occur; then, one hypothesis at least among the subset of hypotheses must be false. The subset of hypotheses runs this risk repeatedly, as each prediction is checked. Glymour (1980) has shown how the author of a case study (Freud, in the Rat Man case) can carry out such a test of a subset of his hypotheses.

Alternative Explanations of the Evidence

An alternative hypothesis expresses some belief about what factors operating during the process of obtaining data, or operating in the situation in which data are obtained, might have led to the occurrence of an event or outcome held to support a test hypothesis (i.e., what factors might account for the data obtained), even if the test hypothesis were false. Similarly, an alternative hypothesis could express some belief about what factors of this kind might have been responsible for the failure of an event or outcome to occur (when its occurrence would have been held to support a test hypothesis), even if the test hypothesis were true.

If different observations favoring a test hypothesis are obtained in very different ways, in the same case study, in different case studies, or in different kinds of studies, it can be argued that it is improbable that these independent observations should all, or in so vast a proportion of instances, be obtained, if the test hypothesis were in fact false. Since the favorable observations were reached by such different routes, in each one of which different alternative explanations were relatively implausible or eliminated, it is not parsimonious to assume that different alternative hypotheses each time, rather than the test hypothesis all through, account for obtaining the observations.

This line of thought leads to the suggestion that the same hypothesis should be tested by carrying out different kinds of studies, including the case study, in each one of which a different alternative explanation for the data cannot be considered to be really plausible or, if plausible, is controlled or eliminated. Replication of a case study, with similar subjects, preferably, for example, by different investigators, working in different settings, and perhaps following somewhat different methods, but testing the same hypothesis, can also be convincing, on much the same kind of reasoning. The case series for this reason continues to influence clinical practice, despite the disrepute in which the case study is held.

The use of various designs in the single case study, which make use, for example, of carefully documented comparisons of phenomena of interest under different conditions over time (including a baseline period of observation), or which enable the investigator to show that the level of some symptom or performance of a subject is over time consistently higher under one condition than under another, and which eliminate various alternative hypotheses as less plausible than a test hypothesis, are well described in such works about developments in the methodology of single-subject research as those by Barlow, Hayes, and Nelson (1984), Campbell and Stanley (1963), Chassan (1979), Edelson (1984, especially Chapters 4 and 11), Kazdin (1982), and Shapiro (1963). (For additional citations, see Edelson 1984.) As these works also make clear, it is possible to perform statistical tests in some single case studies in an attempt to reject a particular kind of alternative hypothesis—the null hypothesis that the apparently favorable outcome probably came about as the result of chance or extraneous factors.

Finally, the appropriate, effective use of "historical controls" (e.g., previous clinical experience) or "external controls" (in general, the use of data from other sources to make relevant comparisons in a case study) in order to eliminate alternative hypotheses from consideration are described by Bailar, Louis, Lavori, and Polansky (1984), Campbell and Stanley (1963), and Lasagna (1982).

Conclusion

The case study does not necessarily imply deviation from canons of scientific method and reasoning. It should not be relegated to the context of discovery. The case studv can be an argument about the relation between hypothesis and evidence. The probative value of a case study (i.e., its ability to test a hypothesis) depends on the validity of the author's use of it to claim that his hypothesis or set of hypotheses has explanatory power, has run some risk of rejection, and is more credible than rival or alternative hypotheses.

The following minimal standards should be met by clinical case studies in that psychoanalytic literature which purports to constitute a scientific body of knowledge.

1. What the author is asserting—his hypothesis, conclusion, or generalization about a case or treatment—is clearly and prominently stated.
2. The author shows how his hypothesis about the case or treatment explains or accounts for the observations he reports; he does not merely juxtapose them.
3. The author is careful to separate facts or observations from his interpretations of them, which is to say, he distinguishes what can be observed without knowledge or use of the theory being tested from interpretations based upon the very theory such observations are being used to test.
4. The author specifies what observations, if they had occurred, he would have accepted as grounds for rejecting his hypothesis.
5. The author reports at least some observations that apparently contradict, or at least are difficult to explain by, his hypothesis, and indicates how he plans to deal with such counterexamples. If he does not reject his hypothesis, he makes clear his grounds for holding on to it or how these counterexamples limit its scope.
6. Since any set of observations can be explained in different ways, the author gives some argument why his observations are better evidence for his hypothesis than for at least one comparable rival or competing hypothesis.
7. Even if the author's observations can be argued to favor his own conjecture over a rival, he considers—as Freud concerned himself, for example, with the problem of suggestion—what factors operating in the situation in which he made his observations could have resulted in his obtaining these favorable data even if his hypothesis were false. (Here, interest in the actions, intrapsychic processes, characteristics, countertransferences of the psychoanalyst must loom large). He presents some argument, even though it be a nondesign argument, for dismissing at least one such factor as a plausible alternative

explanation for his having obtained the data he would like to regard as favorable to his hypothesis.

8. He makes clear to what extent he proposes to generalize his hypothesis about the case or treatment to similar cases or treatments, and presents some argument justifying such a generalization.

A case study meeting these standards would not have to appeal to subjectivity, meaning, and complexity and uniqueness as grounds for abandoning ordinary canons of scientific method and reasoning.

The Evidential Value of the Psychoanalyst's Clinical Data[1]

12

Adolf Grunbaum's critique of the foundations of psychoanalysis (1984) questions both the adequacy of the empirical support for some of its major claims, and the cogency of reasoning about the relation between clinical data and hypotheses. His critique is not destructive in its import or, I believe, in its intentions. It promises rather to increase the interest of psychoanalysts in carrying out needed research, clarifying theory, and enhancing the rigor of the reasoning in the psychoanalytic literature about the relation between hypotheses and clinical data.

For surely any psychoanalyst, after a study of this powerful critique, will think twice before arguing that confirming instances alone are enough to provide support for the credibility of a hypothesis, that the mere concomitance of two events in a case study is enough to establish the causal status of one in relation to the other, or that psychoanalysis is a science with its own unique rules of inference and evidence (whatever these might be).

Reflecting on Grumbaum's critique, a psychoanalyst will come to several realizations. It is unavailing in attempting to support a hypothesis to enumerate however many instances are consistent with it. It is necessary instead to pit rival hypotheses against one another, to collect data that can be argued to favor one hypothesis against its rival or rivals. Suppose that data should then indeed favor a hypothesis against its rival or rivals. Al-

1. This chapter is based on material from Edelson (1986a), and also makes use of material from "A Comment on Adolf Grunbaum's Critique of Freud's Dream Theory," part of which was presented at a Colloquium sponsored by the Center for History and Philosophy of Science at Boston University, April 24, 1985.

ternative hypotheses (e.g., that suggestion it responsible for the favorable outcome) might account for obtaining this favorable data even if the hypothesis (apparently supported by these data) were false. It is therefore necessary to find a way to argue convincingly that these alternative hypotheses have been eliminated or at least rendered highly implausible.

Work in this direction has been and is being attempted, for example, by Lester Luborsky at the University of Pennsylvania, by Robert Wallerstein at University of California at San Francisco, and by Joseph Weiss and Harold Sampson at the Mount Zion Research Center of University of California at San Francisco. I also believe that Freud was methodologically sophisticated enough, as indeed Grunbaum himself has documented, to realize the necessity of such arguments and attempt to provide them, with varying degrees of success, in at least some of his case studies (notably, the so-called Rat Man and Wolf Man cases).[2]

Disagreements Based upon Differences in the Depiction of Psychoanalytic Theory

I shall not attempt to address in this chapter questions about the view of science upon which Grunbaum's critique is based.[3] But I shall discuss briefly an example of the kind of difficulty I have with some of his depictions of psychoanalytic theory. Reasons for such difficulties will be clear to the reader who compares what I have had to say about psychoanalytic theory in Part 1 of this book with what Grunbaum (1984) has said about it.

That such a difficulty exists at all is due to an unhappy current state of affairs. Psychoanalysis has not proceeded very far in systematizing or formalizing its theory, even that bounded by Freud's work. There is little agreement among psychoanalysts about which hypotheses are central and which peripheral, about which are most supported by evidence and which least, and about which hypotheses at this point in history are most accessible to testing and which least. The theory is, I believe, even if one were to take Freud's body of work alone, full of inconsistencies, undeveloped lines of thought, and hypotheses that are contradictory in that at the least they appear to have different empirical consequences.[4]

2. See Chapter 13 for a discussion of Freud's Wolf Man case.
3. However, see Chapters 13 and 15, where I do raise such questions.
4. This state of affairs motivates my conviction that formulating a core theory of psychoanalysis is essential for scientific advance.

I shall take Grunbaum's critique of Freud's dream theory as an example of this kind of difference between us. My response to Grunbaum's critique is based, in part at least, simply on the fact that he and I have read *The Interpretation of Dreams* somewhat differently, have neglected and emphasized different passages, and therefore come to somewhat different conclusions. This is not a very important level on which to carry on discussions of the cogency of scientific reasoning or the justification of hypotheses, and is a sad commentary on how little progress has been made in doing the work that needs to be done by psychoanalysis before it can properly benefit from the kind of critique a distinguished philosopher of science such as Grunbaum may direct to it.

Grunbaum claims that Freud's hypothesis that motives causing a dream, if unconscious, can be justifiably inferred from free associations to the manifest content of the dream is not adequately, if at all, supported by evidence. It is certainly not supported by Freud's account of his interpretation of the Irma Dream. Grunbaum's conclusion is that, in general, the method of free association yields epistemically defective data. It cannot be used to provide empirical warrant for believing Freud's hypotheses.

With regard to the Irma Dream, I believe that Grunbaum has somewhat misconstrued what Freud was attempting to do in presenting his associations to the Irma Dream. I do not read Freud's account of the Irma Dream as either an attempt to vindicate or to legitimize free association. I do see it as an attempt to describe the method and to illustrate, somewhat imperfectly, its use.

Freud does attempt to give presuppositions he believes justify his confidence in the relevance of the products of free association. All mental processes are purposive. (That is, all mental processes are related or oriented to goals or ends.) That one content follows another is directly or indirectly determined by what end governs a sequence of such contents. He adduces two theorems, the "basic pillars" of his technique: "when conscious purposive ideas are abandoned, concealed purposive ideas assume control of the current of ideas, and . . . superficial associations are only substitutes by displacement for suppressed deeper ones" (p. 531). He indicates also in his writings that he appreciates the necessity for providing evidence independent of the method and its use in the psychoanalytic situation to warrant acceptance of these theorems.

Furthermore, the psychoanalyst's selection of associations in making an interpretation is not as arbitrary as Grunbaum depicts it.[5] The

5. Chapter 15 of this book is devoted in large part to supporting this assertion.

wish inferred from some selected associations must fill a gap in the patient's conscious mental processes, provide a missing continuity, and make conscious mental processes otherwise inexplicable explicable.

Grunbaum questions the credentials of free association as a method that, it is claimed, warrants conclusions about causes. According to him, such claims are based on a misextrapolation from the experiences described in *Studies on Hysteria* (Breuer and Freud 1893–1895); the mere appearance of a content in a stream of associations cannot provide warrant for the conclusion that an event has occurred or has causal relevance. Despite Grunbaum's overall appreciative depiction of Freud as sophisticated and knowledgeable about methodological issues (even granting that Freud failed to solve methodological problems of which he was aware), Grunbaum's portrayal of the reasoning Freud used in his method of free association suggests that Freud was here guilty of the most egregious errors in logic and gross causal fallacies. It seems more likely to me that Freud believed his inferences from data obtained by the method of free association were justified by a complex set of premises about the mental apparatus, adumbrated, for example, in *The Interpretation of Dreams* (see Edelson 1984, pp. 134–135 and the previous discussion in this chapter of Freud's two theorems). I would prefer a criticism of Freud that raised questions about the credibility of these premises, rather than one that here suddenly has him become a simpleton.

Because I want to argue for the *potential* evidential value of the psychoanalyst's clinical data, I may pay more attention than it deserves, given the power of Grunbaum's overall performance, to his strange lapse (as he attempts to discredit the method of free association) from his usual high level of fairness, his masterly grasp of the complexity of the theory he critiques, and the care and generosity with which he reads other passages of Freud in the light of his knowledge of the context of the passage and aim of its author. The lapse to which I refer is apparent when, in writing about the Irma Dream as the paradigmatic depiction of free association. Grunbaum takes the associations Freud reports and on the basis of these mocks Freud's claims for free association—to the effect that unconscious infantile wishes can be discovered by free association that have been and are active and which have played a causal role in dream-formation. Grunbaum zeros in on the fact that Freud's reported associations involve recent conscious wishes only. Why, here, in particular, does Grunbaum ignore Freud's many explicit statements that on grounds of discretion he was giving a much edited approximation to the products of free association, an imperfect illustration of what a patient's free associations might sound like—wouldn't Freud have been the first to acknowledge that this "patient" (Freud) was having a good deal of difficulty suspending judgment and conscious control over the direction of his thinking and was reluctant to utter aloud what he was in fact consciously thinking?

Grunbaum should not be blamed for this particular lapse, because Freud's unsolved problem in exposition is in part also conceptual and it has never been solved in the psychoanalytic literature. With respect specifically to the problem of the evidential value of what is produced by free association, Grunbaum, with his customary perspicacity, seems here to have at least touched a nerve. For surely the degree of success with which any procedure is carried out must make a difference in the evidential value of what is obtained through its use. And just what difference does the degree of success make, and how is this difference to be detected and calibrated? About these matters, Freud and others in the psychoanaytic literature have very little to say.

In fact, neither Grunbaum nor the many discussions of free association in the psychoanalytic literature carefully distinguish the procedure itself, and the degree of success with which on a particular occasion it is implemented, from the various kinds of productions obtained when attempts are made to carry out the procedure, and the properties these productions manifest on particular occasions of attempted use of the procedure (especially those properties that bear on their evidential value).

Theoretical presuppositions always underlie a procedure for obtaining data in science. What theoretical presuppositions about the mental apparatus do in fact underlie free association (regarded now solely as an instrument for obtaining data relevant to the assessment of the credibility of psychoanalytic hypotheses)? What empirical support exists independently of the use of the procedure itself for these presuppositions? We psychoanalysts have not even precisely specified what variables (e.g., suspension of conscious purposiveness) constitute the procedure itself; how to estimate the values these variables take (e.g., the degree to which such suspension is achieved); and how to determine just what properties of communications change, and in what way, as these variables take different values. The answers to such questions must be sought by psychoanalysis in the years ahead if it is to refute convincingly Grunbaum's claim that there is no way to differentiate among the mass of clinical material those communications having greater degrees of evidential value from those having lesser degrees.

According to Grunbaum, Freud's hypothesis that all dreams have their genesis in, or are caused by, an unconscious infantile wish is also not adequately, if at all, supported by evidence. "Support," of course, involves more, as previously mentioned, than the enumeration of positive instances.

Freud's causal claims or hypotheses about dream-formation are not nearly so unambiguous or consistent as Grunbaum's critique suggests they are. In particular, the causal status of wishes, in general, and of unconscious infantile wishes, in particular, in dream-construction is not nearly so straightforward as his critique would have it.

It is possible, for example, to argue from *The Interpretation of Dreams* that Freud placed a great deal of reliance on thematic affinity among manifest dream images and various associations in interpreting dreams, but also that he emphasized the lack of thematic affinity among dream images and the latent mental contents from which they have been derived by various transformational mental operations.

I have myself argued (Edelson 1975; and Chapter 2 of this book) that in certain passages in *The Interpretation of Dreams* Freud seems to be proposing that the dream work functions primarily to disguise wishes because these are objectionable to the dreamer. However, in other passages, Freud seems to be proposing something quite different. The dream work involves a set of mental operations for representing "ideas." (Ideas, as Freud uses the term, include wishes.) These operations are activated in certain states of consciousness (e.g., sleep), and function to represent ideas in a certain way whether or not a motive to disguise a wish that is to be represented by the dream exists in the dreamer's mind. Such mental operations, then, on this latter view, only incidentally come to be used to censor or disguise a wish in those particular cases—by no means all cases—when some wish represented by the dream is objectionable to the dreamer.

In this latter reading, dreams are *not* like neurotic symptoms. They do not, for example, necessarily involve conflict. They are simply mental representations constructed in a particular way, whose meanings can be unraveled by means of free association in the psychoanalytic situation. On this reading, Grunbaum's quirky notion that the consequence of Freud's theory (which theory?) is that, when dreams are interpreted and the patient becomes conscious of the wishes represented by his dreams, the patient should have fewer dreams, by analogy to the claim that symptoms will disappear when pathogenic unconscious conflicts become conscious, is somewhat of a non sequitur.

My response to Grunbaum's question about the truth-status of the represented-infantile-wish hypothesis is that, on my reading (which I have described in Chapter 2 of this book), what Freud intended to assert about the causal status of wishes in dream-production is by no means unequivocal. Grunbaum may have oversimplified in trying to decide just what is the nature of the conjecture Freud is making here. Further, again, given my reading of *The Interpretation of Dreams,* finding ways to justify belief in whatever this hypothesis may turn out to be may be even more problematic than Grunbaum's critique suggests.

The Scientific Status of the Psychoanalyst's Clinical Data

Grunbaum asserts that empirical warrant for the credibility of psychoanalytic hypotheses is, given scientific standards of method and reasoning, insufficient. (This is very different from asserting, as some critics of psychoanalysis have been wont to do, that it is justifiable on present knowledge to dismiss psychoanalytic hypotheses as either false or scientifically meaningless; the difference is a sign of the sophistication of Grunbaum's critique.) According to Grunbaum, the inadequacy of the empirical foundation of psychoanalysis is in large part a consequence of the stubbornness with which psychoanalysts have fallaciously argued and mistakenly concluded that their clinical data, obtained in the psychoanalytic situation, provide sufficient and satisfactory empirical support for the scientific credibility of their hypotheses. But in his view such data, because of the clash between essential characteristics of the psychoanalytic situation and essential requirements of scientific method and reasoning, cannot possibly *test* psychoanalytic hypotheses and thereby provide the required support. He challenges psychoanalysis, therefore, to produce instead the extraclinical—epidemiologic and experimental—data that are capable of testing the causal claims made by psychoanalytic hypotheses, for passing tests such as these is necessary if a contemporary psychoanalysis is to build the strong empirical foundation upon which its scientific credibility must ultimately depend.

There can be little question about the cogency of Grunbaum's detailed demonstrations that psychoanalysts have argued fallaciously when they have concluded that data obtained in the psychoanalytic situation provide scientific support for inferences that a particular causal event (whether situational or intrapersonal) has occurred and furthermore (assuming one is willing to grant that this event has occurred) that it bears a causal relation—is causally relevant—to the clinical phenomenon of interest. Examples are his documentation of psychoanalysts' dependence on the tally argument (with its empirically unjustified premises), on "confirmation" of hypotheses by enumeration of mere positive instances of them, and on *post hoc ergo propter hoc* reasoning. Yet, as should be clear from my book *Hypothesis and Evidence in Psychoanalysis* (1984), whereas I do agree with Grunbaum that the canons of eliminative inductivism are the canons of scientific reasoning and method,[6] that the case study investigations of psychoanalysis have involved serious violations of these canons, and that it is the responsibility of contemporary psychoanalysis to bring its investigations into conformity with these canons, I cannot agree with his conclusion that any reliance by psychoanalysis on data obtained from the psychoanalytic situation to support its causal inferences according to these canons is by the very nature of things doomed to failure and that psychoanalysis must turn instead entirely to epidemiologic and experimental research for such support.

Of course, I would have no trouble agreeing with Grunbaum that epidemiologic and experimental studies should be attempted. What one looks for in psychoanalysis as in any science is a convergence of conclusions based on tests of the same hypotheses carried out in different ways—intraclinically, epidemiologically, and experimentally; each way has its advantages and disadvantages with respect both to internal and external validity (Edelson 1984, pp. 125–126). Yet, Grunbaum seems to me to be uncharacteristically ingenuous in recommending carrying out epidemiologic and experimental studies to test psychoanalytic hypotheses (while also maintaining that it is fruitless to turn to clinical data to test these hypotheses), without at the same time emphasizing:

(1) the exceptional difficulties posed even in using such studies to test nonpsychoanalytic hypotheses in the clinical realm;

(2) how much greater the difficulties are in psychoanalysis, where causal variables are often intrapersonal (e.g., fantasies) rather than situational (e.g., parental behavior) and where it is impossible, unfeasible, or unethical to mobilize the phenomena of interest (e.g., intense and, for the subject, highly objectionable sexual or aggressive impulses) in anything like the relevant intensities outside the clinical situation.

This apparent ingenuousness is especially manifest in one rather startling lapse by Grunbaum from the scholarly appreciation he has shown throughout his book for what Freud is up to. Grumbaum seriously proposes that the putative causal link between paranoia and

6. But see, however, Chapters 13 and 15.

unconscious conflicts over homosexuality should be tested by comparing groups of subjects, one group from a society with more and another from a society with less tolerant attitudes toward homosexuality, on the assumption that it follows from psychoanalytic theory that the former group should have a lower incidence of paranoia than the latter. But many psychoanalytic hypptheses cannot be tested in this way just because it may not be a situational variable such as "social attitude" that is held to be causally relevant by psychoanalytic theory. What if the theory holds that unconscious conflicts about homosexuality develop in early childhood, more or less independently of the rationalized social codes of adult life even as these are translated into child-rearing practices; that these conflicts have their origin in the child's instinctual wishes, unfolding and varying in intensity relatively independently of the intrinsic features of actually and accurately experienced situations; and that these wishes are expressed in the child's fantasies about his own and others' bodies, about certain kinds of acts, and about the imagined consequences, both desired and undesired, of carrying out these acts? This is a possible, interesting theory, whether true or not, even if its testability by the methods Grunbaum urges upon psychoanalysis in place of its reliance on clinical data is problematic.

In short, I am not inclined to quarrel with Grunbaum's diagnosis, whatever quibble we may have over details, but I am prepared to question his remedy. I have made by own proposal for dealing with the problems he has so compellingly identified (Edelson 1984, especially pp. 157–160), and this includes in part: (a) clarification of just what the hypotheses are that require testing; (b) exploitation of recent conceptual and methodological developments in single subject research; (c) building upon the directions set by Luborsky (1967, 1973; Luborsky and Mintz 1974), who has used his symptom-context method to test hypotheses according to the canons of eliminative inductivism in single subject studies, and Glymour (1980), who has explicated how a bootstrap strategy may be used to test hypotheses in such a case study as the Rat Man; and (d) exploitation of methods recently developed in the social sciences for making valid causal inferences when data are nonexperimental or qualitative.

Grunbaum has every right (and perhaps from his point of view every reason) to remain skeptical that such a program can or will be carried out and that taking such a direction will ever remedy the defects in the foundation of psychoanalysis he has identified. Here, of course, only time will tell. But, to my colleagues, I suggest, let us make a hopeful beginning. What if we were to adopt the minimal set of standards listed in the concluding section of Chapter 11; and these were the standards to be met by clinical case studies in order to be accepted in our psychoanalytic literature? (I refer here of course only to that literature that purports to constitute a scientific body of knowledge.)[7]

Can we then not imagine a day when Grumbaum, faced just by case studies meeting these standards—we will not now imagine confronting him also with studies involving more complex or sophisticated single-subject research methodologies—finds himself compelled to acknowledge that clinical data not only might but do contribute critically to the assessment of the credibility of psychoanalytic hypotheses?

7. As Chapter 13 makes clear, while conformity to these eight canons is a minimal requirement for testing empirical generalizations, conformity to them will not suffice in attempts to establish the scientific credibility of complex theoretical causal explanations.

13 Causal Explanation in Science and in Psychoanalysis: Implications for Writing a Case Study[1]

In response to those who demand epidemiologic and experimental tests of psychoanalytic theory (e.g., Grunbaum 1984), I have argued that the case study, now held in disrepute, is nevertheless necessary to test psychoanalytic hypotheses and, if properly formulated, can indeed test, not merely generate, hypotheses (Edelson 1984). I have in addition described various kinds of scientific arguments relating hypothesis and data that can be used in case studies to provide support for the credibility of psychoanalytic propositions (Chapter 11 of this book).

In this chapter I do not take for granted, as I have previously, the current standard paradigm of explanation and hypothesis testing. Instead I take as a starting point Freud's own view of what science is all about, and the strategy of explanation he used in his case study of the Wolf Man (1918). There he tells a causal story. He makes use of a causal pattern rather than a covering law model of explanation. He makes use of both etiological explanations, in which a cause is external to the object or system affected by it, and constitutive explanations, in which a cause is a relation among the internal constituents of the object or system affected by it. (Notice that "etiology" here does not mean "cause of pathology," since either an etiological explanation or a constitutive explanation can be given of pathology.) Any evaluation of the validity of one of Freud's explanations should take into account the kind of explanation it is.

I write this chapter to confound psychoanalysts (especially of a hermeneutic bent), philosophers of science,

1. This chapter is a revised version of Edelson (1986b).

methodologists, and other members of the academic community who doubt that the psychoanalytic case study can provide support for the scientific credibility of psychoanalytic theory. (It is also true that I write in part because I believe that, regrettably, most case studies in the psychoanalytic literature are not written in such a way that they meet criteria for such support.)

I argue three points: (1) Freud's intent in his case studies was to provide empirical evidence for causal claims. Although he was not always successful in achieving this objective, there is something to learn from his example even when he ran into difficulty. (2) Freud's strategy of explanation, although it involves reference to psychological entities (mental states), is based on the same conception of the causal structure of the world that informs causal explanations in the natural sciences—notwithstanding that his work does not conform to the paradigm of explanation (and ideas about hypothesis testing that go with this paradigm) dominating contemporary philosophy of science. (3) The psychoanalytic case study, when it successfully meets—as most times it does not—*criteria of adequacy* implied by my analysis of Freud's explanatory intent and strategy, can indeed provide evidence capable of supporting not only the scientific credibility of empirical generalizations in psychoanalysis but the credibility of causal explanations involving theoretical entities as well.

Freud's Strategy of Explanation

The Problem of the Retrospective Study

Freud begins by issuing some disclaimers. He acknowledges that a study of a childhood illness based on a later account of it by an adult patient, although instructive, is not as likely to be as convincing as a study of the child's illness at the time he suffered it. He points out, however, that the study of the child has disadvantages; the material is less rich, especially because words and thoughts have to be supplied to the child.

The issue here has to do with the many problems that have to be addressed in establishing both facts and causal connections retrospectively. It is difficult to argue convincingly from the retrospective reports of an analysand alone that particular childhood events (actual or imagined) such as "the primal scene" really did occur and occurred at particular times. Even if one accepts that an event did occur, it is even more difficult to argue from such reports alone that the event was in fact the cause of the analysand's much later behavior or symptoms. Just how difficult this is in a psychoanalysis, E. Kris (1956a,

1956b) has shown. Freud struggled with these difficulties throughout the Wolf Man's psychoanalysis, and the strain is evident in the case study.

The extent to which these problems are at least partially solved in a retrospective case study determines how credible its conclusions are. Gould (1986), discussing Darwin as a historical methodologist, tells us what kind of scientific arguments can make hypotheses about time-remote events scientifically credible.[2]

The Problem of Presuppositions

Freud makes a somewhat more sweeping disclaimer—one which might be taken as made to evade the consequences of failing to construct a convincing argument. He has not published the case study, he writes, to convince those who are skeptical and refuse to accept the word of the scientific investigator, but has published it only to provide new facts for those already convinced by their own clinical experiences.

What does this conviction Freud mentions concern? He is not explicit. Surely, not his conclusions, which he will argue follow from the data he presents in his case study. Does he mean that the reader for whom he writes is one able to accept that the facts are as the investigator reports them to be? If the facts are the patient's reports, as they are in this case, and not empirical generalizations, then Freud's expectation of his reader does not seem unreasonable. However, questions might arise if a reader were to distrust the investigator's memory or written reconstructions of what the patient has reported.

More probably, Freud expects that his reader will be able to accept, at least provisionally, inferences made in the course of the argument about the patient, when these inferences are based on what the patient reports in conjunction with parts of psychoanalytic theory that are presumed to have previously been shown to be scientifically credible. For example, Freud makes inferences in interpreting the wolf dream that presuppose his theory of dreams.

This expectation remains troublesome because, even if a certain knowledge of psychoanalysis (whether based on clinical experience or not) is necessary, the convictions of a knowledgeable reader too may be groundless. That is, by objective scientific standards the presupposed propositions may not previously have been adequately established as scientifically credible. Furthermore, even a knowledgeable reader can be skeptical about the status of such propo-

2. See Chapter 15 for a further and extended discussion of this point.

sitions. If questions about presupposed propositions are likely to arise in a reader's mind, the author should refer him to some other study or studies in which there is at least some valid argument for the scientific credibility of these propositions—not, as is so often the case, merely an enumeration of anecdotes or vignettes that are consistent with them.

A Causal Story

Freud tells a *causal* story. He links particular causes to particular effects. A particular cause may not generally (in all other cases) produce this particular effect, although psychoanalytic theory may postulate that a particular cause *of this type* may in general produce a particular effect *of this type.*

Those regarding psychoanalysis as a hermeneutic discipline regularly overlook that the story told is a *causal* story, which makes causal claims. These claims require substantiation.

1. *The existence of causes.* The causes must be shown to exist. The noun "cause" refers to an actual event, state of affairs, or entity that possesses the power to produce an effect. The verb "causes" does not refer to a conceptual or logical relation between facts, but rather to a brute existential connection between facts. A causal connection is not part of a thought about what is out there in reality; it is a part of what is out there in reality.

Freud, believing that unconscious mental states are causes, exerts himself to demonstrate that these entities actually exist in the patient's mind; they are not just conceptual fictions or mere figures of speech. The ability of an investigator to intervene and manipulate causes, especially when these causes are hypothetical, and even to use hypothetical entities to produce effects in domains other than those in which they serve an explanatory role, argues for their existence (Hacking 1983). For this reason, the importance of the role of hypnosis in the origins of psychoanalysis cannot be underestimated, for here the investigator is able to implant an idea that clearly has causal efficacy with respect to the subject's subsequent actions, although throughout the course of these actions the subject remains unconscious of it.

Freud is constructing a causal account, not simply arguing for the therapeutic efficacy of a procedure. Therefore, he is rightfully concerned in this case study with his patient's immediate and long-term responses to his interpretations. His interventions are meant to influence the properties and powers of unconscious entities, which actually exist in the patient's mind.

So Freud mobilized his patient's incredulity, and "then had the satisfaction of seeing [the patient's] doubt dwindle away, as in the course of the work his bowel began, like a hysterically affected organ, to 'join in the conversation,' and in a few weeks' time recovered its normal functions after their long impairment" (1918, p. 76). Similarly, Freud comments: "The patient accepted this [construction] . . ., and appeared to confirm it by producing 'transitory symptoms.' A further additional piece which I had proposed . . . had to be dropped. The material of the analysis did not react to it" (Freud 1918, p. 80).

The efficacy of interpretative interventions is a major argument in psychoanalysis for the existence of the unconscious ideas it postulates as causal. Establishing that efficacy in studies of the analysand's responses to interpretation throughout the psychoanalytic process should, therefore, have a high priority in psychoanalytic research.

Freud's treatment of causes in this case study is of major importance to psychoanalytic theory, because here the question of the relative roles of external traumas and fantasies looms large. One of Freud's few discussions of the kinds of criteria that might be used to distinguish memories of actual events from memories based on fantasies occurs in this case study. Following Freud's conjecture that the sudden change in the patient's character after the summer of his third year was in part at least the result of a supposed castration threat by the patient's governess during that summer, the patient reported some dreams, which involved material concerned with aggressive actions against his sister or governess and punishments he received because of these. But the content was vague, and the dreams gave the impression of making use of the material in different ways. Freud concludes: "the correct reading of these ostensible reminiscences became assured: it could only be a question of phantasies, which the dreamer had made on the subject of his childhood at some time or other, probably at the age of puberty, and which had now come to the surface again in this unrecognizable form" (p. 19). On the other hand,

> his seduction by his sister was certainly not a phantasy. Its credibility was increased by some information which had never been forgotten and which dated from a later part of his life, when he was grown up. A cousin who was more than ten years his elder told him in a conversation about his sister that he very well remembered what a forward and sensual little thing she had been: once, when she was a child of four or five, she had sat on his lap and opened his trousers to take hold of his penis. (P. 21)

2. *The time relation between cause and effect.* Causes must be shown to exist prior to or simultaneously with their putative effects.

The relation between cause and effect is not symmetrical as is, in contrast, the relation between variables in generalizations about co-occurrence or correlation. Establishing the facts of chronology is especially important when cause is held to be prior to effect. Just one way to demonstrate chronology is, as in an experiment, to intervene and manipulate the cause, and observe what follows this intervention. This way is, however, not available in the psychoanalytic situation, where a retrospective reconstruction of chronology is at issue.

There can be no question that Freud was well aware of the importance of chronology to a causal account. For example, he questions whether one attitude in his patient existed simultaneously with, or replaced, another.

> Whether these contradictory sorts of attitudes towards animals were really in operation simultaneously, or whether they did not more probably replace one another, but if so in what order and when—to all these questions his memory could offer no decisive reply. He was also unable to say whether his naughty period was *replaced* by a phase of illness or whether it persisted right through the latter. (P. 16)

The incessant attention to chronology that Freud pays throughout the Wolf Man case is not motivated simply by his wish to show that the facts of the case favor his own explanation of phenomena, as caused by sexual experiences and motives stemming from *very early* periods of childhood, over Jung's explanation of the same phenomena, as caused by difficulties in coping with current tasks in *adult* life and especially failures in fulfilling high cultural aspirations. Much more, Freud's attention to chronology is motivated by the necessity to substantiate causal claims.

In his attempt to derive the chronology of remote events from a much later account of them, Freud was not, as one should expect, completely successful. His struggles with the problem are manifested in inconsistencies and contradictions in the way he dealt with it. In the text he writes "*certainly* before his fourth year," whereas the editor notes that "In the editions before 1924 this read '*perhaps* in his sixth year' " (p. 13; my italics). The comment "He *must* have been very small at that time" is followed by this note: "Two and a half years old. It was possible later on to determine almost all the dates with *certainty*" (p. 14; my italics). Freud does not reveal his method for determining any of the dates with such certainty. This omission is serious with respect to his argument that the data of the case study support his causal account, because even the plausibility of much of his causal account—to say nothing of its credibility—so much depends on chronology.

Indeed, any psychoanalyst can confirm how difficult it is in fact to establish chronology with any certainty in the psychoanalytic situation. The difficulty is exemplified by many passages from Freud's account (see pp. 15, 37, 76n). However, his frequent claims of certainty about chronology cannot help but shake the reader's confidence in the accuracy, perhaps even the honesty, of Freud's account of the facts, leaving aside the question of his explanation of them.

"*No doubt* was left in the analysis that these passive trends had made their appearance at the same time as the active-sadistic ones, or very soon after them" (p. 26; my italics).

"Since Christmas Day was also his birthday, it now became possible to establish with *certainty* the date of the dream and of the change in him which proceeded from it. It was immediately before his fourth birthday" (p. 35; my italics).

I do not myself believe that the problem here lies in Freud's character but rather with a limitation in his conceptual preferences and knowledge. He is concerned to maintain a strict determinism, and statistical reasoning and probability, so much a part of contemporary science and so necessary in dealing with uncertainty, does not seem available to him. He rarely relativizes his conclusions to current knowledge: this is what I can assert now, but I do not know what other factors may eventually turn out to make a difference or how much of a difference they might make. That there are irreducibly random processes in the world is, of course, a notion he rejected.

Now we can say that it would have been better if Freud had indicated even roughly the probability that supposed states of affairs had occurred in the way he supposed, and in each case exactly what evidence led him to even an estimation of that probability. The moral for the writer of a case study, if he wishes to retain the reader's confidence, is to avoid Freud's example here, acknowledging uncertainty where it exists and qualifying his conclusions accordingly. (Such a stance is, of course, just what one would expect a psychoanalyst to have as he offers explanations to an analysand in the psychoanalytic situation.)

It is also possible that Freud at times is so positive about a chronological inference on theoretical grounds. He has inferred a chronology from parts of psychoanalytic theory that he regards as not in question in this case study. For example, we have the following passage: "I naturally assumed that these obvious symptoms of an obsessional neurosis belonged to a somewhat later time and stage of development than the signs of anxiety and the cruel treatment of animals" (p. 17).

Freud does not refer the reader to the grounds for his assumption. Even though the reader with some knowledge of the psychoanalytic theory of development may guess what these grounds were, I do not recommend that a writer follow Freud's example here either. Psychoanalysis is a body of knowledge that is important to a wider intellectual and scientific community than that comprised by psychoanalytic practitioners. In its documentation, it should follow the scholarly practices and accept the standards of that community.

3. *Causal powers.* The writer of a case study must show that a cause has the power to produce its effects by virtue of its structure or properties, and also just how—by virtue of what processes or mechanisms—its causal influence is propagated from one space-time locale to another.

Freud attributes properties and causal powers to unconscious mental contents and processes different from those possessed by conscious mental contents and processes.

> Our bewilderment arises only because we are always inclined to treat unconscious mental processes like conscious ones and to forget the profound differences between the two psychical systems. . . . the whole process is characteristic of the way in which the unconscious works. A repression is something very different from a condemning judgement. (Pp. 78ff)

He describes *mental operations* (mechanisms), which his causes set into motion. These mental operations transform psychological entities (mental states, or particular constituents of mental states such as mental representations). Such operations are not exhausted by a list of defense mechanisms. Examples are:

a) Substitution of one object for another as the object of a mental state or event. Examples: the patient replacing his sister with his governess as the object of his anger; replacing his sister with his nurse, and then his nurse with his father, as the object of his sexual wishes; replacing others with himself as the object of his sadistic wishes; replacing himself in his fantasies with unidentified boys or an heir to the throne as the object of his sadistic wishes.

b) Substitution of one kind of mental state or event for another. Examples: replacing one kind of wish (to play with his penis) with another (to torment others); replacing *memories* of himself as a passive victim with *fantasies* of himself as an active aggressor.

c)Altering the intensity or importance of a mental state or event. Example: exaggerating his love for his nurse.

Use of mental operations may be motivated (caused) by a wish to avoid a painful state of affairs—for example, states of affairs in which loss, shame, or guilt are experienced.

4. *Rival explanations.* The investigator must argue that it is more likely, given his data and the way they were obtained, that what he claims produced the effect in which he is interested did in fact cause it, rather than some other factor that might have done so.

Freud argues that the data of his case study favor his explanatory account of all the phenomena over Jung's and others' (1918, pp. 7n, 48–60, 103n). His account gives causal priority to childhood traumas or fantasies occurring well before culture impinges in any direct way on a child and well before the child aspires to realize "higher" cultural values and aims. The rival account, which gives causal priority to cultural influences, postulates that the fantasies a patient reports are formed later in life and merely carried backward to childhood.

Freud also argues that data favor his explanatory account of such specific phenomena as the patient's inveterate idiosyncratic heterosexual object choice and the patient's sudden change of behavior in his third year of life. Adler would have given another explanation of the former (pp. 22f.), and the boy's relatives do give another explanation of the latter (p. 15).

Freud's determination to refute rival views is not, from the point of view of science, an idiosyncrasy or fault. Case studies that have a strong impact, changing the beliefs held about a particular domain, almost always achieve this effect because of a convincing argument that the data reported favor the investigator's conjectures over rival conjectures. An investigator should make explicit what rival conjecture he is attempting to eliminate and what data provide a warrant for dismissing it.

Covering Law Model vs. Causal Model

The telling of a causal story as an explanation bears little resemblance to the received view of scientific explanation, the so-called covering law model. "According to the 'received view,' particular facts are explained by subsuming them under general laws, while general regularities are explained by subsumption under still broader laws" (Salmon 1984, p. 21). Grunbaum contrasts the covering law model of explanation with causal explanation in the following passage:

> It is crucial to realize that while (a more comprehensive law) *G* entails (a less comprehensive law) *L* logically, thereby providing an explanation of *L*, *G* is *not* the 'cause' of *L*. More specifically, laws are explained *not*

by showing the regularities they affirm to be products of the operation of *causes* but rather by recognizing their truth to be special cases of more comprehensive truths. (Grunbaum 1954, p. 14)[3]

In the Wolf Man case, Freud certainly does not seem to be subsuming the particulars of his case under some general law, from which—given particular circumstances ('initial conditions')—they might be deduced, and which might be said then to explain them. He is describing a pattern of cause-effect connections; phenomena are explained by the way they are fitted into this pattern.[4]

In explaining phenomena in the Wolf Man case, Freud moves freely among interrelated but essentially independent theories. These theories are not deducible from each other or from one small set of axioms, postulates, or basic laws. He makes use of the psychoanalytic theory of dreams (e.g., in interpreting the wolf dream as evidence for a primal scene experience or fantasy early in childhood); the psychoanalytic theory of psychosexual development (e.g., in dating the chronology of symptoms); and the psychoanalytic theory of neurosis (e.g., in explaining disturbed behavior as the result of a regression from later to earlier kinds of sexual wishes). Psychoanalytic theory is clearly demonstrated in this case study to be a *concatenated* theory, whose components "enter into a network of relations" and, "typically, converge upon some central point, each specifying one of the factors which plays a part in the phenomenon which the theory is to explain," rather than a *hierarchical* theory, which "is a deductive pyramid, in which we rise to fewer and more general laws as we move from conclusions to the premises which entail them" (Kaplan 1964, pp. 298f.).

The fruits of Freud's strategy of explanation should be evaluated with this distinction in mind. In the received view (covering law model of explanation, hierarchical theory), deducing particular facts from empirical generalizations, and empirical generalizations or less general laws from more general laws, is a major method for testing conjectures. If a more general law is true, what is predicted—because it follows—from it, is expectable (certain or at least highly probable). If the prediction fails, the more general law is falsified. However, deducing predictions is not necessarily the way to test theory if it is concatenated, as psychoanalytic theory is, and if one uses, as Freud does, a causal pattern model of explanation.

3. The covering law model of explanation is trenchantly criticized by Cartwright (1983) and Salmon (1984).

4. Kaplan (1964, pp. 327–336) gives a general account of the pattern model of explanation.

Etiological vs. Constitutive Explanation

Freud uses different kinds of causal explanation in his case study. In some instances (e.g., in detailing the responses of the patient to an external trauma—his sister's seduction of him), Freud makes use of *etiological* causal explanations. In other instances (e.g., in explaining a phobic or obsessive symptom as an expression of an internal constellation of conflicting motives), he makes use of *constitutive* explanations.[5] Salmon (1984, pp. 269–270, 275) gives a general account of this distinction.

Etiological explanations postulate causal connections in which cause and effect are at the same level, and causes are external to and impinge on the system whose properties are to be explained. A given fact is explained "by showing how it came to be the result of antecedent events, processes, and conditions" (Salmon 1984, p. 269). Example: the effect of one moving medium-sized body on the motion of another with which it collides.

Constitutive explanations postulate causal connections in which cause and effect are at different levels, and causes are internal to, or constituents of, the system whose properties are to be explained. Such explanations exhibit the internal causal structure of what is to be explained. Examples: pressure inside a container is caused by the impact of moving molecules on the sides of a vessel (pressure is an expression of such impacts); temperature is a function of, and is caused by, the velocities of molecules (temperature is an expression of such molecular velocities).

One of Freud's explanations—of the patient's marked change in character during his third year—seems to be a mixture of etiological and constitutive explanation. The patient had repeatedly heard during his adult illness that he had exhibited a marked change of character. Before the summer of his third year, he had been good-natured, tractable, quiet, but after that summer during which he was in the care of an English governess he became discontented and irritable; easily offended, he often screamed in rage. The patient remembered becoming upset when he was not given double presents one Christmas, which was also his birthday. He also remembered tormenting his nurse, whom he loved.

Members of the family offered two explanations of this change in him: the governess irritated him; dissension between the governess

5. Grunbaum (1984) evaluates psychoanalysis as if its main causal claims were etiological (i.e., attribute causal efficacy to external events or states of affairs) or, either etiological or constitutive, attribute causal efficacy to time-remote events or states of affairs. In my view, neither characterization is true of the *core* causal claims of psychoanalysis.

and his nurse upset him. Freud implies that these explanations are discredited by the fact that there was no change in the boy's unbearable behavior after the governess was sent away. (This is an example of Freud's use of a datum to exclude rival explanations of a phenomenon.)

Freud explains the boy's change of character as follows. An external trauma (seduction by his sister, who was also his rival) was a state of affairs some of the aspects of which caused pleasure (penis being touched) and some of the aspects of which caused pain (loss of masculine self-esteem resulting from his passivity). The external trauma caused a wish for a state of affairs that would cause the same kind of pleasure without causing the same kind of pain. That wish motivated (caused) a mental operation (substitution of nurse for sister in the state of affairs desired). An attempt to satisfy the transformed wish caused a new external trauma (nurse's rebuff and castration threat). The new external trauma caused a wish to avoid a repetition of this painful state of affairs. This wish in turn motivated (caused) various mental operations: substitution of father for nurse in the state of affairs desired; substitution of one kind of wish for another (wish to torment others for wish to masturbate); substitution of one object for another in the new desired state of affairs (himself for another as the one to be hurt). His naughty behavior had multiple internal causes—principally, his wish to torment his nurse and others; and his wish to provoke his father into beating him. By a chain of such external and internal causal connections, Freud explains the change in the boy's character.

In another explanation—of the patient's phobic symptoms—Freud gives an essentially constitutive explanation. (He tends to use "express" rather than "cause" when a constitutive explanation is given: "the symptoms express an unconscious conflict.") The phobic symptoms were caused, Freud explains, by the patient's retrospective interpretation of a memory or fantasy of seeing, at a very early age, his mother and father having intercourse, in the light of later sexual researches (and his preoccupation with castration); his simultaneous realization that gratification of his wishes for passive pleasure from his father seemed then to entail his being castrated; the fear of his father aroused by this realization; and displacement of this fear to animals.

Psychoanalytic Explanation

It is a common mistake to suppose that because psychoanalysis is concerned with meanings or purposes it is not concerned with causal explanation.[6] Some writers automatically assume that an explanation in terms of wishes and beliefs cannot be a causal explanation; they mistakenly contrast motivational and causal explanation. But, of course, motives are among the causes of psychological phenomena, as this chapter has endeavored to demonstrate.

Often the mistake seems to rest on equating "causal" and "mechanistic" and then "mechanistic" and "machinery." The notion of man as a machine is derided as nineteenth century science and contrasted with a humanistic view of man. But a mechanistic explanation, no matter what the content, is essentially any explanation that attempts to answer the question "How?"—by what steps, involving what causal processes? An explanation that employs such concepts as "mental operations," "shifts in emphasis," and "alterations in representations" to account for how a cause produces an effect is a mechanistic explanation.

A similar mistake, as Freud emphasized over and over, rests on equating "what is mental" with "what is conscious." This equation is usually made because the possibility that there really are unconscious mental entities is rejected. It is assumed instead that talk of unconscious mental entities must really be talk about neural events. The conviction here is that true causes cannot be mental but must be physical. It is the burden of this chapter[7] that this is not so.

To have explanatory knowledge is to have more than descriptive or predictive knowledge. It is to have a knowledge of underlying mechanisms. To understand a cause-effect relation is to know what the causal connection between them is.

The point is Salmon's and it is his account of causality and causal explanation (1984) that informs this chapter. It is true that he refers primarily to the natural sciences in his discussion of scientific explanation. However, his view of science and scientific explanation is in many ways close to the view one may infer from the Wolf Man case that Freud himself held. Both adhere to a causal rather than a covering law model of explanation. Both believe in the reality and the causal efficacy of theoretical (nonobservable) entities. ("Atoms" and "unconscious wishes" are nonobservable entities.) Both are "mechanistic."

A *causal process* is a process that transmits its own structure, its own features, and changes in its own structure from one space-time locale to another, and is therefore capable "of propagating a causal influence from one space-time locale to another" (Salmon 1984, p.

6. See Chapter 11 of this book.
7. And of Chapter 7.

155). One event *produces* or causally influences another event by means of the causal process connecting them. A causal process *propagates* causal influence from one space-time locale to another. There may be successive intermediate events between cause and effect; these are connected by spatiotemporally continuous causal processes. Both an arrow flying from bowman to target and a memory are examples of causal processes.

I maintain that Freud regards psychological entities (mental states) as causal processes. These transmit their own structure or features (for example, their contents), and changes in their own structures or features, through time. The significance of his abandoned Project was that it served him, even after he abandoned it, as a *model,* exemplifying cause-effect relations. His metapsychology was an expression of his commitment to causal explanation. A case study, such as the Wolf Man case, tells a causal story; it is a causal explanation of the Wolf Man's symptoms and their interrelations.

It is especially important to note that causes and effects in a causal story are particulars: "A causal process is an individual entity, and such entities transmit causal influence. An individual process can sustain a causal connection between an individual cause and an individual effect. Statements about such relations need not be construed as disguised generalizations" (Salmon 1984, p. 182).

A *causal interaction* occurs when two causal processes intersect in space-time and modify each other's structures. The modifications in structure are produced by the interaction and propagated by the causal processes after the interaction. Cause and effect here are *simultaneous.* Both two colliding balls that modify each other's motion, and the seduction of the Wolf Man by his sister (each one a causal process), are examples of causal interactions. (Freud, interestingly enough, goes to some length to describe what happened subsequently to the sister as well as to the Wolf Man—how each of them might have been modified by this interaction.)

Simpler interactions include the coming together and *fusion* of two causal processes, resulting in one ongoing causal process; and the *bifurcation* of a single causal process, resulting in two ongoing causal processes. Salmon (1984) gives as an example of fusion a snake swallowing a mouse, and as an example of bifurcation an amoeba dividing to form two amoebas. Just so, *condensation,* a mental process, may be considered an example of causal fusion of mental contents, each one of which before fusion was a separate causal process. And Freud certainly seemed to have some kind of causal bifurcation in mind when he spoke of *displacement* as involving the splitting of a mental content and its associated affect, with the content and the affect then as separate causal processes undergoing different vicissitudes. Similarly, differentiations in psychological development may result from causal bifurcations, and the blurring of boundaries between self representations and object representations in introjections and identifications may involve causal fusions.

Two or more causal processes that are physically independent of one another and do not interact with one another but that arise from the same set of conditions are said to have been produced by a *common cause.* Cause here is *prior* to effect. A causal story or pattern may include such constituents as the following: two independent causes produce a common effect; a common cause produces two causal processes that have a common effect.

Although causal processes and interactions may be governed by laws of nature, the same cannot be said about the origination of two or more independent causal processes by a common cause, which involves de facto conditions and nonlawful facts.

Salmon (1984) gives as examples the coincidence that two students' papers are identical, because they both independently copied from the same paper, and the coincidence that a number of people became ill at the same time, because they all independently ate from the same pot. A similar kind of coincidence, often observed in a psychoanalysis, occurs when an analysand utters exactly the same rather unusual phrase in a number of very different and apparently completely unrelated contexts in the same session, inviting inference to a common cause—for example, a hypothetical unconscious state of the analysand.

What is needed here now is an account of psychoanalytic explanation—however tentative, crude, or inadequate at this time—that will fit Freud's strategy of explanation in his case study, and that at a minimum will at least approximate the kind of explanations given by a psychoanalyst to an analysand. I have given such an account in Chapter 6, and so here, for the sake of the continuity of this discussion, I shall merely remind the reader briefly of the highlights of that account, with some additional comments motivated by this chapter's concerns.

The domain of psychoanalysis, which is an intentional psychology, is comprised by mental states such as wishes and beliefs and some of their properties and relations. In explaining inhibitions, compulsions, representational anomalies, or mistakes or failures—regarded as properties of mental states—psychoanalysis tries to answer two kinds of questions. *Why* does a mental state have such a property? This is a question about *what* motives (other mental states) cause mental states to have such properties. *How* do these motives cause mental states to have such properties? This is a question about causal powers—about mental mechanisms or operations.

In general, psychoanalysis explains a mental state (easily accessible to consciousness) as a substitution for a set of unconscious mental states. Mental operations have effected transformations on a set of unconscious mental states, and this process has produced the mental state(s) to be explained.

1. The psychoanalyst infers the existence of unconscious mental states. The mental activities that are constituents of these unconscious mental states are directed to imagined states of affairs (fantasies).

2. There are two kinds of imagined states of affairs, those the analysand wishes to actualize and those the analysand wishes to avoid. The imagined states of affairs the analysand wishes to actualize are states of affairs in which the analysand is experiencing sexual or sensual pleasure.[8] The imagined states of affairs the analysand wishes to avoid are states of affairs in which the analysand is experiencing disagreeable affects.

3. The psychoanalyst infers the existence of unconscious beliefs that states of affairs the analysand wishes to actualize will inevitably be followed by states of affairs he wishes to avoid, or will make impossible other states of affairs the analysand wishes to actualize. In other words, the psychoanalyst infers the existence of unconscious conflicts and dilemmas.

4. The psychoanalyst infers that the analysand is tempted to actualize a state of affairs that he believes will have one or more of these dread consequences, and that his unconscious realization that he is so tempted generates anxiety. The psychoanalyst infers that the anxiety motivates the analysand to develop and carry out strategies for resolving the conflict or dilemma in such a way that anxiety may be avoided.

5. The psychoanalyst infers the nature of these strategies—what symbolic operations the analysand has used upon what mental activities and representations to achieve this resolution. The mental states to be explained realize a state of affairs that to some extent and in some way yields pleasure, and yields it in such a way that the analysand is able to some extent and in some way to escape the consequences he wishes to avoid.

How does psychoanalysis explain that a particular mental operation has been used by an analysand? Often, it postulates that a relation among mental representations causes a choice of mental

8. In Chapter 6, I give my reasons for omitting mention of aggression here.

operations. Paradigmatic example: this representation reminds the analysand of that representation, because of certain similarities between them, similarities that are relevant to his wishes and fears. Therefore, the mental operation of displacement is used. Furthermore, mental operations are used which fulfill at one and the same time a number of wishes or resolve a number of conflicts or dilemmas. Constraints on the choice of strategies or mental operations include the analysand's psychological capacities (e.g., cognition, language, memory, perception), knowledge of which is presupposed, not generated, by psychoanalysis.

An analysand may maintain eternal vigilance—always prepared to avoid what is feared and to grasp what is desired—in order to avoid anxiety. The analysand can perceive any situation as one involving or at least promising pleasure or pain by focusing on characteristics (imagined or real) of others in his situation that are relevant to his wishes and fears, and by attributing to others in his situation motives, intentions, or attitudes toward himself. Interpreting then and representing every situation in such a way that it yields some pleasure and enables him at the same time to evade an imagined danger becomes a strategy for effecting a resolution of an unconscious conflict or dilemma. So the patient appears compelled to experience the same state of affairs over and over.

Here, of course, is the basis for the importance of transference phenomena in both psychoanalytic therapy and research. Grunbaum (1984) argues that the psychoanalyst is begging the question when he uses here-and-now transference phenomena as evidence for explanations (of current states such as symptoms) that give causal status to long past events. He questions the reasoning by which a psychoanalyst claims, first, that current mental representations of particular past events or fantasies are constitutive causes of current behavior, and then goes on to claim, second, that therefore past actual events or fantasies are causes of the analysand's symptoms. For the characterization of transference phenomena as involving a transfer from the past to the present assumes that the past occurrences existed, and furthermore that their causal influence has been propagated through time by accurate mental representations of them. But the truth of these assumptions is the very question at issue.

Transference phenomena are nevertheless non-question-begging evidence for constitutive explanations. Such explanations involve inferences about causally efficacious psychological entities (unconscious mental states) existing or occurring in the here-and-now (including, e.g., memories or fantasies of the past, in which early important objects play a part in the states of affairs represented). Giving, in addition, an account of the *origin* of such causally efficacious present psychological entities (unconscious mental states) is logically independent of the constitutive explanation.

6. The psychoanalyst may offer a genetic explanation. He may locate the origin of present unconscious wishes and beliefs in the analysand's infantile experiences with his body and with parents and siblings, and attempt to show how the influence of these was transmitted and altered through time by developmental and environmental factors. However, genetic explanations are at a different level from constitutive explanations, in which cause lies in the presence of an unconscious conflict or dilemma in the here-and-now, even if current unconscious wishes for pleasure are postulated to be infantile (the kind of wishes a child might have).

It is true that Freud and others do try to explain the existence of unconscious conflicts and dilemmas in terms of origin in early childhood; as a statement of remote etiology of pathology such formulations are part of psychoanalytic theory. However, the Wolf Man case has features (e.g., difficulties in establishing chronology of remote events) that suggest that the psychoanalytic case study might in general offer stronger evidence for the scientific credibility of constitutive explanations (a current symptom is caused by or expresses a currently existing and active unconscious conflict)—or for etiological explanations that involve very recent, putatively causal events (a recent event, perhaps occurring in the psychoanalytic situation itself, causes a phenomenon such as an intensification or transient appearance of a neurotic symptom)—than it can offer for the scientific credibility of explanations that involve remote, putatively causal events.

7. The psychoanalyst also may show an analysand what consequences the resolution he has attempted has for adaptation. This kind of explanation, too, is at a different level from his constitutive explanation of the analysand's resolution as an effect of an internal complex of unconscious mental states and mental operations, for here the analysand's resolution itself becomes the cause of other effects—his relations with external reality—and these effects in turn may act as causes of additional difficulties for the analysand.

In summary, the phenomena to be explained—inhibitions, compulsions, representational or other anomalies, and mistakes or failures of the kind described—are psychogenic or, to use Wollheim's felicitous term (1971), ideogenic. Their causes are unconscious mental states, and the transformations effected upon the constituents of these states by symbolic operations of a certain kind.[9]

9. This account needs to be supplemented by the discussion of imagination, fantasy, and defense in Chapter 9.

Implications for Psychoanalytic Research: Writing a Psychoanalytic Case Study

A psychoanalytic case study may be written to support an empirical generalization, or to tell a causal story.

Empirical Generalizations

A psychoanalyst might want to establish that certain kinds of observable phenomena co-occur or are correlated, and what the strength of their relationship is.[10] (Freud's Wolf Man case has no such objective.) The requirements for such a case study are:

1. An empirical generalization, which goes beyond the particular observations made, is clearly and prominently stated.

2. The phenomena are, in principle at least, intersubjectively observable, and documented in such a way that different judges can agree independently of each other about what they are. Durable records of a psychoanalyst's and an analysand's utterances meet this requirement.

3. The author specifies what observations would count against the generalization, and recounts the attempts he has made to discover if these exist. If he discovers any such facts, and does not reject the generalization, he uses them to limit its scope—he specifies under just what conditions, in which observations might be made of this case, he would expect the generalization to hold.

4. Hypotheses are underdetermined by data. That is, no set of data can decide that one and only one hypothesis is true. (Many different lines can be drawn through the same sample of points.) Since there are many empirical generalizations that will account for the same observations, the author gives some argument why his observations are better evidence for his empirical generalization than for at least one comparable competing or rival one. For example, he may claim that he makes a higher proportion of correct guesses to incorrect guesses with this generalization than with some other about as yet unobserved phenomena (in later sessions, if the generalization has been made on the basis of data from earlier sessions). To make this argument, he must keep count of his incorrect guesses as well as his correct guesses. Other arguments, many of which are nonquantitative, are also available to an author (Edelson 1984, and Chapter 11 of this book).

10. The studies of Luborsky (1967, 1973) and Luborsky and Mintz (1974) are valuable exemplars of this kind of study in psychoanalysis and they are discussed as such by Edelson (1984).

5. His observations and the way they have been obtained may favor his over a rival generalization. However, he still must consider what factors in the situation of observation could have resulted in his obtaining just these observations even if his generalization were false. He presents some argument—it may be a nondesign argument—for dismissing at least one such factor as a plausible alternative explanation of his having obtained the data he claims favor his empirical generalization over a rival.

6. He makes clear in what other cases or kinds of cases he claims the same generalization will hold, and offers some grounds for this claim.[11]

If the relationship thus established has not in addition been demonstrated to be a causal relationship, it is not an explanation of anything. If it were true that most of the time if an analysand talks about fire and water early in a session, he talks about ambitions later in that session (Meehl 1983), such an empirical generalization (even if it held universally rather than assigned some probability to the later event) would not be an explanation of the fact that an analysand in a particular session talked about ambition later in a session in which he had earlier talked about fire and water. The generalization (together with the occurrence of the antecedent event) entails that the later event will occur (with some degree of probability), but it does not *explain* its occurrence. Nor does the occurrence of the prior event itself explain the occurrence of the later event.

Such an empirical generalization—even if one were able to represent it, which seems unlikely, by a sophisticated mathematical equation describing a functional relationship among quantities—is, at most, a regularity or a pattern in nature to be explained, not by subsuming it under some more comprehensive generalization in a deductive argument, but by discovering the causal nexus in which it is embedded. What the writer of this kind of case study has achieved—it is not a small achievement—is to establish that such a regularity does in fact exist, and that it calls for explanation.

One could well argue that it is the facts of psychoanalysis that are most in question in the scientific community, and that such studies as these are badly needed to establish the scientific credibility of psychoanalytic theory. I have already reviewed recent developments in the methodology of the single case study (1984) that make it feasible to use single case studies to demonstrate empirical generali-

11. That these six requirements can be met in a case study is argued in Edelson (1984) and Chapter 11 of this book. But for reasons to have some reservations about this use of the case study in psychoanalysis, see Danziger (1985).

zations. However, contrary to what some might think, establishing empirical generalizations is not the sine qua non of science. For those who think so, a case study such as Freud's must be baffling indeed.

Writing a Causal Story

1. *What is to be explained. Distinguishing facts and inferences.* The writer organizes a set of facts about a case so that they tell a story about causes and effects. A fact in such a story is neither just something to be explained nor something that explains, but is usually cause in relation to some effect(s) and effect in relation to some cause(s)—a constituent in a pattern of cause and effect relations. Such a pattern may be, but need not be, a time-ordered sequence of events; feedback loops, for example, may also belong to such a pattern.

The writer may begin by specifying the phenomena that he intends to explain, essentially by showing how they are causally interconnected.

> What was the origin of the sudden change in the boy's character? What was the significance of his phobia and of his perversities? How did he arrive at his obsessive piety? And how are all these phenomena interrelated? (Freud 1918, p. 17)
>
> But if we put together as the result of the provisional analysis [of the dream] what can be derived from the material produced by the dreamer, we then find before us for reconstruction some such fragments as these: *A real occurrence—dating from a very early period—looking—immobility—sexual problems—castration—his father—something terrible.* (P. 34)

The difference between facts and inferences is sometimes glossed over in a psychoanalytic case study.[12] As Freud has intimated in his beginning disclaimers, it is in connection with what is presented as fact and what is presented as inference (rather than in connection with the causal story itself) that the reader is most likely to boggle. Freud goes to some length to show why propositions from his theory of dreams conjoined with statements of the analysand's seem to him to entail the inferences underlined in the passage above, which he then treats as facts whose interrelations will be explicated by his causal story. The writer should follow his example, carefully differentiating facts and inferences, and giving some justification for treating what is inferred as fact.

After presenting a set of facts, the writer then tells a causal story

12. See Chapter 11 of this book.

that explains these facts—that is, that places them in a pattern of cause and effect relations. "Story" is an apt term. Cause and effect are events. Each of these occurs in a space-time locale. The process that connects them—producing an effect by propagating causal influence from one space-time locale to another—does so by transmitting its own structure from one space-time locale to another.

2. *Common causes.* Locating or specifying a common cause may be an element in a strategy of causal explanation.

a) If improbable events occur—even more improbably—together, the writer infers to a common cause. He notices a coincidence—the occurrence of two improbable events, which supposedly arise from physically independent causal processes, and whose joint occurrence, given the probability of the occurrence of each one, is therefore highly improbable. He infers—since the probability of their joint occurrence, which has in fact occurred, is evidently not so low—that they must have an antecedent common cause.

In other words, if they have a common cause, then the probability of their joint occurrence is greater than would be the case if they were truly independent. The probability of the joint occurrence of independent events is the probability of the occurrence of one multiplied by the probability of the occurrence of the other. If one is able to count the number of occurrences of each kind of event participating in a series of such coincidences and the number of such coincidences, one can calculate the relevant relative frequencies and show that the probability of joint occurrence is greater than would be expected if the events were truly independent. Statistics in a causal framework are evidence (but surely not the only kind of evidence) for causal relations (Salmon 1984, p. 261).

b) The writer then locates and specifies the common cause, and demonstrates that its structure or features are such as to give rise to causal processes producing the events in question.

c) If the common cause is nonobservable, he justifies the assertion that it is in fact the explanation of the coincidence by describing with as much precision as possible its structure and features, and then using this description to describe how it produces still other causal processes, which lead to still other events—of very different kinds—that can be and are observed.

In general, then, the author should make use of heterogeneous data in writing a causal story—data that are different in kind and that, given background knowledge, there is no reason to assume have any connection with each other. In the Wolf Man case, Freud explains a behavior change; different symptoms, beginning at different ages; a type of heterosexual object choice—and their interconnections.

A strong argument is made when the same inference (that particular unconscious mental states currently exist) can be made from the very different kinds of information available to the psychoanalyst. Examples of such kinds of information include: responses to interpretations, including exacerbation of symptoms or the temporary appearance or reappearance of symptoms; memories, including the analysand's memories of what he has been told about his past by others (what Freud calls "direct tradition"); dreams; parapraxes; what the analysand reports of his responses to extra-analytic intercurrent events; daydreams; and the analysand's responses to aspects of and events occurring in the psychoanalytic situation, including his speculations and fantasies about the psychoanalyst.

Similarly, the author of a case study uses the very different parts of a concatenated theory such as psychoanalytic theory to make inferences about nonobservable entities and to forge causal links. Different (interrelated but independent) parts of the theory sometimes suggest different causal factors and different causal connections as the author attempts to tell a causal story. Different parts of the theory may also converge on the same nonobservable mental states or mental operations (causal processes or modes of propagation of causal influence).

The principle of the common cause can be used to justify claims that nonobservable entities, such as unconscious mental states, exist. If such a claim is true, different independent methods should lead to the same state of affairs or value (the existence of one and the same unconscious mental state). When this in fact happens, that it should be a coincidence is so highly improbable that science proposes instead to accept as credible the existence of the postulated nonobservable entity as the common cause of reaching in each case the same state of affairs or value (Salmon 1984, pp. 211–238).

3. *Causal interactions.* Another element in a case study writer's strategy of explanation, similar to that involving inference to a common cause, involves inferring a causal interaction when modifications in causal processes are observed to be correlated—for example, changes in the Wolf Man and his sister following their interaction (the seduction).

4. *Causal gaps.* Gaps in a causal story, like coincidences or correlations, call for explanation. The writer notes where there are gaps in his causal story, and infers to causal processes that propagate causal influence in order to close such gaps. Freud, of course, inferred to unconscious mental states and processes to account for gaps in the continuity of conscious mental life—gaps, instead of causal relatedness among mental contents in consciousness.

5. *Analogies and models.* Another explanatory strategy a writer may use is analogy. He shows, or tells how it has been shown, that an event of a particular kind causes another event of a particular kind. Then he argues by analogy that, since the event he is explaining is similar to that effect, it is caused by a similar cause.

Salmon (1984, pp. 233f.) shows how causal arguments and argument by analogy are used to argue from observables to nonobservables, arguments similar to those used by Freud in arguing for the existence of unconscious entities.

A special case of an argument by analogy is the use of a *model* (Cartwright 1983).[13] A model is a concrete or abstract exemplification of a set of causal relations. Events in the model are interpreted by or coordinated with events in the case study. Causal processes are interpreted by or coordinated with causal processes in the case study. Outcomes of interventions are predicted from the model. Since the model does not include any de facto conditions or nonlawful facts, the predictions usually turn out to be only approximately correct. However, the model, which is a kind of metaphor, is fertile; it suggests what kind of changes in it will result in more accurate predictions. An imaginative application of a sophisticated model (catastrophe theory) to psychoanalytic clinical case material can be found in Sashin (1985).

A model is an idealization. Statements describing entities and their relations in a model are not strictly true of anything but the model itself. However, a model is useful in organizing facts in a system under observation. It raises new questions. It suggests what kinds of things and what properties of these things are causes, and how causal influence is propagated. Mainly, the ability to predict facts, events, or values with ever increasing accuracy, as the model is modified in various attempts to apply it in specific cases, determines just how useful a model is. The use of a model calls for a series of case studies.

Arguing for the Validity of a Case Study

1. *Distinguishing between different kinds of causal explanations.* The writer is careful to distinguish between etiological explanation and constitutive explanation. In general, the case study can give more powerful arguments for accepting a constitutive explanation, or an etiological explanation assigning causal status to a very recent or

13. The role of models in science has been an important theme in this book. See, in addition to this chapter, especially Chapters 1, 2, 3, 9, and 15. The term "model" has been used in somewhat different ways in different chapters, depending on the context and the objective of the chapter in which it occurs.

concurrent external event or state of affairs, than for accepting an explanation that attributes causal efficacy to a time-remote event or state of affairs. The latter kind of causal explanation will achieve credibility to the extent that evidence for intermediate causal connections can be found, but Freud's strenuous and perhaps unique effort in this direction possibly demonstrates the difficulty in supporting such an explanation with data from a case study alone. There is no reason not to write such a case study if it is clear that acceptance of its conclusions must wait for the convergence by other methods of investigation on the same conclusion.

At the moment, it is not clear what these methods might be, although direct observation of children, despite the drawbacks Freud points out, can surely make a contribution.

Epidemiological and experimental evidence is, of course, welcome on specific points; such evidence will have bearing not on the particulars of the causal story but on the theory that has been used for the purposes already described in constructing it. Glymour's bootstrap method (1980), applied by him in explicating one of Freud's arguments in the Rat Man case, has implications for both kinds of explanation (constitutive explanations, or etiological explanations appealing to proximal external events or states of affairs, on the one hand, and explanations appealing to time-remote inner or external events or states of affairs, on the other hand).

Similarly, the author of a case study will find it difficult to justify explanations of phenomena, where cause or effect is adaptation or maladaptation to extra-analytic reality, using data obtained in the psychoanalytic situation alone. Here, the effect of the analysand's symptoms and character traits on his relation to external reality is at issue, as well as the extent to which his character traits are the result of responses or adaptation to reality (external causes). It is easy to confuse cause and effect in dealing with such problems. Are impaired object-relations, for example, cause or effect?

2. *Rival explanations.* What exactly is the role of rival explanations in justifying a causal story? I do not think that causal stories are underdetermined by data in the same sense that empirical generalizations are. If I tell a causal story in which a bowman shoots an arrow from a taut bow, the arrow flies from one space-time locale to another, and pierces its target, what is the alternative explanation of the effect on the target? It is just the particularities of a causal story, in contrast to mere instances of an empirical generalization, that raise questions in my mind about the underdetermination thesis.

Nevertheless, it is important to ask the following question, at least when there is another *comparable* explanatory account of the same data. What leads to the judgment that an explanation of a phenomenon is better than a rival explanation of that phenomenon? The writer may make one or more of the following arguments.

a) No rival explanation exists, or a rival explanation exists but is not truly comparable. It may, for example, offer a rival explanation of one or another phenomenon, but it does not offer an explanation for the set of phenomena—one that shows the interrelations of all the facts.

It is, of course, always possible to think up different explanations for each part of a causal story. However, to prefer a congeries of rival explanations for different data, all of which are explained by a causal story, violates the principle of parsimony. It also violates the principle of the common cause; that all these different explanations should be correct seems much more improbable than that one causal story is correct.

Freud emphasizes that his objective is not simply to explain the Wolf Man's behavior change, phobia, and obsessive piety. Using a covering law model of explanation, all three of these might be explained by one law, for example, a law involving the causal impact of constitutional defects—or each one of them may be explained as a positive instance of three different laws. But Freud seeks to do something other than that. He seeks to explain the interrelations among the three. He does not explain them as the manifestations of one diagnostic entity or even as three different effects of one cause. He wants to explain how one event, the effect of one set of causes, leads together with additional causes to another event, and this event together with still other causes to another.

For an example of this kind of explanation, see the summary of Freud's explanation of the Wolf Man's symptoms (pp. 63–70, 106–119).

As part of his explanatory strategy here, Freud distinguishes between the cause of a neurosis and the cause of an attack of a neurosis or of an intensification of a neurotic symptom. "The obsessional neurosis ran its course discontinuously; the first attack was the longest and most intense, and others came on when he was eight and ten, following each time upon exciting causes which stood in a clear relationship to the content of the neurosis" (p. 61). Also: "We already know that, apart from its permanent strength, it [the obsessional neurosis] underwent occasional intensifications" (p. 68).

In this attempt to explain the causal interrelations among phenomena, and in particular how these causal interrelations determine the sequence of phenomena, we have the hallmark of dynamic as against nosological explanation.

> The genetic approach in psychoanalysis does not deal only with anamnestic data, nor does it intend to show only "how the past is contained in the present." Genetic propositions describe why, in past situations of conflict, a specific solution was adopted; why the one was retained and the other dropped, and what causal relation exists between these solutions and later developments. (Hartmann and Kris 1945, p. 17)

An explanation that purports to account for phenomena but not for their interrelations is not a comparable rival to an explanation that accounts for both.

> I am of course aware that it is possible to explain the symptoms of this period (the wolf anxiety and the disturbance of appetite) in another and simpler manner, without any reference to sexuality or to a pregenital stage of its organization. Those who like to neglect the indications of neurosis and the *interconnections* between events will prefer this other explanation, and I shall not be able to prevent their doing so. (Freud 1918, p. 107; my italics)

b) A more fertile explanation is preferred to a less fertile one. The author's explanation leads to important and interesting questions—in fact, generates an entire research program.

> On these issues I can venture upon no decision. I must confess, however, that I regard it as greatly to the credit of psycho-analysis that it should even have reached the stage of *raising* such questions as these. (Freud 1918, p. 96)

Freud's explanation in the Wolf Man case leads to such questions—still largely unanswered—as: What determines an agent's choice of mental operations, in general, and with respect to a particular mental state at a particular time? Not only why, but how, does a mental state, or any of its constituents, become unconscious?

c) The author's inferences, unlike those in a rival account, result in increasingly detailed, precise descriptions of the structure and properties of nonobservable entities; such knowledge leads to new findings and/or an improved fit between case and model. Similarly, an author's constitutive explanation may be more complete than a rival account. It makes use of all the properties of mental states that theory suggests are important. It addresses the difference that topographic, economic, dynamic, and structural properties of mental states make to a causal story—that such states are unconscious, that they are ordered on dimensions of priority and intensity, that they are of different kinds, and that they have different and often competing or conflicting conditions of satisfaction.

d) The author's explanation results in the interpolation of causal connections where other accounts leave gaps. Here, Freud's injunction in the Wolf Man case that an interpretation of a dream must account for all details of the dream comes to mind.

e) The author is able to render implausible claims that his data are favorable to his explanation over a rival explanation simply because of factors in the situation or in the way he obtains his data. Although suggestion is one possible such factor often mentioned (for example, by Freud himself throughout his writings and by Grunbaum 1984), it does not seem to me that it is reasonable to offer it as an alternative way to account for the data obtained in a case study without taking note of a careful documentation—if one exists (as, of course, ideally, it should)—of factors that count against the plausibility of such an alternative account. These include: the content and sequence of the psychoanalyst's interventions; the extent of the analysand's knowledge of psychoanalytic theory; the degree to which the psychoanalyst is unprepared for, surprised by, and even unbelieving about what the analysand reports (as Freud shows he is in the Wolf Man case), even when it is consistent with the theory; the very long periods during which the analysand stubbornly and persistently rejects, ignores, forgets, and fights about not simply an explanation but the very reports he himself has given that the psychoanalyst offers in support of it; and the degree to which the psychoanalyst not only pays attention to the analysand's disagreements with him but ends up accepting the analysand's views (as Freud reports he does in the Wolf Man case).

"I could not succeed, as in so many other differences of opinion between us, in convincing him; and in the end the correspondence between the thoughts which he had recollected and the symptoms of which he gave particulars, as well as the way in which the thoughts fitted into his sexual development, compelled me on the contrary to come to believe him" (Freud 1918, p. 62).

It is noteworthy that Freud makes clear that he has confidence in an interpretation just to the extent that the interpretation makes use of, or that the response to the interpretation includes, particular idiosyncratic details *which the psychoanalyst could not have predicted from the theory alone (i.e., from general formulations).* Nothing could make clearer how far Freud is from a covering law model of explanation.

For example, in attempting to explain his patient's recurrent memory of being seized with fear of a butterfly, Freud puts forward "the possibility that the yellow stripes on the butterfly had reminded him [the patient] of similar stripes on a piece of clothing worn by some woman." The patient's associations, however, indicate that the stripes on the butterfly in fact reminded him of a pear, the name of which was the same as the name of a nursery-maid who, it turned out, was important in his life. Freud mentions this "as an illustration to show how inadequate the physician's constructive efforts usually are for clearing up questions that arise, and how unjust it is to attribute the results of analysis to the physician's imagination and suggestion." That "the opening and shutting of the butterfly's wings" had looked to the patient "like a woman opening her legs, and the legs then made the shape of a Roman V, which . . . was the hour at which, in his boyhood, and even up to the time of the treatment, he used to fall into a depressed state of mind" were associations that, Freud declares, he could never have arrived at himself. Freud also mentions deprecatingly his own "facile suspicion that the points or stick-like projections of the butterfly's wings might have had the meaning of genital symbols" (pp. 89ff.).

f) An important point with respect to the validity of explanations is that explanations of phenomena, based on limited data—a limited subset of the total set of relevant data—are less likely to be valid than explanations based on a larger subset of these data. An advantage of the psychoanalytic study is that it makes use of data obtained in the psychoanalytic situation, which are otherwise unavailable. Such data, for example, cannot be obtained in experimental and field-questionnaire studies. Imagine testing hypotheses such as Freud's thesis that "a man's attitude in sexual things has the force of a model to which the rest of his reactions tend to conform" (1909, p. 241) by methods that do not afford access to intimate facts a person is reluctant to communicate even when he is conscious of them. Examples are: the content of masturbation fantasies; and under what circumstances such fantasies are excited. Given a stream of "purposeless" communications, the relation of such intimate information to facts about "nonsexual" areas of life can be inferred from sequences, contexts, or metaphoric and other linguistic linkages.

The argument that the investigator has selected his facts to suit his explanation is feckless. It is just these facts that he intends to show can be fitted into a causal pattern. His challenger must show that these facts—along with others—may better be fitted into some other causal pattern; he gives other factors, kinds of events, or entities, a causal status, or causally relates the same facts in a different way. The investigator, writing with his rival in mind, must accept the obligation to include in his account any facts relevant to his rival's purposes—especially events that are not encompassed by his causal pattern but might well be encompassed by his rival's. He argues strongly, of course, if he can show that he is able to encompass the facts that are most relevant to his rival's purposes but that his rival is not able to encompass those most relevant to his own. An examination of Freud's *Interpretation of Dreams* (1900) will show that Freud so argued against the view that the choice of content and other properties of dreams are determined solely by somatic or external stimuli.

One who thinks in terms of a covering law model of explanation and who intends to use it to explain one phenomenon may argue that all the rich data collected by the psychoanalyst might well be irrelevant for the explanation of a neurotic symptom, for example, and that the psychoanalyst's use of them simply results in a "fairy tale" (an epithet one of the critics of Freud's case studies is especially fond of using). However, these data are not irrelevant if they are produced unexpectedly by the analysand with a kind of particularity and detail that along with other characteristics of the situation tend to render implausible the supposition that they are suggested by the psychoanalyst; if they are relevant to a consideration of the kind of thesis of Freud's exemplified by the one (a central one) quoted above; and if the investigator's theory—or set of concatenated theories—enables him to fit them into a causal pattern that has no comparable rival.

Conclusion

A psychoanalyst will distinguish between using a case study to establish facts (empirical generalizations) and using a case study to tell a causal story (i.e., to give a causal explanation of phenomena and their interrelations). No matter which is his objective, he will pay attention to the necessity of arguing for the scientific credibility of his conclusions. In particular, if he attempts the kind of causal explanation Freud offers in the Wolf Man case, he will not only pay attention to the power of his explanation (how many facts it encompasses and how elegantly it does so), but will include in his case study some appropriate argument, based on theoretically relevant evidence, for the scientific credibility of that explanation.

Some psychoanalysts are fond of claiming that psychoanalysis is different from other sciences and has its own rules of evidence, but they are often at a loss to explicate in just what sense it is different and just what these rules of evidence are. A strong argument can, in fact, be made that scientific methodology (e.g., hypothesis-testing) is not independent of theory (Danziger 1985). Theoretical presuppositions about the structure of the world underlie methodological rules (e.g., the rules of statistical inference). Such presuppositions include that a system of phenomena has the same *structure* as the numerical system, or that relations among phenomena are of the same type as relations among numbers. These presuppositions, therefore, determine what constitutes evidence and which limited class of theories will be considered suitable for testing at all.

It may be true, as Danziger claims, that the rules about what constitutes evidence are neither independent of theory nor fixed forever, and that different methods are required to establish the scientific credibility of different kinds of theories. But this does not necessarily imply there is no general characterization of science in terms of methodology that holds across different disciplines or domains. For example, I know of no theory, no matter how complex, that does not at least involve *classification* of entities, states of affairs, or events, and at a minimum the kind of *measurement* that makes use of dichotomous yes-no, on-off variables. At a minimum, a scientific theory will attempt to explain why an entity has one of at least two values, "has the property" or "does not have the property"; why a state of affairs has one of at least two values, "exists" or "does not exist"; or why an event has one of at least two values, "occurs" or "does not occur."

If we offer a causal explanation such as Freud offers in the Wolf Man case, it is still so that some carefully reasoned (and, with respect to the kind of theory at issue, appropriate) argument must be provided as to just why, given a body of empirical evidence (also, with respect to the kind of theory at issue, appropriate), we should believe this causal explanation has indeed captured at least some of the features of, or in some way reflects, the causal structure of the psychological no less than of the physical world. Such an argument is the irreducible methodological imperative of any scientific enterprise—and psychoanalysis, as Freud well knew, perforce must yield to it.

Some Reflections on Responses to Adolf Grunbaum's Critique of the Empirical Foundations of Psychoanalysis[1]

14

Grunbaum's Critique

Must psychoanalysis accept that the only source of possible evidence for its causal inferences are experimental and epidemiologic studies? The philosopher Adolf Grunbaum (1984) seems to think the answer to that question is "yes." His reasoning goes something like this.

The task psychoanalysis faces is to provide evidence for its causal claim that repression of conflict is a necessary cause of neurosis, necessary in childhood for bringing about neurosis and necessary thereafter for maintaining it. Even if a psychoanalyst does not accept that psychoanalysis makes such a causal claim today, Grunbaum's argument will apply to any causal claim in current psychoanalytic theory that has the form that some particular event is a necessary cause of some other event or some state of affairs—that unconscious infantile wishes are necessary causes of dreams, or that unconscious conflicts are necessary causes of neurotic symptoms, or that psychoanalytic interpretations leading to insight and therefore to making the unconscious conscious are necessary causes of the eradication of neurosis.

What would it mean, in Grunbaum's terms, to provide such evidence? It would mean to show that the postulated cause makes a difference. It would mean to show that whether or not the effect occurs depends on whether or not the cause occurs.

If psychoanalysis claims that repression of conflict is

1. In preparation for a dialogue with A. Grunbaum, midwinter meetings of the American Psychoanalytic Association, December 20, 1986.

a necessary cause of neurosis, it must essentially show that *whenever* repression of conflict does *not* occur in childhood (or repressed conflict is *not* present), neurosis does *not*, or is much less likely to, develop (or is *not*, or is much less likely to be, present). In other words, psychoanalysis must show that there is *no* instance, or relatively few instances, in which both "repression of conflict does not occur in childhood" (or "repressed conflict is not present") and "neurosis develops" (or "neurosis is present") are true.[2]

All of this follows from the very meaning of the term "necessary cause." One could not demonstrate that oxygen is a necessary cause of fire if it were impossible to create conditions in which there were no oxygen and so compare what happens under these conditions with what happens when there is oxygen. If psychoanalysis were to take the position that repression of conflict always occurs in childhood or that repressed conflict is always present, then Grunbaum would point out that that would be tantamount to saying that there is no way of obtaining evidence for the causal claim that repression of conflict is a necessary cause of neurosis, or tantamount to saying that this causal claim has the same status as "life is a necessary cause of neurosis" or "being born is a necessary cause of neurosis."

Even if psychoanalysis does obtain observations consistent with the repression-as-necessary-cause hypothesis, for these observations to constitute evidence in support of the scientific credibility of the hypothesis, psychoanalysis would have to obtain them by controlled inquiry. Controlled inquiry makes it possible to argue that the observations favoring the hypothesis were not produced by extraneous factors (such as suggestion or biased sampling of observations) in the situation of investigation, whatever it is, such that these factors would have produced the observations favorable to the repression-as-necessary-cause hypothesis even if that hypothesis were false.

The essential thing about controlled inquiry is that it uses experiment or statistical-analytical tools to eliminate certain other (from the investigator's point of view, extraneous) factors as accounting for his having obtained the data he thinks favor (or are explained by) his own causal conjecture. Controlled inquiry also attempts to determine the weight to be given or role to be assigned to various factors as causes of the event, pattern, or outcome the investigator is interested in explaining.

Strictly speaking, if an investigator has no evidence that favors his

2. Grunbaum, correctly, does not depict psychoanalysis as claiming that repression of conflict is a sufficient cause of neurosis or a necessary and sufficient cause of neurosis.

conjecture, he does not need to concern himself with eliminating chance or other factors as responsible for the outcome of his investigation—unless he wants to eliminate the possibility that some extraneous factor, and not that his conjecture is *false*, prevents him from obtaining such favorable evidence. An investigator would not be concerned with asking, for example, "Could Chance instead account for this pattern of observations?" unless the pattern of observations favored his conjecture and he wanted to eliminate Chance as a factor in order to argue that his observations constituted support for the scientific credibility of his conjecture. In other words, the null hypothesis is not in relation to the tested hypothesis a rival conjecture. Chance presents itself when an investigator is considering what *extraneous* factors may be operating to account for the pattern of observations he has obtained.

Before continuing with my presentation of Grunbaum's argument, I want to make two points about controlled inquiry.

1. I would add to the usual account of controlled inquiry (which Grunbuam follows) and emphasize that controlled inquiry may also use *nondesign arguments about the features of a nonexperimental situation* to rule out extraneous factors as responsible for obtaining favorable data.

2. I hold to the view that a crucial aspect of controlled inquiry is collecting evidence that will favor one causal conjecture over another. Further, and here Grunbaum and I seem to part company, I hold that rival causal conjectures are not ordinarily from entirely different paradigms, but that they are generated within a particular conceptual frame of reference. For examples from molecular biology (does DNA have one or two strands?), see Platt (1964). In psychoanalysis, preoedipal or oedipal factors as ultimate causes of neurosis are the usual kinds of rival conjectures; behaviorist and psychoanalytic accounts of the causes of neurosis are not what is usually meant in normal science by rival conjectures. They are competing paradigms. Shifts in paradigms do not usually occur through elimination by crucial test. They are more likely to occur as the result of differences over time in a paradigm's success, compared to its rivals, in generating intra-paradigm rival conjectures, and in obtaining evidence that will eliminate one in favor of another, thereby progressively increasing knowledge about and understanding of its domain. The failure of psychoanalysis to proceed in this way will disable it more than any supposed failure to pass tests in which a psychoanalytic hypothesis is challenged by a hypothesis generated by a different paradigm.

The conclusion of Grunbaum's argument is that, since psychoanalysis has not been able to make the comparisons of outcomes under different causal conditions that have to be made to show that repression of conflict is a necessary cause of neurosis, the empirical foundation for one of its key causal conjectures is weak. It is unlikely if not impossible that psychoanalysis will be able to make such comparisons using clinical data obtained by the case study method. That is not the case simply because these data are not obtained by controlled inquiry—that is, are contaminated in unknown ways by extraneous factors. That is the case because the very nature of the psychoanalytic situation and the case study method of inquiry prevents systematic comparisons.

Grunbaum believes that, if psychoanalysis continues to rely on data obtained by such means, rather than carrying out the kind of experimental and epidemiologic studies that enable the investigator to make such comparisons, its empirical foundations cannot be strengthened. It is this argument that is at the center of his critique of the empirical foundations of psychoanalysis. Neither his discussion of the problems that arise in any attempt to use clinical data as evidence for the key causal conjecture of psychoanalysis (that is, that these data are not obtained by controlled inquiry, and so are contaminated in unknown ways by such extraneous factors as suggestion, which cannot then be eliminated as responsible for producing the data), nor his discussion of the problems that arise in any attempt to use therapeutic efficacy as evidence for a key causal conjecture of psychoanalysis (the repression-as-necessary-cause hypothesis), are as fundamental.

Grunbaum believes that, even if it could be shown that the data obtained in the psychoanalytic situation were uncontaminated (that is, suggestion, the analysand's desire to please the psychoanalyst or conform to his expectations, and the psychoanalyst's selection biases, for example, were not responsible for producing them), these data cannot provide evidence for the causal claims of psychoanalysis. He does not deny that such data may be evidence for the *existence* of unconscious conflicts, for example—only that such data cannot justify the belief that these unconscious conflicts are necessary to produce and maintain neurotic symptoms. To demonstrate that something is a necessary cause, one must be able to show that in its absence the effect does not occur (although, since it is not a sufficient cause, in its presence the effect may or may not occur). He does not believe that this can be done in the psychoanalytic situation. For even if clinical data were not contaminated by extraneous factors, Grunbaum does not see any way for the psychoanalyst to make the systematic comparisons in the clinical situation required to demonstrate that repression of conflict is a necessary cause of neurosis—or that anything is a necessary cause of anything else—to demonstrate, in other words, that the putative necessary cause, whatever it might be supposed to be, makes a difference.

My own view of Grunbaum's critique is that, because of the explicitness and utter lucidity of his argument, and the thorough scholarship with which he has documented his depiction of psychoanalysis, the critique can function as a powerful stimulus to hard thinking about the issues he has raised.[3] I do not know for what more one could ask from a philosopher of science.

Responses to Grunbaum's Critique

The critique will not function in this way for those who incorporate it into an apocalyptic vision of the demise of psychoanalysis.

The End is Near

There are those who see the critique as warning that the demise of psychoanalysis is at hand, unless it is promptly taken to heart and its proposals implemented. What is called for is an immediate and too-long-delayed mending of ways. These advocates of Grunbaum's surrender to his critique. Grunbaum is taken uncritically as the one who knows what is necessary to justify rational belief.

I find this kind of response repugnant because it is so unskeptical, especially since it usually comes, ironically enough, from people who pride themselves on the skepticism they have shown previously in insisting on adherence to canons of scientific research in assessing psychoanalytic knowledge. Grunbaum's critique after all is not, being the work of a human intelligence, invulnerable, and makes many assumptions that are at the present time controversial (for example, assumptions about the nature of the scientific enterprise, causality, the basis of rational belief in science, what is core in psychoanalytic theory and indeed what kind of theory psychoanalytic theory is). It does not deserve simple enthusiastic agreement—but only agreement that survives the most searching and critical examination of it.

Psychoanalysis Is Finished

There are those who see Grunbaum's critique as finishing off psychoanalysis once and for all, a too-long-delayed demise finally accomplished. The tone is exultation, more or less ill-concealed:

3. An attempt to deal with the central issue raised in his critique follows in Chapter 15.

"The last nail has been driven into the coffin of psychoanalysis." It is usually nonpsychoanalysts who respond so, but I have sat one whole afternoon while a group of psychoanalysts, discussing the issues raised by the critique, returned again and again to this coffin metaphor, a tinge of despair well-tempered by satisfaction that "what we have been warning for so long will come to pass has now come to pass—maybe now the psychoanalytic establishment will be sorry that it has not listened to us."

I find this kind of response an outrageous misuse of Grunbaum's critique, betraying the spirit in which it is offered by closing out any possibility that it might be taken positively and used constructively by those to whose endeavors it is directed. I have of course no access to Grunbaum's motivations (which conceivably are full of wicked animus toward psychoanalysis). But I regard his motives as irrelevant, as long as his critique is a work of reason, raising important and impossible-to-ignore issues, arguing about them clearly and powerfully (an argument that I am, of course, free to analyze and criticize), and arriving at conclusions (which I am free to present grounds for rejecting or accepting) and sane constructive proposals based on those conclusions (to which, since "should" cannot logically follow from "is," I am free to develop alternatives, even if I were to accept his conclusions).

In particular, those who make this kind of use of Grunbaum's critique fail to make some important distinctions and to take into consideration some important points.

1. *The question of plausible alternatives.* There may be differences of opinion about what constitutes rational grounds for belief in the presence of comparable plausible alternatives challenging that belief compared to what constitutes rational grounds for belief in the absence of comparable plausible alternatives. I am not persuaded that such conceptual systems as learning theory (quite a variety of theories come under this rubric), nonpsychoanalytic personality theories, social psychology theories, cognitive theories, or neuroscientific theories have come up with explanatory schemes capable of dealing with even a fraction of the mental phenomena in the intended domain of psychoanalysis encountered in a psychoanalysis, in everyday life, and in any of Freud's case studies.

Certainly those responding to Grunbaum in this way present nothing that persuades me. As far as making sense of, or explaining, a circumscribed set of mental phenomena that I am able to observe, in which I am interested, and that belong to the domain I regard as the intended domain of psychoanalysis (Edelson 1984; and Chapters 6, 7, and 13 of this book), psychoanalysis is still pretty much the only game in town.[4] Psychoanalysis is not required to abandon what might be admittedly a leaky craft to jump into a sea of conceptual chaos.[5]

2. *Grounds for believing vs. grounds for not believing.* Questioning how strong are the grounds for believing is not the same as asserting grounds for not believing. There is a difference between:

what one might take to be rational grounds for tentatively and provisionally *accepting* a belief, and for being willing to act upon it;

what one might take to be rational grounds for remaining *uncommitted* with respect to the question of the truth or falsity of a belief, and for allowing oneself to decide on the basis of a variety of considerations (including value considerations) as to whether to act on it or not (e.g., the desirability of continuing to test it through one's actions as part of an inquiry into its possible truth);

and what one might take to be rational grounds for *rejecting* a belief as false.

Grunbaum, who knows these distinctions, nowhere argues that he has evidence sufficient to justify rejection of psychoanalytic theory as false or psychoanalytic therapy as ineffective. He cannot cite methodologically adequate studies comparing the relative weight evidence confers on alternatives if these studies have not been done, however much he rightly deplores that they have not been done.

In this domain, experimental studies usually, and perhaps inevitably, produce "weak effects" (that the results were obtained by Chance is plausible), and have had as startling methodological deficiencies as any case study. If and when adequate experimental studies are done, even beginning students in statistics will know the difference between a result compelling one to recognize that one cannot reject the null hypothesis (Chance) as explaining why one obtained favorable data (a kind of "not proven" verdict), and the kind of result compelling one to recognize that a tested hypothesis is false.

Erwin (1986) writes:

> Most of Freud's hypotheses are implausible given our background evidence; that is one reason they are interesting. Most are neither trivial nor self-evident. Of course, they are not all equally implausible, and *some may have indirect support even if the evidence does not yet*

4. I do not say this out of ignorance of other conceptual frames of reference or other kinds of investigations of mental phenomena. P. Meehl (1983) comments similarly in a rigorous, tough-minded paper on the problem of subjectivity in psychoanalytic inference, "if you are interested in learning about your mind, there is no procedure anywhere in the running with psychoanalysis" (p. 352).

5. An assertion that is applicable to the proponents of any theory—see Stegmuller (1976).

> *warrant their acceptance.* Nevertheless, for those that are implausible, all things considered, does not the absence of evidence warrant a harsher verdict than "not proven"? Are we not entitled to say, although tentatively, "there is some reason to think them false"? (P. 236; italics added)

I would say to that question, "No." First, evidence that supports indirectly is not "no evidence." Second, the absence of evidence is not the same as the presence of negative or falsifying evidence. Third, "clinical data" is not "no evidence."

Erwin and I do not agree on the evidential value of the clinical data we already have, though we may agree that such data do not have the same evidential value they would have if we also had experimental data to complement them. We probably also agree about the enhanced evidential value of clinical data obtained by case study methods paying more attention than hitherto to the need for increasing the extent to which inquiry is controlled. I do not believe that the case study method is an inferior method, compared to the experimental method, for obtaining data. I believe that each kind of research method is a different mode of inquiry, and that each has its own particular advantages and disadvantages, strengths and weaknesses (see McGrath 1981).

If someone has given "blind assent" (Crews 1985, p. 33) to a theory, of course, he is likely to be shocked and disillusioned when it does not provide unassailable grounds for believing it. But "benevolent skepticism" (Lieberson 1985, p. 40) is the stance of science toward conjectures. That stance—not blind assent—has always been the appropriate stance toward psychoanalytic conjectures (in or out of the treatment situation). Grounds for tentative provisional acceptance of a belief, especially as more acceptable at some particular time than its rivals, and *not* certainty, are always what science seeks. Psychoanalysis should be held to no stronger aspiration.

3. *The grounds for practical action.* There is a difference between the rational grounds for belief sufficient for a clinician to continue offering a particular treatment such as psychoanalysis to the particular patient who consults him, the rational grounds for belief sufficient for the patient to accept such an offer, and the rational grounds for collective belief in the probable efficacy of that treatment should it be offered by many therapists to a population of patients. A psychoanalyst, in offering such treatment, might well lay his cards on the table, indicate his inability to predict or guarantee particular results, invite the patient's skepticism, commit himself and the patient to a limited trial before proceeding further, and together with

the patient monitor the effects of both that trial and any subsequent work the two may decide to join together in doing. There is nothing about psychoanalysis that implies that a psychoanalyst should adopt an attitude of limitless complacency in the face of a deteriorating clinical course. Nor is "blind assent" part of the contract in a psychoanalysis; I can think of nothing more contrary to the spirit in which, ideally, it is carried out, or more contrary to its rationale or goals.

That my own experience is sufficient for me to accept tentatively and provisionally, and act on, quite a number of beliefs in seeking practical ends need not delude me into believing that my experience alone provides sufficient grounds or a compelling reason for others to accept and act on these beliefs.

The comments of the methodologist D. Campbell, in a paper (1975) in which he retracts earlier animadversions on the shortcomings of the case study method, are apposite.[6]

> The quantitative . . . generalization will contradict such anecdotal, single-case, naturalistic observation at some point, but it will do so only by trusting a much larger body of such anecdotal, single-case, naturalistic observations.
>
> This is not to say that such common-sense naturalistic observation is objective, dependable, or unbiased. But it is all that we have. It is the only route to knowledge—noisy, fallible, and biased though it be. We should be aware of its weaknesses, but must still be able to accept it if we are to go about the process of comparative (or monocultural) social science at all. (Pp. 178–179)

The Critique Is Malevolent or Incompetent

There are those who see Grunbaum's critique as motivated by a wish to bring about the demise of psychoanalysis. They dismiss it as a shabby, shallow argument by one who does not know very much about psychoanalysis, has not done his homework, and does not realize how much and in what ways Freud has changed or how much psychoanalysis has changed.

I find this kind of response defensive and ill-conceived (that is, based on a misconception or inaccurate reading of Grunbaum's work). It is distinguished by the foolish inability to see that he has raised any serious issue for psychoanalysis to consider, or by the destructive illusion that any serious issue raised by him has been or is being satisfactorily settled within psychoanalysis.

It is certainly true that Grunbaum gives his own reading of the

6. Campbell's animadversions on the single case study have been widely read, while his later retraction is hardly known.

psychoanalytic corpus, but he is entitled to do so, especially given that different psychoanalysts have always felt entitled to read and understand Freud's work in their own way. Since Freud was neither consistent in his development nor systematic in his revision of theory, different lines of thought twisting through his work somewhat independently, it does not make sense to impugn someone's scholarship or knowledge of psychoanalysis on the basis of a reading that emphasizes one line of thought over others. Pointing out that Freud changed his mind is usually unavailing, because Freud was notoriously reluctant to abandon a position once taken and is quite ready to assert it later even following an apparent change of mind.

Referring to post-Freud developments in psychoanalysis *as a way of neutralizing Grunbaum* is apposite only if these developments make a difference with respect to the problems he has raised. Grunbaum could surely ask whether those who respond in this way to his critique make much of a case that these developments make that kind of a difference.

Rational Grounds for Believing Clinical Inferences in Psychoanalysis[1]

15

The Problem: Causal Inferences in the Clinical Situation

We do not try to understand everything. We do not try to understand a person. We try to understand something *about* a person. A person reports a dream, makes a slip of the tongue, describes a symptom. From inevitably fragmentary observations—from things the person says about himself, his experiences, his mental states—we formulate a conjecture about the dream, the slip, the symptom. We formulate an explanation of it. These fragmentary observations, no matter how many of them we have, do not by themselves logically entail the conjecture or explanation. In fact, many conjectures or explanations, based on the same observations (things seen or said), are equally possible. So what is it that makes our explanation believable—and more believable than other explanations?[2]

This is our question. *What provides the grounds for our belief that, from what the analysand tells us in the psychoanalytic situation, we can infer the causes of his symptoms, dreams, parapraxes?*

Our problem is to provide grounds, typically in the form of empirical evidence, for believing an explanation. We want evidence that not only favors the credibility of an explanation but favors it over a currently plausible rival conjecture or explanation. Some such plausible

1. A very much abbreviated version of this chapter was presented in a three-hour dialogue with A. Grunbaum at the midwinter meetings of the American Psychoanalytic Association, New York, December 20, 1986.

2. Whether a good causal explanation has rivals in this sense is a question I raise in a later section of this chapter.

rival almost always exists as part of background knowledge at the time the conjecture or explanation is formulated.[3]

Of course, we cannot want or have evidence that favors our own explanation over *all* possible rivals, for some rivals are implausible, given current knowledge, and some rivals will only appear with the advent of new knowledge. So, acceptance of an explanation is always tentative or provisional, never certain, since the process by which it comes to be accepted at a particular time is always dependent on what evidence is then available, what background knowledge exists at that time, and which plausible rivals are around to be eliminated.

I do not consider the problem of providing grounds for belief esoteric—of interest to philosophers of science, theoreticians, and research investigators only. It is a question that is on the minds, or should be, of practicing psychoanalysts as they do clinical work. It is an intrinsic part of that work. A psychoanalyst reflecting on his work seeks to satisfy himself as to his grounds for believing in the accuracy of his interpretations. He wonders just how much at any particular time what he hears and observes in the psychoanalytic situation justifies his believing his interpretations. He is or should be ready to reflect on just what it would take to get him to change his mind about the accuracy of an interpretation, and to revise or altogether reject the interpretation. It cannot be possible that he suspends all such concerns until after a psychoanalysis is over, for psychoanalyst (and analysand for that matter) change their minds throughout a psychoanalysis as they acquire more information. (Of course, similarly, a psychoanalyst may change his mind about a case even years afterward if fresh information comes to light or additional knowledge about the workings of the mind or the psychoanalytic process becomes available.) This changing one's mind is inevitable, given what I have just described—the time-relative provisional nature of even justified belief.

The practicing psychoanalyst is concerned with this problem even if he is concerned only with the grounds for his own beliefs. He may or may not be concerned as well—as at least some psychoanalysts should be—with the question: What kind of evidence in support of my explanation will I need to provide that is sufficient to convince *another* psychoanalyst, or, perhaps a more difficult task, a *non-psychoanalyst,* of its credibility.

I believe that we must think about this problem and develop ways of solving it if we are to resist successfully the view that the clinical

3. But see footnote 2.

data and the case study method of psychoanalysis have no evidential value—have nothing to contribute to attempts to confer scientific credibility on psychoanalytic explanations.

The Justification of Clinical Causal Inferences

In this chapter, I shall focus on a single question. A psychoanalyst's conjecture, clinical inference, or explanation is often a causal one. What problems face a psychoanalyst who tries to provide evidential support for a clinical *causal* inference—e.g., a causal explanation of a parapraxis, dream, symptom—using only information or data obtained in the psychoanalytic situation, even if this information or these data were *not* considered to be problematic?

Grunbaum (1986) argues that, even though the psychoanalyst were to have uncontaminated data, he would still not be able to justify clinical causal inferences using clinical data and the case study method. If that were true, then there would not be much point in worrying about how to mitigate the contamination of clinical data by extraneous factors such as suggestion, the analysand's desire to please the psychoanalyst or conform to his expectations, and the psychoanalyst's selection biases. I shall ignore in what follows questions about the validity or reliability of psychoanalytic data, because worrying about the influence of the kind of extraneous factors I have just mentioned is futile, if in fact, even though the psychoanalyst were to have reliable, valid, clinical data, he would not be able to justify clinical causal inferences using clinical data and the case study method. Of course, I do not intend to imply that questions about the reliability and validity of data are unimportant.

I shall try to make clear in what follows that any solution to the problems encountered in attempting to justify clinical causal inferences will depend on both the psychoanalyst's conception of what kind of theory psychoanalysis is and his conception of causality.[4]

Freud's Arguments

What arguments do psychoanalysts use, in order to provide grounds for believing—in order to justify—their clinical causal inferences in the psychoanalytic situation? I take Freud's arguments to be repre-

4. It will not escape the reader's notice that, in this chapter and in Chapter 13, I have departed somewhat from the view of science espoused in Edelson (1984). In what way will shortly become clear.

sentative. It may be true that the theory and practice of psychoanalysis have changed significantly since Freud wrote his major works. However, for the most part, psychoanalysts rely on the same kinds of arguments he used to justify their clinical causal inferences. A careful critical reading of Freud suggests that at times he gave poor arguments and at times powerful, sophisticated arguments. Any helpful assessment or critique of his work will differentiate between these. I shall try to cite his better arguments.

The Specter of the Causal Inversion Fallacy

In some passages in *The Introductory Lectures,* Freud attempts to justify his inference to a causal explanation of a slip or dream from what a person reports while in a special state of mind (free association). The person holds the slip or dream-element in mind and suspends critical screening of what comes to mind. Then, in this special state of mind in which the train of thought is more governed by primary than by secondary process, it is just this memory of this slip or dream-element, held in mind, that evokes what he reports. The cause of the slip or dream, according to Freud, can be inferred from the total of what is evoked. (Philosophers of science tend to ignore that what the analysand says in "association" to a parapraxis or dream-element is said in a special state of mind in which connections among mental contents are formed very differently than in an ordinary state of consciousness.)

What is evoked—what the person reports under these special conditions—is a tortuous network of ideas, memories, feelings, and images, the interconnections between them produced by the operations of a primary process. Often what he says refers to events, including mental states, occurring prior to or contemporaneously with a slip or dream, for example. But even a reference to an event occurring *after* a slip or dream may lead by a train of thought to memories of, or allusions to, events occurring prior to or simultaneously with the slip or dream. Of course, if what the person says refers *only* to events occurring after the slip or dream, then the use of it to infer a causal explanation of the slip or dream would be problematic.[5]

Perhaps the person's free associations involve a sequence of mental contents, some of which are things or happenings that are similar, or were contiguous in time or space, to a putative cause of the slip, and

5. See Freud's discussion of a premonitory dream (1900, pp. 623–625) for his explanation of how it comes to be that someone has the thought, "I dreamt of this experience last night" (that is, *before* the experience occurred).

thus may be said to allude or point to it. Can this event, or this putative cause, on these grounds alone, be supposed to be the cause? Can we infer that the memory of a parapraxis or dream is necessarily connected by some means to the memory of whatever caused it, such that holding the memory of the effect (for example, a slip or dream-element) in mind will necessarily evoke the memory of what caused it?

The philosopher of science Glymour (1983) has argued, mistakenly in my opinion, that the fact that a memory of a putatively causal event came to a person's mind *after* a dream suggests, if any causal connection exists, it was the dream that caused the memory, not vice versa. He concludes that Freud has fallen prey to a causal inversion fallacy. What comes *after* a dream, an "association" to it, as an effect of the dream, cannot be used to infer the cause of the dream.

Glymour's argument, it seems to me, may involve, or pave the way to, a failure to distinguish between the occurrence of the analysand's *utterance* (the report of his "association") and the occurrence of quite another event—the one to which the utterance refers—and a failure also to distinguish between the memory of an event and the event itself. But defending Freud against the accusation of falling into the causal inversion fallacy does not require demonstrating that the accuser has fallen into these confusions.

It is more relevant to the concern about causal inversion fallacies in psychoanalytic inference that Freud made quite explicit that "new" associations produced during the day *following* a dream may be connected to each other and to the latent dream-thoughts by a set of "paths," "loop-lines," or "circuits," which differs, in both materials and pathways, from the set of interconnections between materials and dream-thoughts that had *previously* entered into the construction of the dream. There is, evidently, more than one road to Rome. A new set of materials and thought-elements, drawn from experiences *after* the dream, may lead, by alternative pathways, to the same latent meanings of the dream (1900, p. 280).[6]

Begging the Question by Speaking of Associations

Freud describes asking someone why he made a slip. That person "gave the explanation with the first thing that occurred to him." Freud's editor is aware of the pitfalls lurking here. He comments that the use of the word "association," as in "the patient's association,"

6. This formulation of Freud's is also discussed in Chapter 2. I shall continue to discuss the causal inversion fallacy later in this chapter in the section on inference to the most likely cause under the heading *Another Look at the Specter of the Causal Inversion Fallacy.*

would be question-begging in this context, for "association" implies "that the something else that has occurred to him is in some way connected with what he was thinking before." But much of Freud's discussion here "turns on whether the second thought is in fact connected (or is necessarily connected) with the original one" (1916–1917, pp. 47–48).

Freud is ambiguous concerning the two alternatives: (1) The first thing that the person gives us is his explanation of the slip. But then why are we justified in believing the person's explanation? (2) The first thing that the person gives us is something on his mind on the basis of which we infer our explanation of the slip. But then why are we justified in believing our own explanation?

Psychic Determinism

Freud gives one answer that is not very good. It is that determinism rules our mental life. In discussing dreams, he argues that it is not *arbitrary* "to assume that the first thing that occurs to the dreamer is bound to bring what we are looking for or to lead us to it. . . ."

> [It is] a fact that *that* is what occurred to the man and nothing else. . . . It can be proved that the idea produced by the man was not arbitrary nor indeterminable nor unconnected with what we were looking for. (1916–1917, p. 106)

But to provide a convincing argument that just this last assertion is scientifically credible (not "proven") is the problem. (Since Freud is most reasonably concerned with presenting grounds for tentative, provisional acceptance of his causal inference, he—or his translator—would have been well-advised in this and other passages to eschew the word "proof," with all its connotations of mathematical deductive certainty.)

The "psychic determinism" argument is a poor one. Determinism in the mental sphere only implies that it is not accidental that what came to his mind came to his mind; something caused it to come to his mind. Determinism does not imply that what came to his mind is necessarily causally connected to the parapraxis or dream that preceded it. We need grounds for believing that. The grounds for belief must be stronger than that thinking of the parapraxis or dream-element seems to have prompted his "associations."

A Poor Analogy

Freud anticipates one objection to his belief that just what a person says first in response to Freud's query about a slip will be important or will lead to contents that will be important in explaining what caused the slip. Since the person was anxious to fulfill the request to explain the slip, naturally "he said the first thing that came into his head which seemed capable of providing such an explanation." But that he said the first thing that came into his head under these conditions does not necessarily provide the information we need to infer the cause of the slip or to justify our causal inferences as to the cause of the slip. Why should we give such weight to this information?

> [It is not] proof that the slip did in fact take place in that way. It *may* have been so, but it may just as well have happened otherwise. And something else might have occurred to him which would have fitted in as well or perhaps even better.

In other words, there is no necessary connection between the slip and what then occurs to him.

In response to this objection, Freud argues, again badly, that it is a "physical fact" that the person said what he said. He compares what-the-person-said-in-response-to-his-[Freud's]-query to the weight of a component of a certain substance discovered through chemical analysis of that substance. He compares the inferences drawn from what the person said to inferences drawn from that particular weight.

> Now do you suppose that it would ever occur to a chemist to criticize those inferences on the ground that the isolated substance might equally have had some other weight? Everyone will bow before the fact that this was the weight and none other and will confidently draw his further inferences from it. But when you are faced with the psychical fact that a particular thing occurred in the mind of the person questioned, you will not allow the fact's validity: something else might have occurred to him! (1916–1917, pp. 47–49)

Yes, we can accept that it is a fact that the person did indeed respond as he did to Freud's query. But the reliability and validity of the data are not importantly at issue here. What is at issue is that we do not know that the person's utterance is relevant in inferring the cause of his parapraxis, for causal factors other than the one(s) causing the parapraxis might have determined that what came to mind came to mind or influenced him to say what he said.

In discussing the same problem in connection with dreams, Freud indicates he knows that just to show something exists—an unconscious conflict or fantasy, for example—is not the same as showing this unconscious conflict or fantasy is connected in any way, much less causally connected, with a dream or dream-element. "[W]hat occurs to the dreamer in response to the dream-element will be determined by the psychical background (unknown to us) of that particular element." But even if "what occurs to the dreamer in response to the dream-element will turn out to be determined by one of the dreamer's complexes," that by itself "does not lead to an understanding of dreams but . . . to a knowledge of these so-called complexes." The question remains: What have these complexes to do with dreams?

The Common Cause of Both a Parapraxis or Dream and the Analysand's Associations

Freud tries to deal with the problem of showing that the complex not only exists, but is also causally connected with both the dream or dream-element and the dreamer's associations to the dream or dream-element, by arguing as follows.

> [In the association-experiments] the single determinant of the reaction—that is, the stimulus-word—is arbitrarily chosen by us. The reaction is in that case an intermediary between the stimulus-word and the complex which has been aroused in the subject. In dreams the stimulus-word is replaced by something that is itself derived from the dreamer's mental life, from sources unknown to him, and may therefore very easily itself be a 'derivative of a complex'. It is therefore not precisely fantastic to suppose that the further associations linked to the dream-elements will be determined by the same complex as that of the element itself and will lead to its discovery. (1916–1917, pp. 106–110)

The plausibility of this supposition seems to depend then on our obtaining associations to specific dream-elements, or perhaps to the dream itself, as a basis for making a causal inference. Does it become less plausible as we get further away from this model of free association, and regard all of what an analysand says, whether or not it follows closely on the dream report or not, as a basis for explaining the dream or as evidence for such an explanation once we have it?

It is probably not correct to construe Freud as arguing that he is justified in inferring a causal connection between what-a-person-refers-to (even in his immediate associations to a parapraxis or dream-element) and the parapraxis or dream-element, merely from the fact, assuming we may take it as a fact, that the *thought* of the parapraxis or dream-element caused what-he-now-refers-to-in-what-he-says to come to mind. Freud seems rather here to believe that there is a *common cause* of both parapraxis or dream-element and associations, and that this cause is a "complex" or, as we might say now, an unconscious conflict or fantasy.

"The last nail has been driven into the coffin of psychoanalysis." It is usually nonpsychoanalysts who respond so, but I have sat one whole afternoon while a group of psychoanalysts, discussing the issues raised by the critique, returned again and again to this coffin metaphor, a tinge of despair well-tempered by satisfaction that "what we have been warning for so long will come to pass has now come to pass—maybe now the psychoanalytic establishment will be sorry that it has not listened to us."

I find this kind of response an outrageous misuse of Grunbaum's critique, betraying the spirit in which it is offered by closing out any possibility that it might be taken positively and used constructively by those to whose endeavors it is directed. I have of course no access to Grunbaum's motivations (which conceivably are full of wicked animus toward psychoanalysis). But I regard his motives as irrelevant, as long as his critique is a work of reason, raising important and impossible-to-ignore issues, arguing about them clearly and powerfully (an argument that I am, of course, free to analyze and criticize), and arriving at conclusions (which I am free to present grounds for rejecting or accepting) and sane constructive proposals based on those conclusions (to which, since "should" cannot logically follow from "is," I am free to develop alternatives, even if I were to accept his conclusions).

In particular, those who make this kind of use of Grunbaum's critique fail to make some important distinctions and to take into consideration some important points.

1. *The question of plausible alternatives.* There may be differences of opinion about what constitutes rational grounds for belief in the presence of comparable plausible alternatives challenging that belief compared to what constitutes rational grounds for belief in the absence of comparable plausible alternatives. I am not persuaded that such conceptual systems as learning theory (quite a variety of theories come under this rubric), nonpsychoanalytic personality theories, social psychology theories, cognitive theories, or neuroscientific theories have come up with explanatory schemes capable of dealing with even a fraction of the mental phenomena in the intended domain of psychoanalysis encountered in a psychoanalysis, in everyday life, and in any of Freud's case studies.

Certainly those responding to Grunbaum in this way present nothing that persuades me. As far as making sense of, or explaining, a circumscribed set of mental phenomena that I am able to observe, in which I am interested, and that belong to the domain I regard as the intended domain of psychoanalysis (Edelson 1984; and Chapters 6, 7, and 13 of this book), psychoanalysis is still pretty much the only that anything is a necessary cause of anything else—to demonstrate, in other words, that the putative necessary cause, whatever it might be supposed to be, makes a difference.

My own view of Grunbaum's critique is that, because of the explicitness and utter lucidity of his argument, and the thorough scholarship with which he has documented his depiction of psychoanalysis, the critique can function as a powerful stimulus to hard thinking about the issues he has raised.[3] I do not know for what more one could ask from a philosopher of science.

Responses to Grunbaum's Critique

The critique will not function in this way for those who incorporate it into an apocalyptic vision of the demise of psychoanalysis.

The End is Near

There are those who see the critique as warning that the demise of psychoanalysis is at hand, unless it is promptly taken to heart and its proposals implemented. What is called for is an immediate and too-long-delayed mending of ways. These advocates of Grunbaum's surrender to his critique. Grunbaum is taken uncritically as the one who knows what is necessary to justify rational belief.

I find this kind of response repugnant because it is so unskeptical, especially since it usually comes, ironically enough, from people who pride themselves on the skepticism they have shown previously in insisting on adherence to canons of scientific research in assessing psychoanalytic knowledge. Grunbaum's critique after all is not, being the work of a human intelligence, invulnerable, and makes many assumptions that are at the present time controversial (for example, assumptions about the nature of the scientific enterprise, causality, the basis of rational belief in science, what is core in psychoanalytic theory and indeed what kind of theory psychoanalytic theory is). It does not deserve simple enthusiastic agreement—but only agreement that survives the most searching and critical examination of it.

Psychoanalysis Is Finished

There are those who see Grunbaum's critique as finishing off psychoanalysis once and for all, a too-long-delayed demise finally accomplished. The tone is exultation, more or less ill-concealed:

3. An attempt to deal with the central issue raised in his critique follows in Chapter 15.

Furthermore, the parapraxis or dream-elements and the associations are interconnected in a complicated weblike network, not linearly, by mental operations belonging to primary process rather than secondary process. Freud's argument that clinical data obtained in the psychoanalytic situation can serve as evidence supporting the conjecture that this common cause not only exists but is also causally connected to both dream-element or parapraxis and the analysand's associations depends heavily on the demonstration that mental operations belonging to primary process (or dream work) exist and serve as causal mechanisms or processes. For an illustration of Freud's reliance on primary process or dream work in constructing and justifying his explanations, see his account of the dream of the botanical monograph (1900).

The Question about Clinical Causal Inferences Repeated: Three Ways of Arriving at a Satisfactory Answer to It

Does what occurs to a person or what he says in his free associations *necessarily* or *accidentally* follow his report of a parapraxis or dream? ("Accidentally" would apply if what occurs to him or what he says is determined by something other than the cause(s) of the parapraxis or dream.) Does what he says refer or not to something preceding or contemporaneous with the parapraxis or dream that was causally (necessarily) connected to it? Freud is arguing that, given a certain state of mind (required for free association), a person is bound to say something that has relevance for inferring a causal explanation of the parapraxis or dream. Freud is claiming, in other words, that he can answer the question: Why did just *that* come to the person's mind under these circumstances? What caused just *these* mental contents he reports?

I shall describe in what follows the kinds of arguments Freud used, explicitly or implicitly, on my reading of his works, to justify such clinical causal inferences. I believe that these arguments, which psychoanalysts still use, successfully justify clinical inferences from the mental contents reported by an analysand to the causes of the parapraxis, dream, or symptom the psychoanalyst is trying to explain.

There are three major arguments: one involves *analogy*, another *consilience of inductions* (convergence of inferences on the same conclusion), and the third a *concept of causal powers* and an *elucidation of causal mechanisms.*

Freud uses argument by analogy to what is confidently known from commonsense psychology in order to justify his belief that, from what the analysand tells him in the psychoanalytic situation, he can infer the causes of symptoms, dreams, and parapraxes.

Freud justifies clinical causal inferences by pointing out the variety of phenomena such inferences are capable of explaining as effects of the same cause.

Freud justifies clinical causal inferences, especially in such works as the Rat Man and Wolf Man cases, by giving not just some explanation but a *good* (that is, a successful) *causal* explanation.[7]

Analogy

Analogy and Models in Science

Discovery and justification, rather than being sharply separate phases in scientific activity, come together in the use of models and metaphors in science (Diesing 1971; Harre 1970, 1972). The model or metaphor acts as a vehicle of discovery. When we compare the realm we attempt to understand to the model we do understand, we are led by the model to postulate and discover evidence for the kinds of features theoretical entities must have if they are to perform their causal role in the realm we are exploring. A model is fertile when disanalogies lead us to modify our theory in such a way that its explanatory power is increased. The model gives us epistemic access to theoretical entities, for it enables us to name them, to refer to them consistently throughout a series of changes in theory and to know we are still referring to the same things, and to understand what kinds of features these entities ought to have (Boyd 1979).

We are familiar with the model and understand its workings. This knowledge gives us confidence in the causal inferences we make in the realm with which we are not familiar, which we are trying to understand, but which we argue is importantly similar to the model.

How Freud Argues by Analogy

It is important to note, in any attempt to understand how Freud justified his clinical causal inferences, that he repeatedly argues by analogy. (The importance of models and metaphors to Freud's scientific thought has been documented by J. Edelson [1983].) From the way Freud argues by analogy in such relevant contexts as the following, we can construct his largely, but not entirely, implicit argument for justifying his clinical causal inferences.

Freud moves along the continuum of parapraxes, from those in which a person knows what intruding intention caused a parapraxis, to those in which a person knows but is reluctant to tell someone else

7. What characterizes a good causal explanation will be described later in this chapter.

what intruding intention caused a parapraxis, to those (the cases of particular interest to psychoanalysis) in which a person is mystified by his own parapraxis.

Freud moves along the continuum of dreams, from those in which a person has no difficulty in identifying the wish represented by the dream as fulfilled (dreams of hunger, thirst, sexual excitement); to those dreams in which a person is able to identify the wish, by reflecting upon the events immediately prior to the dream that clearly instigated it (thematic affinity enters here), but is reluctant to tell someone else what the wish is (for example, the ambitions to which Freud confesses in his reflections on the Irma Dream); to those dreams (the cases of particular interest to psychoanalysis) in which the dreamer is mystified by his own dream (there may be little, if any, thematic affinity between the manifest dream and the latent dream-thoughts).

In passing, I note that those who reject Freud's emphasis on wish-fulfillment as having priority in explaining dreams seem never to mention the existence of erections during REM sleep or the explicit sexual scenes and excitement in wet dreams.

Freud is not simply arguing for the universality of an explanation as covering the entire range of cases along the continuum. He uses what can be confidently known from cases at one end of the continuum, with the modifications introduced by cases in the middle of the continuum, in order to explain the cases at the other end of the continuum.

Similarly, he also argues that certain actions in infancy and childhood, hitherto regarded by many as "innocent," are manifestations of sexual excitement or attempts to gratify sexual wishes, because the same actions are so clearly manifestations of sexual excitement or attempts to gratify sexual wishes, not only peripherally in normal adult sexual life, but—taking center stage—in the sexual life of those with perversions.

Commonsense Intentional Psychology as a Homely Model

It is sometimes suggested, from one kind of reading of Freud's work, that, in order to justify his clinical causal inferences, he used natural science models—a system of hydraulic pipes, or a quasi-neurological network (his *Project for a Scientific Psychology*). Another kind of reading of his work suggests, interestingly enough considering the frequent characterization of psychoanalysis as violating common sense, that his major model probably derives from commonsense knowledge of psychology. (In a somewhat related vein, Campbell [1975] has characterized "common-sense naturalistic observation" as our "only route to knowledge.") What is meant by "common-sense knowledge of psychology" will shortly become clear.

To be sure, commonsense knowledge of psychology—which includes knowledge of the way wishes and beliefs accessible to awareness conjoin to produce intentions to act—provides a homely model. Darwin also used a homely model to justify his inference about the process by which variation in nature takes place over eons of time. Darwin's homely model for the mechanism of *natural selection,* which incidentally is not a hypothesis that can be tested in a laboratory, was the selective breeding of domestic plants and animals to bring about desired changes.

It will be difficult for many readers to accept that psychoanalytic theory is an extension of commonsense psychology, or makes use of commonsense psychology as a model. Grunbaum, for example, in a personal communication, writes:

> I no more think that psychoanalytic theory is an *extension* of commonsense psychology than I think theoretical physics is an extension of common-sense "physics." What common-sense man believes a table is mostly *empty space* between particles?? That is why Eddington spoke of two tables.
>
> If psychoanalytic theory were the extension of common sense you depict, why did it encounter so much disbelief? And why was it so counterintuitive that the "ego is not master in its own house"? It is *utterly* incredible common-sensically that horror dreams should be wish-fulfilling.

Perhaps Grunbaum confuses the characteristics of the model with which a scientist begins with the theory that results as theory changes in order to account for what in terms of the model would constitute counterintuitive facts. That wishes bring about what appear to be unpleasurable effects is accounted for by postulating a new class of wishes—unconscious wishes—and a new set of mechanisms (mental operations characteristic of primary process) by which such wishes produce these effects.

In the same way, Darwin did not conclude from his model—breeders breeding selectively to bring about desired outcomes—that there was a Breeder in the sky operating over eons of time to bring about the products of evolution. He postulated a purely physical mechanism—natural selection—operating in a way analogous to that of his homely breeder. It was this theoretical emendation of his initial model—the idea of a purely physical mechanism—and not the homely model itself that aroused so much opposition and disbelief.

It is the idea of unconscious wishes and primary process mental operations, theoretical concepts devised to account for phenomena that are difficult to comprehend given our ordinary commonsense knowledge, that arouse disbelief. But surely the idea of unconscious wishes is an extension of the idea of wishes that is part of ordinary commonsense intentional psychology, which appeals to the causal efficacy of such entities as wishes and beliefs.

Reasons Justifying Acceptance of a Person's Causal Explanation of His Own Parapraxis In the Unproblematic Case

Here is an example of the kind of model Freud used in the way Darwin used domestic breeding. There are slips a person, on being asked, is able to explain. A person says, "This meeting is closed," rather than, what he intended to say, "This meeting is opened." Asked why he made the mistake, he reports a memory. Just before the meeting began, he was aware of a reluctance about beginning it; he did not want the meeting to take place. He wished it were already at an end.

Here is our model—an unproblematic case in which a person's own explanation of a parapraxis is credible. We shall use this model in explaining mystifying parapraxes. Therefore, we should be clear about the reasons we are warranted in accepting a person's causal explanation (unless we have reason not to accept it) in the model situation.

First, from much experience, we have reason to believe that *rational actions,* such as closing a meeting, are generally caused by conjunctions of desire (I wish the meeting were at an end) and belief (I believe that announcing the meeting is closed will result in closing it).

A Second reason to accept the causal explanation of the-meeting-is-closed slip is that we can see the clear connection here, the *thematic affinity,* between what was desired and believed, on the one hand, and what took place, on the other.

Explanations of rational action depend on thematic affinity in tracing the causal connections between certain mental states (desires and beliefs), on the one hand, and intentions-to-act or actions, on the other. An explanation of rational action might proceed as follows. Let *X* and *Y* be distinct states of affairs. John has a mental representation of *X,* and he wants *X.* John has a mental representation of *Y,* and he believes that *X* will occur only if *Y* occurs, or he believes that if *Y* occurs, then *X* will occur. The conjunction of John's wish and belief, which are identified by the contents *X* and *Y* symbolically represented in these mental states, causes his intention to bring about *Y.* We accept the conjunction of wish and belief as the cause of that intention, because of the thematic affinity of the contents of wish and the contents of belief (a representation of *X* appears in both), and the thematic affinity of the contents of wish, belief, and intention (a representation of *X* is shared by the former pair and of *Y* by the latter pair).

When Freud says "a man's attitude on sexual things has the force of a model to which the rest of his reactions tend to conform" (1909, p. 241), so that, for example, ambivalence and indecision in love will manifest itself in ambivalence and indecision in many other areas of life as well, his evidence depends, in part, on the argument by thematic affinity. Similarly, the psychoanalyst's reasoning, in part, depends on appeal to thematic affinity, when he adduces, as evidence for the existence and causal status of an oral, anal, or phallic wish or fantasy, the confluence in a single session or series of sessions of phrases metaphorically or literally describing a variety of kinds of activities as "taking in," "merging with," "swallowing," "sucking in," "spitting," or "biting"; or as "hiding away," "keeping secret," or "storing in a dark place"; or as "penetrating," or "being penetrated."

Reasoning by appeal to thematic affinity is not only central in the explanatory strategies of psychoanalysis. It plays an important role in the definition of the domain of psychoanalysis. For psychoanalysis, distinctively, asks questions about mental contents—objects and states of affairs. These objects and states of affairs are represented internally by persons. These persons relate in a variety of ways (wishes that, believes that, perceives that, remembers that, thinks that, feels that) to these mental representations.

Psychoanalysis wants to explain mental contents, in particular those that are mystifying to the persons in whose mental representations they appear. Not *how* someone is able to *believe,* but *why* does he believe just *that?* Not *how* someone is able to *remember,* but *why* does he remember just *that?* Not *how* is he able to form or carry out *intentions,* but *why* does he intend doing, or why does he carry out the intention to do, just *that?*

By analogy to explanations of rational action, which appeal to the contents of wishes and beliefs as causal processes linking one mental state to another, psychoanalysis explains the contents of some mental states in terms of the contents of other mental states. These contents link one mental state to another and they may propagate causal influence through space and time as, for example, the contents of memories do (Salmon 1984).

A Third reason to accept the causal explanation of the-meeting-is-closed slip is that, as a matter of commonsense psychology, we believe that, in some cases at least, a person may have *direct access to the causes of his actions.* We recognize that, in some cases at least, the "phenomenology" of the mental state a person is in (what it is like to be in such a mental state) is *enough* (not necessary, not sufficient, but enough) to explain his actions.[8]

If I ask you why you ran, and you state that you ran because you were in a state of terror, I accept that explanation, even though I do not know the cause of your state of terror. Some philosophers will regard me as naive to do so. I realize that a philosopher can think up cases that demonstrate that I might be mistaken in accepting that the state of terror caused the running, but I do not accept the invention of what I consider to be fringe or marginal cases to be a convincing reason for not accepting both the word of a person that he was in a particular state and that this state caused his action. I conclude rather that the excessively skeptical philosopher does not know what being in a state of terror is like.

I know that when I speak before an audience, I see that audience because the presence of the audience, which is really there, causes me to see it. I see no compelling reason for a psychoanalyst to subscribe to the philosophical fiction that I only know patches of sensation and not objects, and that I must infer an audience as cause from these patches of sensation. Nor do I see any reason to be dissuaded by the citation of fringe phenomena (for example, hallucinations) from my belief that I know it is the existence and presence of an object that causes me to see it.

However, if a person says that he cannot do a certain kind of work because his father mistreated him as a child, I require more grounds than that the person offers me this causal explanation to believe it. To the extent that a postulated cause is remote from the effect, and when there is no obvious connection of contents (thematic affinity) associated with postulated cause and effect, we require more grounds than the person's stated explanation of his actions or difficulties to accept the causal inference. But there is a defect in any conception of causality that, because of the problems posed by this last case, can admit of *no* direct access on the part of a person to the causes of his actions. This is skepticism in the service of explanatory nihilism rather than inquiry.

Is the notion of "necessary cause" as relevant as Grunbaum (1984, 1986) assumes to a consideration of the problems of justifying a psychoanalytic clinical causal inference? The analogical use by psycho-

8. I am indebted to Wolheim (1984) for his discussion of the phenomenology of mental states, and to Robert Shope (personal communication) for the idea of "enough," which he attributes to Anscombe.

analytic explanations of features of the explanations of rational action (described in the following paragraphs) raises this question, as did my description of the "phenomenology of a mental state" as *enough*.

A person's wishes and beliefs (for example, "I want to kill my father" and "I believe feeding him a particular dose of arsenic is sufficient to kill him") clearly make a difference in, and clearly are causally connected to, that person forming and carrying out the intention to feed arsenic to his father. Yet this conjunction of wish and belief, while *enough* to cause forming and carrying out such an intention, is not a necessary cause. A person's wish to preserve his father's life and belief that this is just the dose of arsenic required medicinally to accomplish that will cause the same intention.

Nor is this conjunction of wish and desire a sufficient cause for forming and carrying out such an intention. The person may have the wish and belief, but his belief about the dose may be mistaken, or he may be prevented from forming and carrying out the intention by internal and external obstacles, including lack of means, opportunities, and conflicting wishes.

The conjunction of wish and belief is neither a necessary nor a sufficient cause, but if it occurs, and the intention is formed and carried out, we have no question that it was *enough* to cause that outcome. Grunbaum (1984) himself has argued persuasively, and I think decisively, that it won't do to start talking about "reasons" in a case like this, with the idea that "reasons" are not "causes." So we can't get out of this difficulty about "necessary cause" that way.

Grunbaum may still argue that the claim requiring justification is that the conjunction of wish and belief is *causally relevant*. It makes a difference. How do we know? He might say we know because a comparison of a group of those having such a wish and belief with a group of those having no such wish and belief will show that in the former group there is a higher incidence than in the latter group of forming and carrying out the intention to feed arsenic to father.

However, even if we were not to obtain such a result, and even if we knew only one person with such a conjunction of wish and belief who forms and carries out such an intention, we would still have the same confidence in the causal inference that the conjunction of wish and belief, if it were to exist, would be *enough* (not necessary or sufficient) to produce or cause the intention. After all, we are interested in the sense of causality that is involved when an active agent produces fire by virtue of its causal powers, not interested merely in the kind of causal relevance oxygen has (it must be present, although it is not an active agent with causal powers).

All this suggests that some revision and clarification of our ideas about causality, as used in psychoanalytic explanation and in the case study, may be in order. Perhaps there are philosophers who can help us with this task.

Fourth, we assume the person in such cases as the closing-the-meeting case can and will tell us directly, *if nothing acts to prevent him,* what conjunction of desires and beliefs, in conflict with other desires and beliefs, triumphed over these others to cause the mistake. But from the model provided by commonsense psychology, we expect that if he feels ashamed of the operative desires and beliefs, guilty about having them, or fearful of the consequences of someone else knowing what they are, he will remain silent before a query, or answer evasively or misleadingly. We know that we must persuade him by some means to override his reluctance and tell us what he knows, and that our conviction that he then tells us the truth depends on our grounds for believing that we have provided an incentive for him to tell us the truth that is greater than his reluctance to do so, or that we have successfully mitigated the shame, guilt, or anxiety that caused him to be reluctant. (A good deal of practical clinical wisdom used in psychoanalytic treatment, and discussed under the headings "analyzability" and "technique," is an extension of this knowledge.)

This model for obtaining information from which we can validly infer the causes of action, inferences we accept without question, is one that has over and over in our practical actions demonstrated its usefulness. So far, it owes nothing to psychoanalysis. As a first step, Freud invites us to trust the psychoanalyst's causal inferences, because they are based on the same kind of reasoning we have found has "worked" in explaining actions by commonsense psychology.

What Is Different about the Psychoanalyst's Method of Inquiry and Causal Inferences?

It is that they are adjusted to encompass those cases in which a person is mystified by his parapraxis, dream, symptom. In these cases, there is no apparent conjunction of wish and belief leading to the formation of an intention to act, and no thematic affinity between those wishes and beliefs of which the person is aware and that which mystifies him.

First, we reason analogically that the mystifying parapraxis, dream, or symptom will have a cause similar to the one it had in the case in which the person was able to tell us its cause.

Second, we reason analogically that, if inquiry and the person's response to it leads to obtaining information about the cause (or information from which we can infer the cause) in the unproblematic case, then inquiry and the person's response to it will do so in the problematic case as well—but only if the method of inquiry is suitably adjusted to take into account complicating, interfering, or negative causes. Such adjustment is, or course, the raison d'etre of the method of free association.

The principle of reasoning by analogy, which operates here, may be stated: "similar effects, similar causes" (in the first argument by analogy), and "similar causes, similar effects" (in the second argument by analogy). Trust in the results of reasoning by analogy is justified by the successes science has had using models or metaphors to explain the unfamiliar by the familiar, or what we do not understand by what we do understand (suitably modified to take complicating, interfering, or negative causes into account). Such trust is also justified by the successes science has had in adhering to the criterion of parsimony, which might be paraphrased in this context: "Make all the use you can of the kind of explanation that has already worked in analogous situations. Turn to another kind of explanation only if repeated and uncorrectable failure of the kind of explanation you have been using forces this move upon you."

Arguing by analogy to commonsense psychology, Freud explains a parapraxis, a dream, a symptom when these are just the kinds of cases that constitute the distinctive domain of psychoanalysis: the cases when a person has no access to the causes of his actions. He is mystified by his parapraxis, his dream, his symptom. Freud reasons that, if he could tell us the cause, then, as in the instance when he has access to it, he would.

Freud satisfies himself that the person is not simply reluctant to reveal the causes. The person is indeed unaware of any states that would account for the slip, the dream, the symptom. They make no sense to him; that is, they are unrelated to wishes and beliefs that are accessible to consciousness.

Freud assumes there must be some other *causal process* operating to prevent a person not just from telling us, but from being aware of, the cause of his parapraxis, dream, or symptom. This negative causal process acts against the expression of, or access to, the cause of the phenomena of interest. It acts, in other words, to prevent or distort what would occur if the cause of the prapraxis, dream, or symptom were producing its effects unimpeded. This negative causal process is something analogous to the actions a person takes to conceal or obscure wishes and beliefs when he is knowingly reluctant to reveal that it is these that are the causes of his actions—and analogous, too, to the actions he takes to deceive himself about objectionable causes of his actions. This negative causal process now prevents him not just from telling but from being aware of the causes of his actions.

Such a negative causal process produces its effects by means of a set of mental operations, which the person carries out apparently unknowingly. These mental operations are analogous to—have features similar to—the workings of censorship in society, the operations of editing a text, and the evasive strategies all people, including ourselves, use (this we know without the help of psychoanalysis) when they feel ashamed of, guilty about, or afraid to reveal their wishes and beliefs. These mental operations (e.g., repression) have the power to deny wishes and beliefs access to consciousness, as a passage in text is erased by an editor or blacked out by a censor. These mental operations (e.g., displacement, condensation) are responsible for producing phenomena, the contents of which have no apparent thematic affinity with the contents of the unconscious wishes and beliefs that are their putative causes.

Justifying the Assumption that the Method of Free Association "Works"

Assuming that a person wants to be aware of and will reveal the causes of what mystifies him, if he can (thus the locution "the unconscious mental states strive for expression"), and that free association (which is designed to mitigate what prevents him from doing so) will make it possible for him to do so, Freud expects this method to yield information from which we can infer these causes (unconscious mental states). How does Freud justify his expectation that the method of free association *works*—that by using it, he will indeed overcome internal obstacles and succeed in getting the information from which he can infer unconscious mental states and mental operations, which, by virture of their causal powers, produce parapraxes, dreams, symptoms. He justifies this expectation in five ways.

First, by analogy to the effects of interventions we carry out in practical actions, guided by commonsense psychology, to reduce the reluctance of those we wish to communicate their secrets to us.

Second, and for Freud very importantly, by analogy to the effects of interventions in hypnosis, where something like an experiment occurs. The hypnotist is able to introduce an idea into a person's mind, which subsequently is clearly the cause of his actions, which he cannot help carrying out, although he is unaware at the time of these actions of its existence or causal role. The hypnotist is also able to remove that idea from the person's mind, and observe that following this intervention its effects also disappear. Under hypnosis, that is, in a special state of mind, a person is able to provide information about mental contents and to achieve access to the phenomenology of mental states of which he is not aware when not under hypnosis.

Third, by analogy to the effects of "chimney sweeping" or "the talking cure" reported in *Studies on Hysteria* (Breuer and Freud 1893–1895). A train of thought leads to the recovery of a traumatic memory (such as the memory of nausea brought on by seeing a dog lapping at one's dish), which is related by thematic affinity to a symptom, persistent nausea, which disappears when the memory is recovered, thereby testifying to the causal powers the memory possessed while unconscious.

This latter sequence of events is not, contra Grunbaum, vitiated as a model because there is no "cure"—that is, just because the patient's remission is followed by the return of symptoms or the generation of new symptoms.

No one has shown that invariably it is just this particular symptom—the persistent nausea itself—that returns.

No one has eliminated the possibility that, if it is this symptom that returns, it does so as a result of the patient's enduring capacity to respond again with nausea to some *new* experience. More precisely, psychoanalytic theory now proposes that the symptom can return as a manifestation of a persistent unconscious fantasy, which some new experience again evokes. This is not essentially different from supposing that a persistent viral infection can be recurrently manifested, exacerbated from time to time by a change in circumstances or the state of an organism. The psychoanalyst must, of course, present other evidence for the existence and causal efficacy of such enduring mental dispositions, so that an appeal to this possibility does not merely beg the question.

No one should any longer emphasize that it is the recovery of a memory (of a past external event or of a fantasy in the past) that is decisive in the model of the "talking cure." What is decisive in fact is the recovery of the phenomenology of a mental state, such that a person again experiences what it is like to be in that state, and thus gains direct access to the causes of his actions. The intense *conscious* experience in the transference, as part of the transference neurosis, of mental states that have hitherto been *inferred* as unconscious entities but now exist and exercise causal powers in the here-and-now, is important as evidence for psychoanalytic inferences. The recovery (in the transference) of the phenomenology of a mental state, which the person now directly experiences as the cause of his actions, is also important in understanding the therapeutic action of psychoanalysis.

Freud's fourth justification of his belief that free association works is based on an analogy to word-association experiments, the results of which Freud thought provided reliable evidence for the existence, at least, of unconscious "complexes" (what we would call unconscious conflicts or fantasies).

Fifth, just as hypnosis and word-association experiments provide independent evidence for the existence of unconscious causally efficacious mental states, so studies of art, literature, autobiography (the Schreber Case), and jokes provide independent evidence for the existence of the kinds of mental operations Freud postulated in his clinical causal inferences. This is the raison d'etre of the great interest psychoanalysis has in what is called "applied psychoanalysis," especially in the study of works of art and literature. (See, e.g., Edelson 1975, and Chapter 8 of this book.)

Now none of these justifications is unassailable by itself, but together they provide strong warrant for the credibility of clinical causal inferences in psychoanalysis.

Darwin as a Methodologist of Historical Science

Freud's thought has important parallels in the thought of Darwin. So much so, that psychoanalysis may claim that it is entitled to the same scientific status granted the theory of evolution. As previously noted, Darwin also makes use of a homely model.

> Darwin did not know what were the processes by which change in the animals and plants of nature came about, so he constructed a model. He knew very well that there is change in domestic animals and plants and he knew that that change is due to the fact that the breeder *selects* those plants and animals from which he wishes to breed, which are more suited to whatever purpose he has in mind, and that after several repetitions of selection a quite different-appearing creature can be derived from appropriately chosen individuals solely by breeding. There is a variation in nature, and Darwin conceived of a process analogous to domestic selection which could be a model of whatever process was really taking place in nature. He called this process, modelled on domestic selection, *natural selection.* (Harre 1972, pp. 176–177)

Gould (1986) also, in acclaiming Darwin as a methodologist who showed how history could be scientific, mentions two themes in his work that have to do with justifying the causal inferences a historian makes about the past. The historian faces a problem similar to the one Freud as methodologist faced in justifying causal inferences from clinical data.

The first theme (an argument by *analogy*) is the uniformitarian argument. The assumption is that change takes place gradually and in the same way over eons of time. Therefore, one is warranted in extrapolating from current small-scale phenomena that can be seen and investigated (for example, the observed selection by breeders) to what has occurred in the remote past.

The second theme (an argument by *consilience,* or the convergence of inferences from different types of evidence) is "the establishment of a graded set of methods for inferring history when only large-scale results [in the present] are available for study" (p. 61). We can make an inference about history by ordering temporally coexisting phenomena ("constructing, for example, a sequence leading from variation within a population, to small-scale geographic differentiation of races, to separate species, to the origin of major groups and key innovations in morphology"). Another (also nonexperimental, let it be noted) source of evidence, converging on the conclusion of evolution by natural selection through eons of time, are the cases of imperfect design ("vestigial organs" and "adaptations as contrivances jerry-rigged from parts available"). These imperfections eliminate the creationist hypothesis, which supposes a Perfect Engineer. Note that examples of impressive adaptation, which are consistent with both hypotheses, cannot eliminate the creationist hypothesis. Also belonging to the class of "oddities" that have evidential value are "odd biogeographic distributions made sensible only as products of history" (Gould 1986, pp. 63–64).

Consilience: Convergence of Inferences Upon the Same Conclusions

Each time we infer a particular unconscious mental state or conflict as a cause, we make a risky inference. Risky, because we have not ruled out other possible causes of what we are attempting to explain, or other causes of a person's giving us the information from which we infer the causes of what we are trying to explain. Risky, because we do not know if, and to what extent, an analogy holds in a particular instance in which we are relying on it. Our confidence in each such inference may be qualified. The evidence for it may be taken as relatively weak. But if a number of such risky inferences converge on the same conclusion, we may feel a greater degree of confidence in the conclusion, and regard the evidence *taken as a whole* to be quite strong.

Grunbaum (1984) has argued that when the method (e.g., the method of free association) of arriving at the same conclusion does not vary, the *same* risks may be present each time an inference is made. Therefore, since it is possible that the same kinds of errors in inference may be multiplied many times over, we are not justified in feeling greater confidence because many inferences converge on the same conclusion.

I would argue in return that our confidence is justifiably increased by the fact that the inferences are from very different kinds of information (different types of evidence)—therefore, in some relevant sense, they are independent and involve different risks. The range or variety of information a psychoanalyst has, and makes use of, in making inferences surely contributes to his confidence in the validity of these inferences. Compare our confidence in our inferences about a person when we know one thing about him with our confidence when we know many, and many differet kinds of, things about him. Indeed, things we would have to guess in the first case, we might have direct information about in the second.

Different inferences in the psychoanalytic situation are from:

descriptions of early life experiences;

reports by others to the analysand about early life experiences;

descriptions of current life experiences;

observable vicissitudes of the analysand's efforts to follow the method of free association (in what context, following what content, are there gaps, discontinuities in subject matter, pauses, missing affects, mismatch of content and affect, conscious reluctance to report thoughts? when and how does he justify deviating from the method? when does he out-and-out abandon the effort to follow the method althogether?);

patterns—contiguities and similarities—in the contents of the analysand's associations;

descriptions of the analysand's mental states as they occur in the psychoanalytic situation (sudden images, sensations, eruptions of feeling, fantasies);

what is internally evoked in the psychoanalyst;

and the analysand's images of and reactions to the psychoanalyst and events in the psychoanalytic situation.

Images of and reactions to the psychoanalyst and events in the psychoanalytic situation are especially important because of the analysand's access to the direct causes of his actions in the here-and-now—the phenomenology of mental states. He appreciates what it is like to love or hate someone, to be jealous, to be envious, to want to hurt or be hurt, and to experience the power of these states to produce or influence other mental states or his actions. This, to put it simply, is why, in psychoanalytic treatment, so much emphasis is placed on creating the conditions that will make the development of a transference neurosis possible, and on the importance of the transference neurosis in the curative process.

I would also argue in return that our confidence is justifiably increased by the fact that the inferences from parapraxes, dreams, and symptoms, which converge, are based on analogy to different models, and so justified in different ways, in some relevant sense independently, by what we know confidently from these different models. Parapraxes, dreams, and symptoms are unlike each other in important ways. Inferences about the causes of mystifying parapraxes take as their model what we know from commonsense psychology about instances in which there is no doubt that a mistake is the outcome of the intrusion of a checked but insistent intention. Inferences about the causes of mystifying dreams take as their model what we know from commonsense psychology about instances of unquestionable wish-fulfillment. Inferences about the causes of mystifying neurotic symptoms take as their model what we know from commonsense psychology about the way in which compromises are formed as a person seeks to solve the problem of gratifying a wish under conditions of threat.

Furthermore, there was no knowledge prior to psychoanalysis suggesting that parapraxes, dreams, and neurotic symptoms were other than independent phenomena, roughly grouped with a variety of other kinds of phenomena thought to be caused by somatic dysfunction or deterioration. No theory prior to psychoanalysis would have picked them out as in some way like each other and different from these other kinds of phenomena. It is only their susceptibility to psychoanalytic explanation that links them.

The philosopher of science Salmon (1959) summarizes these points in the following passage:

> It is necessary that there be independent evidence for the existence of [a] psychic mechanism, apart from the specific item it is supposed to explain. Other parts of psychoanalytic theory indicate what the independent evidence is. The theory gives a limited list of inferred entities such as unconscious feelings, desires, impulses, conflicts, and defense mechanisms. In some cases, at least, the theory states that such entities are created (with a high degree of probability) under certain specifiable conditions. The occurrence of such conditions constitutes independent inductive evidence for the existence of the entity. Furthermore, according to the theory, if one of these unconscious psychic entities exists, it is possible under specifiable conditions to elicit a certain kind of conscious entity (which may go under the same name without the qualification "unconscious"). Free association, hypnosis, and narcosynthesis are ways of eliciting the unconscious entity. It is not that the subject becomes aware of an unconscious entity—there is a sense in which this is impossible by definition. Rather, according to the theory, the occurrence of the conscious entity (or the report of it if one insists upon excluding introspective evidence) under the specified conditions constitutes inductive evidence for the existence of the inferred entity at an earlier time. Other items of

> behavior such as slips, dreams, and neurotic symptoms constitute further inductive evidence for the existence of the inferred entity. It may be, and often is, the case that none of these items of evidence is by itself very conclusive, but we must keep in mind that inductive inference often involves a concatenation of evidence each item of which is quite inconclusive. Nevertheless, the whole body of such evidence may well be conclusive. (Salmon 1959, pp. 258–259)

Those who use probabilistic reasoning and statistical inference argue, similarly, that having a body of different items of evidence, each of which although consistent with a hypothesis is by itself "weak," makes a difference. For example, one does a study that has an outcome favorable to a particular conclusion. One cannot, however, reject the possibility that this outcome was obtained by chance. The probability of obtaining the outcome by chance is calculated to be quite high, .10, let us say. One is not justified in rejecting the hypothesis that the outcome was obtained by chance alone, unless the probability of its occurrence by chance alone is low, around .02, let us say.

But suppose one has a series of such studies, all showing the same trend, that is, suggesting the same conclusion. Suppose the probability of obtaining this favorable outcome by chance is high in each study (possibly, .10); in no study is one justified in rejecting the hypothesis that the outcome occurred by chance. However, the probability of obtaining favorable data by chance alone in a series of independent studies is much lower than .10; depending on the number of studies, it might be as low as .02. Therefore, one is justified in rejecting the hypothesis that the outcome of the *series* of studies was obtained by chance alone, and in accepting the conclusion on which all the studies converge. (See, for example, Chassan 1979).

Campbell and Stanley (1963, pp. 36–37) argue that if a number of independent kinds of studies are tests of a particular hypothesis, and each study eliminates even one, but a different, alternative explanation of how data came to be obtained in each study that favored the hypothesis, then it is more parsimonious to assume that the truth of the hypothesis accounts for this outcome in each study than to assume instead that (a really improbable coincidence) in each study a different alternative explanation accounts for it.

But the most important parallel again is to Darwin's work in developing a methodology for making causal inferences about history. Gould (1986) argues that historical science is not to be rejected because it is not experimental or predictive.

> First, it is not true that standard techniques of controlled experimentation, prediction, and repeatability cannot be applied to complex histories. Uniqueness exists in toto, but 'nomothetic undertones' [common themes of impact] . . . can always be factored out. . . . Nature, moreover, presents us with experiments aplenty, imperfectly controlled compared with the best laboratory standards, but having other virtues (temporal extent, for example) not attainable with human designs.
>
> Second . . . Darwin labored for a lifetime to meet history head on, and to establish rigorous methods for inference about its singularities. . . . We do not attempt to predict the future. . . . But we can postdict about the past. . . . Darwin, so keenly aware of both the strengths and limits of history, argued that iterated pattern, based on types of evidence so numerous and so diverse that no other coordinating interpretation could stand—even though any item, taken separately, could not provide conclusive proof—must be the criterion for evolutionary inference. (The great philosopher of science William Whewell had called this historical method "consilience of inductions."). (Gould 1986, pp. 64–65)

Gould (p. 65) goes on to quote Darwin on consilience. Darwin, facing the same problems Freud faced in justifying causal inferences nonexperimentally, argues: "Change of species cannot be directly proved. . . . The doctrine must sink or swim according as it groups and explains phenomena."

And again:

> This hypothesis [natural selection] may be tested—and this seems to me the only fair and legitimate manner of considering the whole question—by trying whether it explains several large and independent classes of facts; such as the geological succession of organic beings, their distribution in past and present times, and their mutual affinities and homologies. If the principle of natural selection does explain these and other large bodies of facts, it ought to be received.

Psychoanalysis also "explains several large and independent classes of facts." For example:

many different kinds of people's mental states and actions;

many different kinds of mental states and actions (dreams, mistakes or faulty functioning in the absence of physical pathology, neurotic symptoms);

some features of, and some oddities found in, cultural artifacts, such as jokes, art, and literature;

observations of child behavior and the vicissitudes of human development;

sociological phenomena, such as result from the use of social institutions as defenses against anxiety;

and cross-cultural anthropological phenomena.

Psychoanalysis, therefore, may also argue that "iterated pattern, based on types of evidence so numerous and so diverse that no other coordinating interpretation could stand," is a criterion to which its explanations conform.

Psychoanalysts, for the most part, are not being unreasonable in recognizing the limitations of the experimental methods of "hard" science for testing psychoanalytic conjectures about persons. They would do well to commit themselves also to a thoughtful, rigorous study and use of the methods Darwin championed. Psychoanalysis can reasonably aspire to justify its clinical causal inferences to the same degree Darwin was successful in justifying his historical causal inferences.

Gould's collections of essays (1977, 1980, 1983, 1985), his reflections in natural history, are casebooks on the case study. They should be studied both by psychoanalysts who aspire to write case studies having evidential value and by those who denigrate the evidential value of nonexperimental data and the case study. Each essay describes a usually naturalistically observed phenomenon and shows how it bears evidentially on a general hypothesis—and, very often, how it operates to eliminate a plausible rival hypothesis in favor of a hypothesis of interest.

Inference to the Most Likely Cause: Causal Powers and Causal Mechanisms

Successionist and Generative Conceptions of Causality

Psychoanalytic explanations clearly embody what Harre (1972) has called a *generative* rather than a *successionist* conception of causality. I paraphrase his descriptions of these two conceptions in what follows.

Neither of these conceptions is wrong; they both have their uses in science. The successionist conception of causality is especially useful in generalizations about the way in which one moving body, hitting another, may be expected to change its motion, or in making predictions about, given one event, what other event will follow. The generative conception of causality is especially useful in formulating theories that will explain illness in terms of the existence and nature of viruses, differences between materials in terms of the existence and nature of atoms, the growth and development of plants and animals in terms of the unfolding of an innate epigenetic plan, and linguistic performances in terms of an innate competence or knoweldge of the rules of grammar.

In a *successionist conception of causality,* a cause comes regularly before an event or state. We come to have reason to expect that the effect will follow the cause; we can predict it.

Knowledge is of correlations among types of events. Laws describe functional relations between empirically observable variables (most typically, quantities). Theoretical terms (such as force) are calculated from observational terms, and are eliminable. (The laws of mechanics can be reformulated without using the concept "force.")

Explanations account for the occurrence of an event by showing that it is an instance of, or follows from, a universal generalization. Less general empirical regularities and ultimately singular events can be deduced from a few very general laws. Each general law is the explanation of a less general law, which can be deduced from it, and each less general law is the explanation of still less general laws, or singular events, which can be deduced from it.

It is important to note that psychoanalysis is not this kind of hierarchical theory, nor does Freud use the covering-law model of explanation.[9]

Guided by the successionist conception of causality, we may discover that two kinds of observable events co-vary or co-occur. These two kinds of events might, for all we know, occur quite independently of each other. The characteristics of one kind of event can certainly be given without any reference to the other kind of event. We satisfy ourselves that one kind of event occurs prior to the other. We satisfy ourselves, for example, by preventing the prior event from occurring, that, if that event does not occur, the other event will not occur. We call the prior event the necessary cause of the other event.

A cause is necessary but not sufficient if the occurrence of its effect is expected only if it occurs, but its occurrence does not guarantee the occurrence of its effect. One may infer from the absence of the cause the absence of the effect, or from the presence of the effect the presence of the cause—but may not infer from the presence of the cause the presence of the effect, or from the absence of the effect the absence of the cause.

A cause is sufficient but not necessary if, whenever it occurs, its effect may be expected to occur, but its absence does not guarantee that the effect will not occur. One may infer from the presence of the cause the presence of the effect, or from the absence of the effect the absence of the cause—but may not infer from the absence of the

9. This is argued to be the case in Chapter 13.

cause the absence of the effect, or from the presence of the effect the presence of the cause.

In the successionist view of causality, the relation "causes" between cause and effect is external. There is nothing in nature that corresponds to the relation. It is a conceptual relation—our way of organizing our experience. A cause can be characterized without reference to its possible effects. An effect can be characterized without reference to its possible causes.

Things are passive. An effect is what happens *to* a thing because something outside itself has acted on it or influenced it.

My impression is that Grunbaum's critique of psychoanalysis (1984), and other critiques as well (e.g., Glymour 1983)—especially their criticisms of Freud's arguments and method as involving causal fallacy—are grounded in a successionist conception of causality.

In a *generative conception of causality,* in contrast to the successionist conception, things have causal powers that can be evoked in suitable circumstances. These causal powers are not under all circumstances manifested, nor are they always manifested by the same effects.

Knowledge is of structure—the mode of organization of elementary parts, the structural relations among elementary individuals, what can be made from what—rather than of correlations among types of events. The scientist imagines and constructs models rather than laws, and in developing his theory judges analogy and disanalogy. The model is successful when in that model specifiable relations of internal constituents produce, by means of the operation of specifiable causal mechanisms or processes, certain outcomes that are similar to the properties or features of the system under investigation.

Theoretical terms refer to hypothetical entities, some of which are supposed to exist; their existence may come to be demonstrated under appropriate conditions, usually by pointing to their effects. General existential statements ("there are such-and-such things") are important in theory, more important perhaps than universal generalizations.

Explanation is not by subsumption under a "covering law." Explanation describes a causal mechanism that is capable of producing something like the event or pattern of happenings to be explained.

Cause is internal. There is a real connection—a causal process or mechanism—existing between, and connecting, cause and effect. The cause stimulates the causal mechanism, which produces the effect. A causal mechanism (not necessarily "machine-mechanical") is any kind of connection through which causes bring about effects. A cause is not a thing independent from its effect. It is connected to it. It cannot be described without referring to the effects it is capable of generating or producing. That capacity is part of its nature. The effect could not happen without the cause and would not be just what it is if it were caused by something else or in some other way.

Things are active, not passive. They have the power to bring about effects, and respond to influences and stimuli in the way they do because they are constituted the way they are. Materials and things possess causal powers by virtue of their internal constituents and the relations among these. When some condition varies, against a background of otherwise stable conditions, changes in internal constituents or in the relations among them set into motion generative or causal mechanisms, which produce the observable effects.

Another Look at the Causal Inversion Fallacy

Philosophers such as Glymour (1983) and Grunbaum (1984, 1986), perhaps because they work primarily with a successionist rather than a generative conception of causality, are concerned, as I have previously noted, with what they regard as a causal fallacy in Freud's argument—what Grunbaum calls a causal inversion fallacy.[10] Surely, the mention of an occurrence (an independent event in this conception of causality) *after* a parapraxis or dream in the free associations of an analysand is *caused* by the parapraxis or dream (another independent event in this conception of causality), and can surely not be used therefore as evidence for the *cause* of it.

This misdiagnosis of causal fallacy is based on at least three misunderstandings.

First, even from the point of view of a successionist conception of causality, it should be noted that the free associations, which *follow* the phenomena to be explained, *refer to* events that have frequently occurred *before* or *concomitant* with the phenomena to be explained.

Second, the free associations are indices or expressions of a causal entity, a complex of mental states, which is an *enduring* disposition or propensity (not an isolated independent event). This causal entity has, in other words, the power to produce or generate manifestations, depending on circumstances impinging on a psychological system from outside of it and the state of the system itself. The parapraxis. dream, or symptom *and* the free associations are *connected* to that causal entity by causal processes. Disposition, parapraxis, dream, sympton, and free associations are all part of the same network of

10. Already discussed earlier in this chapter in the section on Freud's arguments under the heading *The Specter of the Causal Inversion Fallacy.*

connections. These connections are forged not by rules of cognitive inference (as in rational thought) but by the operations of the dream work or primary process (including such operations as condensation, displacement, and representation by images, rather than words, and by iconic symbolizations). The cause has the effect it does because that effect is a manifestation of the power the cause has, by virtue of its very nature, to produce just that kind of effect. Cause and effect are not independent events that have no particular relation other than that the preceding one has the conceptual relation *causes* to the subsequent one.

Third, free associations are the products of a method that is intended to mitigate the negative causal process referred to above. So the cause of parapraxis, dream, or sympton is freed to a greater or lesser degree depending on the extent to which the method is successfully implemented to manifest itself more directly, to reveal its nature, to become more apparent than in the presence of the unmitigated negative causal process.

Glymour and Grunbaum may certainly question whether the evidence is adequate to support the theory that such dispositional causes exist, that they have the causal powers to produce the manifestations attributed to them, and that they do so, and by means of the causal mechanisms postulated. In the same way critics of evolution have raised questions about the adequacy of the evidence, mostly nonexperimental, for the existence of evolution and, especially, for the postulated nature of the mechanisms which it is thought produce its effects. But Freud's account is theoretically sound, in the sense that it is not infected, as these critics suppose, by causal fallacy.

The Domain of Psychoanalysis

The wishes and beliefs denied access to consciousness are theoretical entities, whose existence is inferred from what an analysand says as he attempts to follow the procedure of free association. These theoretical entities (together with conscious mental states that are inexplicable to the one possessing them and observable acts that do not make sense to the one who acts) constitute the distinctive domain of psychoanalysis. The putative causal powers of the postulated theoretical entities (dispositional mental states) are analogous to those possessed by conscious mental states by virtue of their phenomenology (what it is like to be in that state). More precisely, the causal powers of dispositional mental states have to do with the causal powers of occurrent mental states that are the manifestations of such dispositions (Wollheim 1984).

In free association, a person attempts to say whatever comes to mind, suspending criticism or selection of, and control according to conscious purposes over, these mental contents. The method of free association is the means by which the operations that censor or obscure, which are motivated by shame, guilt, and anxiety, are exposed or rendered inoperative, so that the relevant causal dispositions may manifest themselves. These mental operations are theoretical causal mechanisms or processes, which are inferred to account for the particular paths free association takes—the sequence of mental contents produced by an analysand, how one content leads to the other. These mental operations, in general, are inferred to account for how one gets by transformative operations from the contents of the mental states that are the putative causes of mystifying parapraxes, dreams, and symptoms, to the contents of the parapraxes, dreams, and symptoms, when there is no thematic affinity between them.

Such theoretical causal operations, mechanisms, or processes—together with manifest phenomena that do not make sense and inferred unconscious mental states (theoretical entities)—constitute the distinctive domain of psychoanalysis.

I shall continue to indicate the difference a generative conception of causality makes by reviewing (1) the study of viruses in biology; (2) a criticism of Grunbaum's critique of the empirical foundations of psychoanalysis based on the study of syphilis in medicine; and (3) a philosopher of science's view of causality in natural science (especially physics).

Then, building on discussions in Chapter 6, 11, and 13, I shall conclude by indicating the extent to which Freud's strategy of explanation can only be understood and assessed from within the frame of reference provided by such a conception of causality.

Viruses

How much more psychoanalysis is involved in inferring causes that are like viruses rather than billiard balls is suggested by the problems medical science has had in establishing criteria for accepting a belief that a particular agent is the cause of a disease. The story is told by Evans (1978).

The *classical criteria* were:

1) "the parasite must be present in every case of the disease under appropriate circumstances";

2) "it should occur in no other disease as a fortuitous and nonpathogenic parasite";

3) "it must be isolated from the body in pure culture, repeatedly passed, and induce the disease anew" (p. 250).

Many causes of illness did not fulfill the criteria: "typhoid fever, diphtheria, leprosy, relapsing fever and Asiatic cholera" (p. 250). The second criterion especially became a problem when the carrier state was discovered for diphtheria and typhoid.

The limitations of these criteria became evident from the study of virus diseases:

> 1) More than one agent might be needed to produce a given disease;
> 2) viruses could not be grown on lifeless media but required living cells in contrast to bacteria; and
> 3) asymptomatic carriers existed. (p. 250)

The following criteria for establishing etiology were added to take into account the nature of viruses:

> 1) the reproduction of the disease experimentally should exclude the possibility that the laboratory animal was infected with some laboratory virus and include appropriate controls; and 2) . . . antibody to the disease should appear during the course of the illness. (Pp. 250–251)

The addition of the host's response to the parasite as an indication of the causative status of the parasite is very much in the tradition of the generative conception of causality.

Then, it was discovered that many viruses were simultaneously present in people who are ill and in people who are healthy. Some infections were caused by multiple viruses. Viruses could produce chronic illness or carrier states. Viral reactivation occurred: the primary exposure occurs years earlier (when, for example, host resistance is weak) than the manifestation of chronic disease (which may also result if the immune system of the host later on is weakened by the disease or by drugs used in treating disease).

Establishing the presence of the virus no longer could establish causation. Longitudinal and cross-sectional studies were both required for establishing causation. The response of the host was again emphasized—in an argument from therapeutic efficacy—when the fact that the prevention of the disease could be achieved by a specific vaccine became an important link in the chain of evidence. This emphasis was enhanced by the growing "recognition that the same clinical picture or syndrome could be produced by a variety of different agents, that different agents predominated under different epidemiologic circumstances, and that the host response to a given virus would vary from one setting to another" (p. 251). That co-factors and host susceptibility determined whether or not the clinical illness was manifest made it clear that the parasite was not a sufficient cause of illness, and that the nature of its interaction with the host (which depends in turn upon the state of the host) is decisive in determining whether contact (that is, infection) results in manifest illness.

Parasites such as the Epstein-Barr virus, causally related to infectious mononucleosis, "could not be grown in pure culture nor did their injection reproduce the disease in experimental animals" (p. 251). So, evidence for causal status came to include:

1) "that antibody to the agent is regularly absent prior to the disease and that it appears regularly during illness";

2) that the "absence of the antibody should indicate susceptibility to infection and its presence indicate immunity"; and

3) that "antibody to no other agent should be similarly associated with the disease unless it is a co-factor" (p. 251).

The causal status of the Epstein-Barr virus was established without fulfilling even one of the three *classical criteria!*

Then, slow viruses were discovered with long incubation periods, which were associated with chronic neurologic diseases, and which could not be isolated in tissue culture in the laboratory, did not produce an immune response, and were resistant to many physical and chemical agents. Causal relationship, then, had to be established by: "1) the consistency in animal transmission of the agent[;] 2) the fact that the agent should be serially transmissible in experimental animals with filtered material[; and] 3) the fact that similar results should not be obtained from normal tissues" (p. 252).

That viruses can be present in normal tissues is well known, creating a problem with respect to this last criterion.

Problems in establishing criteria of causation for the relation beween *viruses and cancer* "include the long incubation period between exposure to the putative agent and the cancer and the probability that disease results from reactivation of the virus rather than from a primary infection" (p. 252).

> The agents most commonly recognized as the best oncogenic candidates are the herpes viruses. But proof of causation is difficult because they are common and ubiquitous viruses, probably require cofactors, and there are difficulties in reproduction of the cancer in animals. There is also the probability that the cancer may have different causes in different geographic areas or under different epidemiologic settings. . . . My own view of the relationship of viruses to cancer is that the host plays a critical role in the development of the malignancy. Such factors as the age at the time of infection, the

> immunologic state of the host, the presence of co-factors such as concurrent infection, and the genetic attributes as expressed by human leukocyte antigens, all play a role in the pathogenesis of the malignant process. (Pp. 252–253)

The consideration of causation in *chronic disease* led to a new set of criteria for accepting the belief that an agent was causally responsible for an illness:

> . . . 1) the suspected characteristic must be found more frequently in persons with the disease than in persons without, 2) . . . persons possessing the characteristic should develop the disease more frequently than those not possessing it, . . . 3) this association must be tested for its validity by investigating the relationship of the characteristic to other diseases in order to establish the specificity of the characteristic and the disease[,] . . . [4)] the incidence of the disease should increase in relation to the duration and intensity of the suspected factor, [5)] its distribution should parallel that of the disease in all relevant aspects, [6)] a spectrum of illness should be related to the exposure to this suspected factor, [7)] reduction or removal of the factor should reduce or eliminate the disease[, and 8)] human populations exposed to the factor in controlled or even in natural experiments should develop the disease more commonly than those not exposed. (Pp. 253–254)

Evans concludes with his own eight criteria for establishing a causal relation in both infectious and noninfectious diseases, but comments that these, too, "should be taken only as guidelines, subject to our changing knowledge of technology and causation" (p. 254).

I have quoted from this story at some length, because it has an important moral for those who wish to advise psychoanalysis concerning the path its research should take if its causal claims are ever to find empirical support.

1. Some understanding of the nature and behavior of the causal agent determines what criteria we use for establishing causal claims. The criteria will change through time, as our knowledge of the possibly quite different natures and behaviors of variants of the same kind of agent, or of different agents, changes. Psychoanalysis will have to establish criteria for establishing its causal claims that bear some relation to the nature of the causal agents it postulates.[11] These criteria will not necessarily correspond to those used in establishing the causal status of a virus. On the other hand, there may be some

11. Paul Ecker called my attention to the Evans paper with just some point as this in mind.

correspondence. Evans thought there might be some correspondence between the links in a chain of evidence establishing guilt or innocence in a court of law and those establishing the cause of a disease (pp. 254–255). Certainly, the concept of circumstantial evidence seems to correspond in some way to the concept of consilience.

2. The effects of causes depend heavily on the nature of the objects on which they impinge. What effects occur depends heavily on the nature of the host, and changes in the nature of the host. The events (inner or outer) that impinge on fantasy systems will have effects, if any, that depend heavily on the nature of the fantasy systems and their relation to one another, and *changes* in their nature and relation to one another.

Syphilis

Ferguson (1985), in his critique of Grunbaum's critique, makes a somewhat similar point. He questions Grunbaum's demand that psychoanalysis should show that its success as therapy compares favorably with "spontaneous remission rate."

1. Grunbaum focuses on remission of symptoms (what in this chapter I have been calling effects or manifestations of dispositions), rather than on alteration in character ("character" as used here is probably related to what in this chapter I have called the capacity or propensity to form symptoms). Specifically, Ferguson comments on what is lost by grouping people for such studies as Grunbaum proposes by their symptom-syndromes rather than by their propensities to respond in certain ways to particular kinds of events. His example: recurrent depression in response to vicissitudes in a woman's relation to her son.

> But, with the *insight* that these recurring depressions always begin subsequent to the return of the woman's son, with whom she has a destructive, pathological relationship, we begin to get some indication of the causes of this woman's depression and an indication how further treatment should go, i.e., suggest modifications in her lifestyle to avoid the precipitating causes of the depression; explore the pathological nature of her relationship to her son and attempt to find the solution through that route; or concentrate on the specific narcissistic vulnerabilities of the woman which make her susceptible to depressive attacks under conditions accompanying her son's presence. (P. 336)

2. The kind of experiment Grunbaum proposes would lead to very misleading results if the causal agent were anything like syphilis and we ignored the structure of the causal agent, its behavior, the nature of its relation to the host, and the mechanisms by which, in interac-

tion with the host, it produces its effects. Remission of the primary stages of syphilis occurs without treatment. The secondary symptoms appear twelve to twenty-four weeks later. In 100 percent of cases, signs and symptons in all untreated patients will then clear within one to three months. Four to five years after onset, two-thirds of patients enter into a latent noninfectious stage, and will not present further signs and symptoms. One-third will progress to tertiary syphilis, the time of onset of which will vary from three-to-four years to twenty-to-thirty years later. Since any organ or system of the body can be affected, symptoms or manifestations of the disease vary enormously.

> If medical treatments are evaluated solely on the basis of symptom occurrence as opposed to the spontaneous remission rate, then in many cases (cases where disease is chronic, or has a complex evolutionary development), the outcome of the experiment, i.e., the results vis-a-vis the value of the treatment will depend on the point in the course of the illness when the experiment is conducted, the simplicity of the causal factors underlying the symptoms, as well as the illness having characteristic features at all stages of its development that allow it to be clinically identified simply by looking at the presenting symptoms (not to mention whether the treatment is any good or not). Illnesses with extended or variable latency periods will not lend themselves well to followup studies. (P. 339)

Inference to the Best Cause

Cartwright (1983) has given us an account of causal explanation (paraphrased in the following paragraphs) that makes sense of the kind of explanation Freud is attempting in his case studies.[12] Although she gives considerable weight to the role of experiment, I do not believe that application of her view of causality and causal explanation in science is dependent on the possibility of experiment.

In her view, based largely on examples from physics (she does not regard explanation in physics and biology to be fundamentally different), good causal explanations probably will not have rivals. (She means here plausible substantive *rival* explanations, not *alternative* explanations of outcomes favorable to a hypothesis, which attribute these outcomes to the fact that extraneous factors or "chance" are operating in the situation of investigation.) Eliminating rivals is not the issue in establishing credibility. To establish the scientific credibility of a postulated cause, one must justify belief in the existence

12. Salmon (1984) gives a similar account of causality and causal explanation, as I have indicated in Chapter 13.

and nature of causal entities, which have causal powers, as well as in the existence and nature of causal mechanisms, by means of which causal entities transmit their influence.

> Suppose we describe the concrete causal process by which a phenomenon is brought about. That kind of explanation succeeds only if the process described actually occurs. To the extent that we find the causal explanation acceptable, we must believe in the causes described. (Pp. 4–5)
>
> [W]e have a satisfactory causal account, and so we have good reason to believe in the entities, processes, and properties in question.
>
> Causal reasoning provides good grounds for our beliefs in causal entities. Given our general knowledge about what kinds of conditions and happenings are possible in the circumstances, we reason backwards from the detailed structure of the effects to exactly what characteristics the causes must have in order to bring them about. . . . We must have reason to think that this cause, and no other, is the only practical possibility, and it should take a great deal of critical experience to convince us of this.
>
> We make our best causal inferences in very special situations—situations where our general view of the world makes us insist that a known phenomenon has a cause; where the cause we cite is the kind of thing that could bring about the effect and there is an appropriate process connecting the cause and the effect; and where the likelihood of other causes is ruled out. This is why controlled experiments are so important in finding out about entities and processes which we cannot observe. (p. 6)

Laws may express functional relations among quantities. These noncausal laws do not explain. Laws may be universal generalizations, and it may be supposed that deducing an instance from such a universal generalization explains it. Such an instance may be deduced from many other rival generalizations. These laws do not explain. Laws may express covariation or co-occurrence among variables. Such a covariation or co-occurrence may be thought to be a causal relation. There are many other reasons for obtaining such covariations and co-occurrences than that there exists a causal relation between the variables involved. These laws do not explain.

> [T]he propositions to which we commit ourselves when we accept a causal explanation are highly detailed causal principles and concrete phenomenological laws, specific to the situation at hand, not the abstract equations of a fundamental theory. (p. 8)

A causal law describes a concrete connection between a cause and an effect. It describes the nature, organization, or structure of a cause (in terms, for example, of the nature and relations among its internal constituents). The nature of the cause accounts for its powers, its capacity to influence things or events. This power to influence is not always manifest. When a stimulus, happening, or change in some condition impinges on the cause, the cause exerts its influence by setting a causal mechanism—possibly, a set of operations—into motion. This mechanism or set of operations terminates in the effect, and is therefore a concrete connection between cause and effect.

A causal law commits itself to the existence of a cause, and is a hypothesis about its nature. It commits itself to the existence of a causal mechanism, and is a hypothesis about its nature. A causal law is rarely true, except of a model or in the laboratory situation. It is an idealization concerning what would happen if only a particular cause, and no other causes, were exerting an influence. But in most cases, of course, we are interested in explaining effects produced by many causes operating at once.

In most cases, certainly outside the laboratory, a number of causal powers, often of very different kinds (that is, appearing in very different causal laws, which may obtain in quite different domains or hold true in quite different models), are all operating at once. Some causes may not succeed in exercising their powers at all, and what is actually produced is usually a compromise among different kinds of causal powers.

There are very few theories that specify the procedure by which to combine the various causes to produce an effect. However, after we have seen what occurs in a specific case, we are often able to understand how various causes contributed to bringing it about. That is why the psychoanalyst can understand, after many observations of a single case, what causes contributed to bringing about the effects he wants to explain. A causal explanation, such as Freud's in the Rat Man and Wolf Man cases, tells us how *different* causal powers *working together* produce effects.

Cartwright gives an example illustrating these points about inference to the best explanation. Camellias do well in rich soil, but are adversely affected by heat. There is no way to predict the outcome of planting camellias in composted manure, which is hot. If, after giving the camellias excellent care, they die, there is no problem in explaining what went wrong. "The camellias die because they were planted in hot soil."

> This is surely the right explanation to give. Of course, I cannot be absolutely certain that this explanation is the correct one. Some other factor may have been responsible, nitrogen deficiency or some genetic defect in the plants, a factor that I did not notice, or may not even have

> known to be relevant. But this uncertainty is not peculiar to cases of explanation. It is just the uncertainty that besets all of our judgements about matters of fact. We must allow for oversight; still, since I made a reasonable effort to eliminate other menaces to my camellias, we may have some confidence that this is the right explanation.
>
> So we have an explanation for the death of my camellias. But it is not an explanation from any true covering law. There is no law that says that camellias just like mine, planted in soil which is both hot and rich, die. To the contrary, they do not all die. Some thrive; and probably those that do, do so *because* of the richness of the soil they are planted in. We may insist that there must be some differentiating factor which brings the case under a covering law: in soil which is rich and hot, camellias of one kind die; those of another thrive. I will not deny that there may be such a covering law. I merely repeat that our ability to give this humdrum explanation precedes our knowledge of that law. On the Day of Judgment, when all laws are known, these may suffice to explain all phenomena. But in the meantime we do give explanations; and it is the job of science to tell us what kinds of explanations are admissible.
>
> In fact I want to urge a stronger thesis. If, as is possible, the world is not a tidy deterministic system, this job of telling how we are to explain will be a job which is still left when the descriptive task of science is complete. Imagine for example (what I suppose actually to be the case) that the facts about camellias are irreducibly statistical. Then it is possible to know all the general nomological facts about camellias which there are to know—for example, that 62 per cent of all camellias in just the circumstances of my camellias die, and 38 per cent survive. But one would not thereby know how to explain what happened in my garden. You would still have to look to the *Sunset Garden Book* to learn that the *heat* of the soil explains the perishing, and the *richness* explains the plants that thrive. (Pp. 51–52)

It may be even that, as in Freud's Wolf Man case, different causal mechanisms come into play in sequence.[13] Different causal laws apply to each step in the process whereby an initial event or cause leads to a terminal effect. Different nonlawful or ad hoc happenings trigger different causal powers along the way. *Although a different causal law may govern each step, there may be no causal law relating the initial cause and the terminal effect.*

A causal explanation, if it succeeds in explaining an effect, has very few, if any, plausible rivals—assuming that our confidence in the causal laws to which we appeal in the explanation is rationally justified, and that there are rational grounds for believing in the existence of the entities and causal mechanisms involved in the explanation.

13. This is described in Chapter 13.

Effects are just what they are because of the specific nature of the causal entities and causal processes that have generated them. Therefore, a scientist is justified in inferring from a detailed description of effects their causes (for whose existence he has evidence) and the causal mechanisms (for whose existence he has evidence) that have produced them.

Causes and Causal Mechanisms in Psychoanalysis

Given a generative conception of causality, how does a psychoanalyst proceed?

1. He postulates the existence of hypothetical entities. In psychoanalysis, these entities are here conveniently referred to as fantasy systems. The causal powers of these fantasy systems depend on their nature and the relations among them. The nature of the fantasy systems is determined by their own internal structure. Their internal constituents are wishes and beliefs, especially the contents of these wishes and beliefs, and the interrelations (conflict, for example) among them.

The causal powers of these fantasy systems are not always manifest. These causal powers may be evoked by various ad hoc or nonlawful external events or states of affairs—as these are interpreted by an analysand. These causal powers set off—their influence is transmitted by means of—concrete causal mechanisms or processes. While a cause—one or more fantasy systems, for example—may have the power to produce an effect, it does not always succeed in exercising that power. The effect actually produced depends on what other powers are at work. What is usually achieved is a compromise among different kinds of causal powers.

A set of dispositional mental states may exist for periods of varying length without there being any manifestation of it, as a stick of dynamite may sit for an indefinite period of time without manifesting its capacity to explode. (Unlike the stick of dynamite, a set of dispositional mental states may "spontaneously" manifest itself repeatedly or periodically.)

Under different circumstances, a set of dispositional mental states (wishes and beliefs) may manifest itself in very different ways. How a set of such dispositions manifests itself depends on how these circumstances are interpreted—classified, given certain significance—by the person who has the wishes and beliefs. The person's interpretation itself may be a manifestation of the set of dispositional mental states—an example of how an effect (the response of a person to an event, for example) can only be characterized in terms of the nature of the person (his wishes and beliefs) upon whom that event impinges. How the set of dispositional mental states manifests itself depends on the way in which the circumstances so interpreted connect with, activate, or evoke the wishes and beliefs (internal constituents), and differentially excite, activate, support, or reinforce, or undermine, interfere with, or weaken, one or another of them. What manifestations occur depends on what the specific content of these wishes and beliefs, and what the nature of the relations among them, are.

2. The psychoanalyst conceives the causal mechanisms such fantasy systems set off, which produce, and therefore account for, the effects he wants to explain. The causal mechanisms he hypothesizes have to be plausible and possible; they are like something about which he already has a great deal of knowledge.

The mechanisms postulated by psychoanalysis are essentially mental operations on symbolic representations. These operations have been designated by such terms as "primary process," "dream work,' and "defense mechanisms." These causal mechanisms produce, and therefore account for, the effects (for example, symptoms) to be explained. We are familiar with the analogues of these mental operations from our knowledge of commonsense psychology and of acts carried out on information represented and transmitted in various ways in social systems.

3. The psychoanalyst looks to consilience and the outcomes of interventions to support the scientific credibility of these hypotheses. For example, the results of different ways of demonstrating the existence of these fantasy systems by detecting their effects converge on the conclusion that they exist. The strength of our beliefs that particular fantasy systems exist, and that they produce their effects by setting off particular mental operations, is justified insofar as, when circumstances change, or interventions such as interpretations or the inculcation of a transference neurosis occur, our expectations about what effects will be produced are fulfilled.

Psychoanalytic theory resembles the viral theories of pathogenesis in biology and Chomsky's linguistic theory of competence more than it does the fundamental laws of physics.

It resembles viral theories in postulating a real causal entity, having causal powers. A virus (like a fantasy system) may be present without manifestations of disease; and may produce manifestations long after its introduction into the host as a result of changes in the host. More than one virus (as more than one fantasy system) may be required to produce manifestations. Evidence of the host's response to the presence of the virus (antibodies, for example) may be necessary in some cases to establish its causal role (just as manifestations of resistance in psychoanalysis are evidence for the causal role of fantasy systems). Most importantly, the criteria for establishing the causal role of a virus has changed throughout the years as more has come to be known about different viruses. One imagines that psychoanalysis too will have to establish criteria for establishing its causal claims (its own "Koch-Henle postulates") that bear some relation to the nature of the causal agents it postulates.

Chomsky's linguistic theory also describes a hypothetical dispositional mental state—knowledge of the rules of grammar or linguistic competence. Such knowledge or competence, when evoked by a variety of ad hoc and nonlawful stimuli or happenings, produces linguistic performances with certain properties, which can be explained by the properties (the particular rules of grammar) attributed to the dispositional state. Belief in such a theory is justified not by experiments but by the theory's ability to account for a wide variety of properties of linguistic objects (sentences) and their interrelations.

In his notes on the Rat Man case (1909), Freud writes that he told the patient that the "attempt to deny the reality of his father's death is the basis of his whole neurosis" (p. 300). His explanation of the neurosis, as given in the case study, is somewhat more detailed. In the words of Wollheim's insightful reconstruction (1971):

> [T]he Rat Man can be seen as the victim of two very general conflicts: the first between his father, or his father's wishes, and his lady; the second between love and hatred—a conflict which qualified his relations with both the major figures of his life. However, the two conflicts readily associate themselves, in that (supremely) the hatred of the father couples with love for the lady. (P. 143)

Why hatred of the father rather than fear of the father?

> First, the Rat Man's hatred of his father was in origin tied to the . . . phantasy . . . of his father as an interferer in his sensual desires. . . . Secondly, there was the Rat Man's own violence of character, which invariably led him from awareness of his father's anger (which might have been the object of his fear) to anger or hatred on his own part in response to his father's. (Pp. 143–144).

In the Wolf Man case, the patient's boyhood phobic symptoms are explained by Freud as caused by a retrospective interpretation, in the light of later sexual researches and preoccupation with castration, of a memory or a fantasy of seeing, at a very early age, his mother and father having intercourse. What he realized, as he reflected in imagination on the scene of intercourse between his father and mother, was that for him to receive passive pleasure from father entailed his being castrated. This realization caused fear of his father—intensified by his own passive wishes toward his father. His displacement of this fear from his father to animals caused his phobic symptoms.

What was the origin of the passive wishes toward his father? In response to fear of women (his masculine self-esteem had been hurt by his sister's seduction of him, and his nurse had rebuffed his advances and threatened him with castration), regression to wishes having their origin in an earlier anal-sadistic phase of development had occurred. Anxiety about his sadistic wishes led him to imagine them as belonging to his father not to himself, and therefore to imagine himself as the object rather than the agent in his sadistic fantasies. The wish to be beaten by his father combined with and reinforced the wish for passive pleasure from his father. Abandoning active phallic wishes in relation to a woman and substituting for them passive wishes in relation to a man had become his protection against the threat from women like the nurse.

What kind of explanations are these? They postulate the existence of dispositional mental states (wishes and beliefs), which, by virtue of their nature (for example, their contents) and the relations among them (for example, conflict), have causal powers. They constitute capacities or propensities to generate other mental states or to bring about kinds of actions—to form symptoms, for example.

Freud believes his clinical causal inferences in the Rat Man and Wolf Man cases justified, because they constitute causal explanations—and good (that is, successful) causal explanations. A good causal explanation postulates real causal entities, which by virtue of their nature have causal powers that are of the right kind to produce the effects of interest, by means of concrete causal mechanisms or processes that connect cause and effect, and determine the detailed character of the effect produced. Good causal explanations do not necessarily involve universal generalizations, and certainly do not involve laws such as the fundamental laws we find in physics—expressing functional relations among variables in mathematical equations. If a causal explanation succeeds in explaining an effect, it can have very few, if any, plausible rivals—unlike laws that merely express covariations. Freud points out that others may explain this or that phenomenon, a particular symptom, for example, but his causal explanation is better than such partial explanations, because it accounts not only for the entire range of phenomena but for the *interconnections* among them.

In Freud's case studies, we see an attempt at causal explanation grounded in the generative view of causality. If psychoanalytic clinical causal inferences are to be denied scientific credibility, such denial must be warranted by pointing to failures of methodology that are perceived as such within this rather than the successionist view of causality.

Conclusion

Psychoanalysis relies on inference to the best explanation. But what counts as the best explanation?

First, the best explanation is one that is simple, not necessarily by any logical criteria, but in that it draws upon established knowledge, upon what is already known or familiar. In other words, it makes use of a model, about which a great deal is already securely known. The grounds for belief in the explanation is an argument by analogy: "similar causes, similar effects"—as well as an appeal in so arguing to the principle of parsimony.

Second, the best explanation is one that explains a variety of kinds of instances. Inferences from a variety of kinds of data, information, or facts converge on the same explanation. In other words, the use of the model leads to an explanation of a variety of kinds of facts. We prefer to believe that the explanation is probably true than to regard its explanatory power as an improbable coincidence—a miracle.

Third, the best explanation is an explanation that makes probable or expected even a singular fact (a particular observation, a pattern, a co-occurrence, or a covariation) that is improbable, unexpected, or surprising, given our background knowledge. This is reason to suspect the explanation is true (Peirce 1901).

An even better reason is provided by Hacking's premise of comparative support (1965), which is available as grounds, when it is possible to calculate how much more probable a fact is, given the explanation, than it would be, given some rival explanation.

Similarly, when two or more apparently independent events seem improbably to coincide, we are incited to infer to a common cause that makes their co-occurrence probable or expected. We prefer to look for a common cause, and believe that such an explanation may be true, than to believe that a random improbable coincidence has occurred (Salmon 1984).

Fourth, the best explanation is one that, once inferred, has multiple implications or consequences, facts not used in inferring it. Successful predictions of hitherto unknown facts provide more reason to accept the explanation than simply showing that a large number of different kinds of facts, already known before but not used in inferring the explanation, can be explained by it (Campbell 1975).

Fifth, the best explanation provides more than an empirical generalization, which "explains" facts simply because they are instances of it. The best explanation usually refers to nonobservable, theoretical entities. It provides an account of the causal powers, the causal mechanisms, the causal processes, by virtue of which such entities produce their effects. It is likely to draw upon a model for evidence for the plausibility, and the existence, of the theoretical entities and causal mechanisms it postulates.

Sixth, the best explanation is one that has no comparable plausible rival when it comes to explaining facts in its domain, or is clearly favored by the evidence over some comparable plausible rival explanation.

I state here, in summary, and offering no argument, the following beliefs.

One, if psychoanalytic explanations meet even one of these criteria, they are plausible and worthy of interest and continued attention.

Two, psychoanalytic explanations already have met to some reasonable extent the first, second, fifth, and sixth criteria.

Three, psychoanalysis is capable, even if it should limit itself to data obtained in the psychoanalytic situation (and there is no reason it must so limit itself), of meeting the third and fourth criteria more effectively than is currently the case (Edelson 1984).

Bibliography

Arlow, J. (1969a). "Unconscious Fantasy and Disturbances of Conscious Experience." *Psychoanalytic Quarterly* 38 (1):1–27. (218)

——— (1969b). "Fantasy, Memory, and Reality Testing," *Psychoanalytic Quarterly* 38(1):28–51. (218)

Bailar, J.; Louis, T.; Lavori, P.; & Polansky, M. (1984). "Studies without Internal Controls." *New England Journal of Medicine* 311:156–162. (265)

Barlow, D.; Hayes, S.; & Nelson, R. (1984). *The Scientist Practitioner.* New York: Pergamon Press. (265)

Bibring, G.; Dwyer, T.; Huntington, D.; & Valenstein, A. (1961). "A Study of the Psychological Processes in Pregnancy and of the Earliest Mother-Child Relationship. (Appendix: Glossary of Defenses.)" *Psychoanalytic Study of the Child* 16:9–72. (177,185)

Borst, C. V., ed. (1970). *The Mind-Brain Identity Theory.* New York: St. Martin's Press. (124)

Boyd, R. (1979). "Metaphor and Theory Change: What is 'Metaphor?'" Pp. 356–408 in A. Ortony, ed., *Metaphor and Thought.* New York: Cambridge University Press, 1979. (328)

Breuer, J., & Freud, S. (1893–1895). *Studies on Hysteria. S.E.*, 2. London: Hogarth Press, 1955. (80,173,271,338)

Buchler, J., ed. (1955). "Logic as Semiotic: The Theory of Signs." Pp. 98–119 in *Philosophical Writings of Peirce.* New York: Dover (66)

Bunge, M. (1980). *The Mind-Body Problem.* New York: Pergamon Press. (124,127)

Campbell, D. (1975). " 'Degrees of Freedom' and the Case Study." *Comparative Political Studies* 8(2):178–193. (317,329,364)

———, & Stanley, J. (1963). *Experimental and Quasi-experimental Designs for Research.* Chicago: Rand McNally. (101,265,343)

Cartwright, N. (1983). *How the Laws of Physics Lie.* New York: Oxford University Press. (287,301,355–359)

Note: Numbers in parentheses after each entry refer to page number of text where entry appears. *S.E.* = *The Standard Edition of the Complete Psychological Works of Sigmund Freud* (London: Hogarth Press).

Cassirer, E. (1923–1929). *The Philosophy of Symbolic Forms.* Vol. 1: *Language.* Vol. 2: *Mythical Thought.* Vol. 3: *The Phenomenology of Knowledge.* New Haven, Conn.: Yale University Press. (45)

——— (1944). *An Essay on Man.* New Haven, Conn.: Yale University Press. (45)

——— (1946). *Language and Myth.* New York: Dover, 1953. (45)

Chassan, J. (1979). *Research Design in Clinical Psychology and Psychiatry.* 2d ed. enl. New York: Irvington. (101,265,343)

Chasseguet-Smirgel, J. (1984). *Creativity and Perversion.* New York: Norton. (212)

Chomsky, N. (1957). *Syntactic Structures.* The Hague: Mouton. (22)

——— (1959). "Review of B. F. Skinner's *Verbal Behavior.*" Pp. 547–578 in J. Fodor & J. Katz, eds., *The Structure of Language.* Englewood Cliffs, N.J.: Prentice-Hall, 1964. (22,26,30,66)

——— (1965). *Aspects of the Theory of Syntax.* Cambridge, Mass.: M.I.T. Press. (22,54)

——— (1966a). *Cartesian Linguistics.* New York: Harper & Row. (22)

——— (1966b). *Topics in the Theory of Generative Grammar.* The Hague: Mouton. (22)

——— (1967). "The Formal Nature of Language." In E. Lenneberg, *Biological Foundations of Language.* New York: Wiley. (22)

——— (1968). *Language and Mind.* Enl. ed.: New York: Harcourt, Brace, 1972. (22,71)

Clark, H. (1970). "Word Associations and Linguistic Theory." Pp. 271–286 in J. Lyons, ed., *New Horizons in Linguistics.* Baltimore: Penguin Books. (52)

Cook, T., & Campbell, D. (1979). *Quasi-Experimentation.* Boston: Houghton-Mifflin. (101)

Copi, I. (1967). *Symbolic Logic.* 3d ed. New York: Macmillan. (73)

Costa, E. (1985). "Benzodiazepine-GABA Interactions." Pp. 27–52 in Tuma & Maser. (95)

Crews, F. (1985). "The Future of an Illusion." *New Republic* 653(3):28–33. (316)

Danziger, K. (1985). "The Methodological Imperative in Psychology." *Philosophy of the Social Sciences* 15:1–13. (297,307)

Dennett, D. C. (1978). *Brainstorms.* Montgomery, Vt.: Bradford Books. (124,181)

Diesing, P. (1971). *Patterns of Discovery in the Social Sciences.* New York: Aldine. (328)

Edelheit, H. (1969). "Speech and Psychic Structure." *Journal of the American Psychoanalytic Association* 17:381–412. (61)

Edelson, J. (1983). "Freud's Use of Metaphor." *Psychoanalytic Study of the Child* 38:17–59. (328)

Edelson, M. (1954). "The Science of Psychology and the Concept of Energy." Ph.D. diss., University of Chicago. (75)

——— (1964). *Ego Psychology, Group Dynamics and the Therapeutic Community.* New York: Grune & Stratton. (63,72)

——— (1970a). *Sociotherapy and Psychotherapy.* Chicago: University of Chicago Press. (63,72)

——— (1970b). *The Practice of Sociotherapy: A Case Study.* New Haven, Conn.: Yale University Press. (63,72)

——— (1970c). "Toward a Study of Interpretation in Psychoanalysis." Pp. 151–181 in J. Loubser, R. Baum, A. Effrat, & V. Lidz, eds., *Explorations in General Theory in Social Science.* New York: Free Press, 1976. (3,78)

——— (1971). *The Idea of a Mental Illness.* New Haven, Conn.: Yale University Press. (xxxiii,33,58,60,66)

——— (1972). "Language and Dreams: *The Interpretation of Dreams* Revisited." *Psychoanalytic Study of the Child* 27:203–282. (21,82)

——— (1975). *Language and Interpretation in Psychoanalysis.* Reprint: Chicago: University of Chicago Press, 1984. (xxxiii,33,66,73, 81,82,247,273,339)

——— (1977). "Psychoanalysis as Science." *Journal of Nervous and Mental Disease,* 165(1):1–28. (62,78,188)

——— (1978). "What Is The Psychoanalyst Talking About ?" Pp. 99–170 in J. Smith, ed., *Psychoanalysis and Language: Psychiatry and the Humanities.* Vol. 3. New Haven, Conn.: Yale University Press. (78)

——— (1984). *Hypothesis and Evidence in Psychoanalysis.* Chicago: University of Chicago Press, 1985 (paperback). (xxxiii, 71,74,91,98, 99, 101, 102, 107, 113, 115, 116, 124, 130, 149, 162, 233, 238, 242, 255, 256,265,271,275,276,278,296,297,314,321,364)

——— (1985a). "Psychoanalysis, Anxiety and the Anxiety Disorders." Pp. 633–644 in Tuma & Maser. (89)

——— (1985b). "The Hermeneutic Turn and the Single Case Study in Psychoanalysis." *Psychoanalysis and Contemporary Thought* 8(4):567–614. Shorter version appeared on pp. 71–104 in D. Berg & K. Smith, eds., *Exploring Clinical Methods for Social Research.* Beverly Hills, Calif.: Sage Publications. (231)

——— (1986a). "The Evidential Value of the Psychoanalyst's Clinical Data." *Behavioral and Brain Sciences* 9(2):232–234. (268)

——— (1986b). "Causal Explanation in Science and in Psychoanalysis: Implications for Writing a Case Study." *Psychoanalytic Study of the Child* 41:89–127. (102,278)

——— (1986c). "Heinz Hartmann's Influence on Psychoanalysis as a Science." *Psychoanalytic Inquiry* 6(4):575–600. (102)

——— (1986d). "The Convergence of Psychoanalysis and Neuroscience: Illusion and Reality." *Contemporary Psychoanalysis* 22(4):479–519. (102,122)

Erdelyi, M. (1985). *Psychoanalysis: Freud's Cognitive Psychology.* New York: W. H. Freeman. (175,177,184)

Erwin, E. (1986). "Defending Freudianism." *Behavioral and Brain Sciences* 9(2):235–236. (315)

Evans, A. (1978). "Causation and Disease." *American Journal of Epidemiology* 108(4):249–258. (350–354)

Ferenczi, S. (1909). "Introjection and Transference." Pp. 35–93 in *First Contributions to Psycho-Analysis.* London: Maresfield Reprints, 1980. (226)

——— (1911). "On Obscene Words." Pp. 132–153 in *First Contributions to Psycho-Analysis.* London: Maresfield Reprints, 1980. (206)

Ferguson, M. (1985). "A Critique of Grunbaum on Psychoanalysis." *Journal of the American Academy of Psychoanalysis* 13(3):329–345. (354–355)

Fodor, J. A. (1968). *Psychological Explanation.* New York: Random House. (124,180,181)

——— (1975). *The Language of Thought.* New York: Thomas Y. Crowell. (124,180,181)

——— (1981a). "The Mind-Body Problem." *Scientific American* 244(1):114–123. (124,125)

——— (1981b). *Representations.* Cambridge, Mass.: M.I.T. Press, 1983 (paperback). (124,130,135,140,141,143,180,181)

——— (1983). *The Modularity of Mind.* Cambridge, Mass.: M.I.T. Press. (104,124,147,148,180,181)

——— (1987). *Psychosemantics: The Problem of Meaning in the Philosophy of Mind.* Cambridge, Mass.: M.I.T. Press. (199)

———; Bever, T.; & Garrett, M. (1974). *The Psychology of Language.* New York: McGraw-Hill. (60)

Freedman, A.; Kaplan, H.; & Sadock, B. (1975). *Comprehensive Textbook of Psychiatry—II.* Vol. 1. 2d ed. Baltimore, Md.: Williams and Wilkins. (199)

Freud, S. (1892–1899). "Extracts from the Fliess Papers." *S.E.,* 1:175–280. London: Hogarth, 1966. (3)

——— (1895). "Project for a Scientific Psychology." *S.E.,* 1:283–398. London: Hogarth, 1966. (3,121,291,329)

——— (1900). *The Interpretation of Dreams. S.E.,* 4–5. London: Hogarth, 1960. (xix,4,21–61,64,65,76,84,100,167,173,177,178,198, 240,254,270,271,273,274,307,322,323,327)

——— (1905). "Three Essays on the Theory of Sexuality." *S.E.,* 7:130–243. London: Hogarth, 1953. (xix,64,193,198–199, 207)

——— (1909). "Notes upon a Case of Obsessional Neurosis." *S.E.*, 10:153–318. London: Hogarth, 1955. (xix,101,306,332,361)
——— (1910). "The Antithetical Meaning of Primal Words." *S.E.*, 11:153–161. London: Hogarth, 1957. (52)
——— (1911a). "Psycho-Analytic Notes on an Autobiographical Account of a Case of Paranoia." *S.E.*, 12:9–82. London: Hogarth, 1958. (15–16,50,188,339)
——— (1911b). "Formulations on the Two Principles of Mental Functioning." *S.E.*, 12:218–226. London: Hogarth, 1958. (xix,173,178)
——— (1913). "Totem and Taboo." *S.E.*, 13:1–161. London: Hogarth, 1955. (3,4)
——— (1914a). "On the History of the Psycho-Analytic Movement." *S.E.*, 14:7–66. London: Hogarth, 1957. (4,149)
——— (1914b). "On Narcissism: An Introduction." *S.E.*, 14:73–102. London: Hogarth, 1957 (8)
——— (1915a). "Instincts and Their Vicissitudes." *S.E.*, 14:117–140. London: Hogarth, 1957. (8,12)
——— (1915b). "The Unconscious." *S.E.*, 14:166–215. London: Hogarth, 1957. (7–8)
——— (1916–1917). *Introductory Lectures on Psycho-Analysis. S.E.*, 15–16. London: Hogarth, 1963. (4,5,6,8,322,324–326)
——— (1918). "From the History of an Infantile Neurosis." *S.E.*, 17:3–122. London: Hogarth, 1955. (xix,278–308)
——— (1919). "'A Child Is Being Beaten': A Contribution to the Study of the Origin of Sexual Perversions." *S.E.*, 17: 179–204. London: Hogarth, 1955. (xix,188–189)
——— (1924). "A Short Account of Psycho-analysis." *S.E.*, 19:191–209. London: Hogarth, 1961. (150)
——— (1925). "An Autobiographical Study." *S.E.*, 20: 7–74. London: Hogarth, 1959. (6)
——— (1926). "The Question of Lay Analysis." *S.E.*, 20:183–258. London: Hogarth, 1959. (144)
——— (1931). "Female Sexuality." *S.E.*, 21:225–245. London: Hogarth, 1961. (4)
——— (1937). "Constructions in Analysis." *S.E.*, 23: 257–269. London: Hogarth, 1964. (6)
——— (1939). "Moses and Monotheism." *S.E.*, 23:6–137. London: Hogarth, 1964. (6)
Gabbard, K., & Gabbard, G. (1987). *Psychiatry and the Cinema.* Chicago: University of Chicago Press. (xii)
Gill, M. (1963). "Topography and Systems in Psychoanalytic Theory." *Psychological Issues*. Monograph 10. New York: International Universities Press. (185)

Glymour, C. (1980). *Theory and Evidence.* Princeton, N.J.: Princeton University Press. (101,264,276,302)
——— (1983). "The Theory of Your Dreams." Pp. 57–71 in R. Cohen & L. Laudan, eds., *Physics, Philosophy and Psychoanalysis.* Boston: D. Reidel, 1983. (38,323,347,348)
Gould, S. J. (1977). *Ever since Darwin.* New York: Norton, 1979 (paperback). (189,345)
——— (1980). *The Panda's Thumb.* New York: Norton, 1982 (paperback). (189,345)
——— (1983). *Hen's Teeth and Horse's Toes.* New York: Norton, 1984 (paperback). (189,345)
——— (1985). *The Flamingo's Smile.* New York: Norton. (345)
——— (1986). "Evolution and the Triumph of Homology, or Why History Matters." *American Scientist* 74:60–69. (280,339–340,343–344)
Gray, J. (1985). "Issues in the Neuropsychology of Anxiety." Pp. 5–26 in Tuma & Maser. (95)
Greenberg, J., & Mitchell, S. (1983). *Object-Relations in Psychoanalytic Theory.* Cambridge, Mass.: Harvard University Press. (224)
Grunbaum, A. (1954). "Science and Ideology." *Science Monthly* 79:13–19. (287)
——— (1984). *The Foundations of Psychoanalysis.* Berkeley: University of California Press. (38, 114, 115, 121, 223, 233, 268, 269, 278, 288, 294,305,309,333–334,340,347,348)
——— (1986). Precis of *The Foundations of Psychoanalysis: A Philosophical Critique.* With commentary. *Behavioral and Brain Sciences* 9(2):217–284. (223,321,333,348)
Hacking, I. (1965). *Logic of Statistical Inference.* Cambridge: Cambridge University Press. (100,363)
——— (1983). *Representing and Intervening.* Cambridge: Cambridge University Press. (108,118,281)
Harre, R. (1970). *The Principles of Scientific Thinking.* Chicago: University of Chicago Press. (217,249,328)
——— (1972). *The Philosophies of Science.* New York: Oxford University Press. (217,328,339,345)
Hartmann, E. (1967). *The Biology of Dreaming.* Springfield, Ill.: Thomas. (61)
Hartmann, H. (1958). *Ego Psychology and the Problem of Adaptation.* New York: International Universities Press. (13)
——— (1964). *Essays on Ego Psychology.* New York: International Universities Press. (11,13,116)
———, & Kris, E. (1945). "The Genetic Approach in Psychoanalysis." *Psychoanalytic Study of the Child* 1:11–30. (304)
Hartshorne, C., & Weiss, P., eds. (1932). *Collected Papers of Charles*

Sanders Peirce. Vol. 2. Cambridge, Mass.: Harvard University Press. (66)
Hebb, D. O. (1980). *Essay on Mind.* Hillsdale, N.J.: Lawrence Erlbaum Associates. (124,134)
Herbst, P. (1970). *Behavioural Worlds.* London: Tavistock. (159)
Hersen, M., & Barlow, D. (1976). *Single-Case Experimental Designs.* New York: Pergamon Press. (101)
Hesse, M. (1966). *Models and Analogies in Science.* Notre Dame, Ind.: University of Notre Dame Press. (77)
Holland, N. (1973). "Defence, Displacement, and the Ego's Algebra." *International Journal of Psychoanalysis* 54:247–257. (186)
Holt, R. (1962). "Individuality and Generalization in the Psychology of Personality." *Journal of Personality* 30: 377–402. (253)
Holzman, P. (1976). "The Future of Psychoanalysis and Its Institutes." *Psychoanalytic Quarterly* 45:250–273. (63)
Jakobson, R. (1964). "Towards a Linguistic Typology of Aphasic Impairment." Pp. 21–46 in A. de Reuck & M. O'Connor, eds., *Disorders of Language.* London: Churchill. (61)
Jones, E. (1953–1957). *The Life and Work of Sigmund Freud.* 3 vols. New York: Basic Books. (7)
Kaplan, A. (1964). *The Conduct of Inquiry.* New York: Harper & Row. (287)
Katz, J. (1964). "Semi-Sentences." Pp. 400–416 in J. Fodor & J. Katz, eds.,*The Structure of Language.* Englewood Cliffs, N.J.: Prentice-Hall. (54)
Kazdin, A. (1980). *Research Design in Clinical Psychology.* New York: Harper & Row. (255)
——— (1981). "Drawing Valid Inferences from Case Studies." *Journal of Consulting and Clinical Psychology* 49:183–192. (101)
——— (1982). *Single-Case Research Designs.* New York: Oxford University Press. (101,265)
Kernberg, O. (1986). "A Conceptual Model of Male Perversion." Pp. 152–180 in G. Fogel, F. Lane, & R. Liebert, eds.,*The Psychology of Men.* New York: Basic Books. (212)
Kris, A. (1982). *Free Association.* New Haven, Conn.: Yale University Press. (163)
Kris, E. (1952). *Psychoanalytic Explorations in Art.* New York: Schocken Books, 1964. (35,45,56)
——— (1956a). "The Personal Myth." *Journal of the American Psychoanalytic Association* 4:653–681. (16,55,279)
——— (1956b). "The Recovery of Childhood Memories in Psychoanalysis." *Psychoanalytic Study of the Child* 11:54–88. (16,28,55,280)

Langer, S. (1953). *Feeling and Form.* New York: Charles Scribner's Sons. (11,15,28)

——— (1957). *Problems of Art.* New York: Charles Scribner's Sons. (28)

——— (1962). *Philosophical Sketches.* New York: Mentor Book, 1964. (28)

——— (1967). *Mind: An Essay on Human Feeling.* Baltimore, Md.: Johns Hopkins Press. (11)

Laplanche, J., & Pontalis, J.-B. (1968). "Fantasy and the Origins of Sexuality." *International Journal of Psycho-Analysis* 49(1):1–18. (209)

Lasagna, L. (1982). "Historical Controls." *New England Journal of Medicine* 307:1339–1340. (265)

Leplin, J., ed. (1984). *Scientific Realism.* Berkeley: University of California Press. (108,118)

Levi-Strauss, C. (1958). *Structural Anthropology.* New York: Anchor Books, 1967. (46)

——— (1962). *The Savage Mind.* Chicago: University of Chicago Press. (25,46)

——— (1964). *The Raw and the Cooked.* New York: Harper Torchbook, 1970. (46)

Lewin, K. (1936). *Principles of Topological Psychology.* New York: McGraw-Hill. (76)

Lieberson, J. (1985). "Letter." *New York Review of Books* 32(4):40. (316)

Liebert, R. (1986). "The History of Male Homosexuality from Ancient Greece through the Renaissance: Implications for Psychoanalytic Theory." Pp. 181–210 in G. Fogel, F. Lane, & R. Liebert, eds., *The Psychology of Men.* New York: Basic Books. (213)

Lipton, S. D. (1977). "The Advantages of Freud's Technique as Shown in His Analysis of the Rat Man." *International Journal of Psychoanalysis* 58:255–273. (119)

Luborsky, L. (1967). "Momentary Forgetting during Psychotherapy and Psychoanalysis." Pp. 177–217 in R. Holt, ed., *Motives and Thoughts. Psychological Issues.* Monograph 18/19. New York: International Universities Press. (101,241,254,276,296)

——— (1973). "Forgetting and Remembering (Momentary Forgetting) during Psychotherapy." Pp. 29–55 in M. Mayman, ed., *Psychoanalytic Research. Psychological Issues.* Monograph 30. New York: International Universities Press. (101,241,276,296)

———, & Mintz, J. (1974). "What Sets Off Momentary Forgetting during a Psychoanalysis?" *Psychoanalysis and Contemporary Science* 3:233–268. (101,241,276,296)

———; Bachrach, H.; Graff, H.; Pulver, S.; & Christoph, P. (1979). "Preconditions and Consequences of Transference Interpretations." *Journal of Nervous and Mental Disease* 167:391–401. (101)

McGrath, J. (1981). "Dilemmatics: The Study of Research Choices and Dilemmas." *American Behavioral Scientist* 25(2):179–210. (316)

Meehl, P. E. (1983). "Subjectivity in Psychoanalytic Inference." Pp. 349–411 in J. Earman, ed., *Testing Scientific Theories.* Minneapolis: University of Minnesota Press. (119,244, 297,315)

Michels, R.; Frances, A.; & Shear, M. (1985). "Psychodynamic Models of Anxiety." Pp. 595–618 in Tuma & Maser. (97,99)

Mineka, S. (1985). "Animal Models of Anxiety-based Disorders." Pp. 199–245 in Tuma & Maser. (95)

Nagel, E. (1979). *Teleology Revisited.* New York: Columbia University Press. (124,129,130)

Parsons, T. (1937). *The Structure of Social Action.* New York: Free Press. (75)

Peirce, C. (1901). "Abduction and Induction." Pp. 150–156 in J. Buchler, ed., *Philosophical Writings of Peirce.* New York: Dover, 1955. (363)

Piaget, J. (1945). *Play, Dreams and Imitation in Childhood.* New York: Norton, 1962. (53,61)

——— (1970). *Structuralism.* New York: Basic Books. (46)

Platt, J. (1964). "Strong Inference." *Science* 146: 347–353. (100,311)

Popper, K. R., & Eccles, J. C. (1977). *The Self and Its Brain.* London: Routledge & Kegan Paul, 1983 (paperback). (126)

Pylyshyn, Z. (1985). *Computation and Cognition: Toward a Foundation for Cognitive Science.* Cambridge, Mass.: M.I.T. Press. (147,180,181,184)

Rapaport, D. (1959). "The Structure of Psychoanalytic Theory." *Psychological Issues.* Monograph 6. New York: International Universities Press, 1960. (98,99–100)

———, & Gill, M. M. (1959). "The Points of View and Assumptions of Metapsychology." *International Journal of Psychoanalysis* 40:153–162. (114)

Redmond, D. (1985). "Neurochemical Basis for Anxiety and Anxiety Disorders." Pp. 533–556 in Tuma & Maser. (91)

Reiser, M. F. (1984). *Mind, Brain, Body: Toward a Convergence of Psychoanalysis and Neurobiology.* New York: Basic Books. (130,149)

Rosenthal, D. M., ed. (1971). *Materialism and the Mind- Body Problem.* Englewood Cliffs, N.J.: Prentice-Hall. (124)

Rozeboom, W. (1961). "Ontological Induction and the Logical Typology of Scientific Variables." *Philosophy of Science* 28:337–377. (159)

Rubinstein, B. B. (1965). "Psychoanalytic Theory and the Mind-Body Problem." Pp. 35–56 in N. S. Greenfeld & W. C. Lewis, eds., *Psychoanalysis and Current Biological Thought.* Madison: University of Wisconsin Press. (146)

——— (1967). "Explanation and Mere Description." Pp. 18–77 in R. Holt, ed., *Motives and Thought. Psychological Issues.* Monograph 18/19. New York: International Universities Press. (146)

——— (1976). "On the Possibility of a Strictly Clinical Psychoanalytic Theory." Pp. 229–264 in M. Gill & P. Holzman, eds., *Psychology versus Metapsychology. Psychological Issues.* Monograph 36. New York: International Universities Press. (146)

Runyan, W. M. (1982). *Life Histories and Psychobiography.* New York: Oxford University Press, 1984 (paperback). (170)

Russell, B. (1956). *Logic and Knowledge.* New York: G. P. Putnam's Sons. (65)

Salmon, W. (1959). "Psychoanalytic Theory and Evidence." Pp. 252–267 in S. Hook, ed., *Psychoanalysis, Scientific Method and Philosophy.* New York: Grove Press, 1960. (342)

——— (1984). *Scientific Explanation and the Causal Structure of the World.* Princeton, N.J.: Princeton University Press. (106,118, 127,286,287,288,290–292,299,300,301,332,355,363)

Sapir, E. (1921). *Language.* New York: Harvest Book, 1970. (61)

Sashin, J. (1985). "Affect Tolerance: A Model of Affect-Response Using Catastrophe Theory." *Journal of Social and Biological Structure* 8:175–202. (301)

Schafer, R. (1968a). *Aspects of Internalization.* New York: International Universities Press. (39)

——— (1968b). "The Mechanisms of Defence." *International Journal of Psychoanalysis* 49:49–62. (174,177,186)

——— (1970). "An Overview of Heinz Hartmann's Contributions to Psychoanalysis." *International Journal of Psychoanalysis* 51:425–446. (116)

Schur, M. (1966). *The Id and the Regulatory Principles of Mental Functioning.* New York: International Universities Press. (39)

Searle, J. R. (1983). *Intentionality.* Cambridge: Cambridge University Press. (103,124,132,133,145,181,242)

Shapiro, M. (1961). "A Method of Measuring Psychological Changes Specific to the Individual Psychiatric Patient." *British Journal of Medical Psychology* 34:151–155. (101)

——— (1963). "A Clinical Approach to Fundamental Research with Special Reference to the Study of the Single Patient." Pp. 123–149

in P. Sainsbury & N. Kreitman, eds., *Methods of Psychiatric Research*. New York: Oxford University Press. (101,265)
——— (1966). "The Single Case in Clinical-Psychological Research." *Journal of General Psychology* 74:3–23. (101)
Sherwood, M. (1969). *The Logic of Explanation in Psychoanalysis*. New York: Academic Press. (246,248)
Shope, R. (1970). "Dispositional Treatment of Psychoanalytic Motivation Terms. "*Journal of Philosophy* 67:195–208. (217)
——— (1973). "Freud's Concepts of Meaning." *Psychoanalysis and Contemporary Science* 2:276–303. (246,248)
Sidman, M. (1960). *Tactics of Scientific Research*. New York: Basic Books. (101)
Stegmuller, W. (1976). *The Structure and Dynamics of Theories*. New York: Springer-Verlag. (315)
Stevens, W. (1951). *The Necessary Angel*. New York: Vintage Books. (17)
——— (1961). *The Collected Poems of Wallace Stevens*. New York: Knopf. (7,9,20,80)
Suppes, P. (1957). *Introduction to Logic*. New York: D. Van Nostrand. (73,75)
———, & Warren, H. (1975). "On the Generation and Classification of Defence Mechanisms." *International Journal of Psychoanalysis* 56:405–414. (186,188)
Trilling, L. (1950). "Freud and Literature." *The Liberal Imagination*. New York: Viking Press, 1951. (17)
Tuma, A., & Maser, J., eds., (1985). *Anxiety and the Anxiety Disorders*. Hillsdale, N.J.: Lawrence Erlbaum. (89)
Von Eckhardt, B. (1982). "Why Freud's Research Methodology Was Unscientific." *Psychoanalysis and Contemporary Thought* 5: 549–574. (256)
Waelder, R. (1960). *Basic Theory of Psychoanalysis*. New York: International Universities Press. (110,149,177)
Wiener, P., ed., (1966). "Letters to Lady Welby." Pp. 380–432 in *Charles S. Peirce: Selected Writings*. New York: Dover. (66)
Wollheim, R. (1969). "The Mind and the Mind's Image of Itself." Pp. 31–53 in *On Art and the Mind*, 1974. (160,173,182,194,219)
——— (1971). *Sigmund Freud*. New York: Viking Press. Reprint: Cambridge: Cambridge University Press, 1981 (paperback). (115,160,179,180,185,194,195,219,295,361)
——— (1974a). "Imagination and Identification." Pp. 54–83 in *On Art and the Mind*. (160,173,194,219)
——— (1974b). *On Art and the Mind*. Cambridge, Mass.: Harvard University Press. (160,194,219)

——— (1974c). "Identification and Imagination." Pp. 172–195 in R. Wollheim, ed., *Freud: A Collection of Critical Essays.* New York: Anchor Press (paperback). (160,173,190,194,219)

——— (1979). "Wish-Fulfilment." Pp. 47–60 in R. Harrison, ed., *Rational Action.* Cambridge: Cambridge University Press. (107,160,173,178,179,182,194,219)

———(1982). "The Bodily Ego." Pp. 124–138 in R. Wollheim & J. Hopkins, eds., *Philosophical Essays on Freud.* Cambridge: Cambridge University Press. (160,173,182,194,219)

———(1984). *The Thread of Life.* Cambridge, Mass.: Harvard University Press. (160,173,194,219,333,349)

Index

Note: Names of persons, cases, and books appear in this Index when the person, case, or book is mentioned in the text without any bibliographic citation. In the Bibliography, numbers in parentheses after each entry refer to page numbers of text where entry appears; these references to authors and their works or cases are not repeated under names in this Index.